The Eyes of the World

The Eyes of the World

Mining the Digital Age
in the Eastern DR Congo

JAMES H. SMITH

The University of Chicago Press
Chicago and London

The University of Chicago Press, Chicago 60637
The University of Chicago Press, Ltd., London
© 2022 by The University of Chicago
Published 2022
Printed in the United States of America

31 30 29 28 27 26 25 24 23 22   1 2 3 4 5

ISBN-13: 978-0-226-77435-0 (cloth)
ISBN-13: 978-0-226-81606-7 (paper)
ISBN-13: 978-0-226-81605-0 (e-book)
DOI: https://doi.org/10.7208/chicago/9780226816050.001.0001

Library of Congress Cataloging-in-Publication Data

Names: Smith, James H., 1970– author.
Title: The eyes of the world : mining the digital age in the Eastern
    DR Congo / James H. Smith.
Description: Chicago : University of Chicago Press, 2022. | Includes
    bibliographical references and index.
Identifiers: LCCN 2021031024 | ISBN 9780226774350 (cloth) |
    ISBN 9780226816067 (paperback) | ISBN 9780226816050 (ebook)
Subjects: LCSH: Small-scale mining—Congo (Democratic Republic) |
    Artisanal miners—Congo (Democratic Republic) | Mineral industries—
    Congo (Democratic Republic) | Tantalum industry—Congo (Democratic
    Republic)
Classification: LCC HD9506.C752 S65 2022 | DDC 338.2096751/5—dc23
LC record available at https://lccn.loc.gov/2021031024

CONTENTS

PART ONE

Orientations

An Introduction to the Personal, Methodological, and Spatiotemporal Scales of the Project

This prologue is intended to give a broad overview of the project—mainly, the general topic and the changing historical context in which this longitudinal research took place (carried out mostly in three-month periods between 2006 and 2018). It consists of a brief narrative regarding the motivation for the research and how it was conducted, followed by some relevant (no doubt inadequately represented) history that is punctuated by a revealing fieldwork vignette quickly capturing some important themes that framed the research context. A more detailed discussion of the book's main arguments is found in chapter 1.

The Eyes of the World explores the worlds of artisanal mining in the eastern Democratic Republic of Congo (DR Congo), examining how decades of violent colonial and postcolonial war and exclusion have, in producing dispossession, also encouraged the emergence of a multidimensional extractive economy oriented around fostering movement and collaboration among very different actors, elements, arrangements, and modes of life. The main focus is on the work, lives, and concepts of miners and traders of what I sometimes refer to as "digital minerals" and what Congolese in the Kivu provinces and Maniema province call the "black minerals" (*minéraux noire*) and sometimes "coltan" (though coltan is also the Congolese term for a specific ore). The collaborative work of people involved in this trade connects Congo with places like Silicon Valley in ways that are counterintuitive and concealed (there are currently an estimated two million artisanal miners in Congo, and it has been estimated that ten million people depend on artisanal mining, although not all of these are mining the black minerals; see Garrett and Mitchell 2009).

Coltan is the colloquial term for a silicate essential to the capacitors used in all digital devices, including mobile phones (Nest 2011). It contains tantalum and niobium, elements named after mythological Greek Titans who were punished for their defiant hubris against the gods—an ironic naming, given that the technologies these minerals enable have helped give rise to a putatively "posthuman age" in which people strive to take on some of the characteristics of superior beings, including immortality and omniscience (Hayles 1999).[1] Their material density enables these elements to hold a high electrical charge, allowing for ever-smaller digital devices (as we will see, most Euro-Americans see smallness as a product of the ingenuity of Silicon Valley engineers, or the sui generis evolution of technology itself, rather than as substances and the work of extracting and moving them). Coltan is similar in appearance to and, in Congo, is often found alongside other "black minerals" that are equally crucial for digital technologies, especially wolframite, the ore from which tungsten is derived, and cassiterite, from which tin is derived. Tungsten is used to make computer screens, and it also enables cell phones to vibrate; tin is used in wiring, among many other processes. Eastern Congolese diggers don't specialize in these substances but move among them based on a number of different factors, including what seems to be in demand at any given time. International nongovernmental organizations have come to call these minerals the "3Ts" (tantalum, tin, and tungsten); in recent years, they have been the target of a great deal of humanitarian discourse and intervention.

A Brief Biography of This Project

If I were to look deeply into myself, I suppose part of what first intrigued me about the mining of Congolese coltan was my deep, emic familiarity with and dislike of techno-utopianism and technosolutionism. I grew up in the 1980s near a Massachusetts tech corridor, and all of my high school friends—even the poets—eventually became computer programmers or software developers. They did this partly because that's what people who were deemed to be smart did then and there and because computerization and technology seemed to be where the future lay—it was world transformative, and that was exciting to people, especially young people (and, I suspect, especially males) who felt smart enough to ride this wave into the future. Computers also allowed young men who were not encouraged to practice humanistic arts to be creative; and it was where the money seemed to be. This wouldn't have been so bad in and of itself, but even then, I found that the embrace of technology as the engine of the future and the solution to all the world's problems tended to go hand in hand with a derisive or dismissive attitude toward the concepts and ways of life of those who were deemed to be less technologically "developed." This attitude of dismissiveness usually also extended to everything deemed to be somehow "old."

Later on, I witnessed a vernacular version of this techno-utopianism emerging in Kenya during the 1990s and early 2000s. It was directed against the old men of Kenyan politics, with their focus on what, for urban elites, were outmoded issues like land, and it envisioned a future that moved beyond what some Kenyans would eventually call "analogue politics" (Poggiali 2016, 2017; Nyabola 2018). Eventually, I moved to Northern California, settling in the interstices between the self-congratulatory techno-hub of the Bay Area and the agri-capitalist Central Valley, and witnessed techno-utopianism and its attendant dismissiveness toward that which lay outside of it on a whole new level, with different inflections. Always there was what anthropologist Johannes Fabian referred to as the "denial of coevalness"— the unwillingness to see that which lay outside of the explicitly "technological" as being as equally contemporary as the technological and as part of the same historical totality that produced the technological in the first place (1986).

When I found out that coltan, an ore found in Congo that seemed to be somehow related to the conflicts there, was essential to all digital devices, it was revelatory for me, as it was for many; I unpack some of the implications of this revelation in chapter 1 and delve into some of the nuances of the war below and in chapter 2. It was also familiar because it spoke to a more global disjunction between the self-presentation of capitalist "progress" and its conditions of possibility. Though it was a radically different context, when I was a youth, anhydrous ammonia sales was our family business (one time, an ammonia tank burst open in my father's face, suffocating him and leaving him with third-degree internal and external burns that nearly killed him); before my father, it had been bottom-of-the-chain coal extraction and transportation. And one of my siblings started his career in oil extraction at pretty much the lowest level of the process: painting and repairing oil rigs in the Gulf of Mexico in his late teens. So, like many, I always knew something about the power of the effectively invisible, sometimes widely unknown, substances that make "modernity" possible and how they could change people's realities and social situations quickly and dramatically, often with great—even potentially fatal—attendant risk.

I had heard about the devastating Congo wars and the war economy that included Congolese coltan while I was still in grad school. But my first visit to Congo was by myself in 2003, a year after receiving my PhD.

I took the one-hour flight from Kenya to Rwanda, followed by a two-hour bus ride to Goma, North Kivu, which at the time was still occupied by a Rwandan-backed militia, the Rally for Congolese Democracy (RCD); crossing the border meant dealing with soldiers in tents. It was weirdly serendipitous that, en route to the border from Kigali, I happened to be seated next to a Congolese coltan trader (I will call him Michael). Michael became my friend for a while, showing me Goma (which had recently been burned to the ground by the nearby volcano) and Bukavu and introducing me to a lot of people. At that time, the Second Congo War, or Great War, was over in name only (it officially ended while I was in Goma, in July), and it was almost impossible for us to leave the cities of Bukavu and Goma (local transport would not go, and we did not do it). While it was hard for me to learn very much during that month-long visit, I did come to understand that the coltan supply chain was complicated, that it involved a lot of different kinds of actors, and that it would be good to do this work with another anthropologist. I also realized that, in part because of the collapsed infrastructure and the dollarization of the economy, Congo was quite expensive, and I would need a lot more funding to actually carry out research there. (I was happy to learn that I would be able to do the research using my Swahili and that, in the mining areas, Swahili was better than French, of which I had very little.)

Even before that first visit to Congo, I had talked with my friend and colleague Jeff Mantz, who did research on the social life of things, about the prospect of our doing something like what the anthropologist Sidney Mintz (1986) had done for sugar and modernity but with coltan and postmodernity, or the "postmodern" sensibilities that digital technologies helped to produce. (Mintz had brilliantly showcased the interconnectedness of seemingly disconnected parts of the world by following the commodity sugar from the Caribbean to Europe.) After my return, we applied for and received a National Science Foundation (NSF) High-Risk Research grant to conduct preliminary fieldwork on coltan mining in the eastern Congo (the "risk" referred to epistemological rather than bodily risk and was funding to ascertain whether the project was "doable"; see our co-authored piece, Smith and Mantz 2006; also Mintz 1985). In 2006, after spending a summer visiting mines near Goma and Bukavu (the mines near Numbi and Nyabibwe), we decided to divide up research in the following way: I would travel to and work near rural and forest mines, focusing on the extractive work of artisanal miners and low-level traders, while Jeff Mantz would conduct ethnography "higher up" on the supply chain with higher-level *négociants* (middlepersons) and *comptoirs* (buying houses) in the cities of Goma and Bukavu. With this division of labor in mind, we applied for another NSF grant, a collaborative one, and set about work. Over time, this developed into two separate research projects.[2] Some years later, in 2015, I was awarded another NSF grant to study the impact on artisanal miners of internationally imposed "conflict minerals" regulatory efforts.

Once Jeff Mantz and I had established our division of labor, I set out to do research in Congolese mines and realized that I wasn't sure exactly where to go or how to do it. I wanted to get a sense of the range of mines that were out there because I was worried about generalizing based on a single place, and my thought was to visit several different locations and then focus in on a few. The only person I had to introduce me to people and places and help me navigate state officials, soldiers, and other situations was Michael, the coltan trader I had met on the bus in 2003. But he was a businessman/trader, and his presence certainly didn't help persuade people that I was an academic researcher, something people had a hard time with anyway. Moreover, Michael was always trying to redirect me to gold mines because he wanted me to buy gold with him, which at the time was fetching a better price than coltan. So in 2009, at what may have been his suggestion, and certainly with his enthusiastic support, I brought my Kenyan friend, assistant, and colleague Ngeti Mwadime with me to Congo because we were used to working together and I trusted him.[3] That worked well enough (we made it to the rainforest mine of Bisie, for one example), and Ngeti was as helpful as he could possibly be, but it became very clear that the main obstacles to doing research in rural Congo were the constant bureaucratic shenanigans from state officials "selling papers," and Ngeti's presence didn't help with that; in fact, because he was also a foreigner, it just compounded the bureaucracy and cost too much money.

In 2011, I was back with Michael again, but by this time he was clearly tired of my not buying gold, and I began to sense that he was trying to set me up to be robbed. (I used to wake up in the morning to see him perched by the side of the bed, staring at me with a pensive look I didn't like.) On one occasion, we were in a truck together with some former Mai Mai friends (former combatants in an indigenous militia discussed later in this book) of his whom I didn't know, headed out of Goma, and he asked to be dropped off long before we arrived at our destination, an insecure mining town governed by armed actors. It felt like a setup (all transactions in Congo were in cash, and for a long time, I had no bank account, and even when I had one, there was no way of transferring significant money from one place to another. As a result, I was always carrying large sums of cash, sometimes thousands of dollars, and Michael knew this). But there was another Congolese fellow in the truck with me, who I could tell was just along for the ride. I struck up a conversation with him, and we became fast friends over the course of an hour or so. When the men in the truck were preparing to leave me on my own, I asked him how he'd feel about staying with me for a day or so until I got on my feet. Our suspicions grew when the men in the truck protested, and he insisted on staying (later he confided in me that he stayed because he was concerned for my well-being in this new town). The few days he was to spend with me turned into weeks, then months and years.

The young man in the truck was Raymond Mwafrika (Raymond African, his actual name), and he became my friend and assistant; over the years, we handled many difficult situations, dealt with demanding state figures and "customary authorities," took cargo planes, and drove many hundreds, even thousands, of miles across Congolese roads in a beater jeep. Much more than an assistant, Raymond was a colleague, because he had a lot of experience with mining, having been involved in local politics in his hometown of Luhwindja, a major artisanal gold-mining town where the Canadian mining company Banro was beginning to extract gold industrially and coming into conflict with the diggers. With a bachelor's degree in political science from the University of Bukavu, Raymond also had a good deal of experience working for NGOs and academics. Raymond's influence, ideas, and interests are strongly reflected in this text. Over time, I helped him grow some more academic connections, and he has assisted other international academics conducting research in Congo.

I didn't realize it at the time, but through these different moments, I was learning a lot about what I refer to, broadly, as the eastern Congolese practice of "collaborative friendship" that had become so important for people, partly because of the danger and destruction brought about by the war (although this practice was not brought about solely by the war). This practice consisted in forging mutually beneficial friendships with people who were formerly strangers, usually on the fly in moments of crisis, and building enduring networks out of these happenstance alliances. While it seems simple enough, it is a crucial tactic and also a kind of art.

When Raymond and I visited the Institut Supérieur de Développement Rural (ISDR) in Kindu, Maniema, the director introduced us to one of their instructors, Joseph Nyembo, who was also studying the relationship between mining and conflict (he had earned a master's degree from the University of Brussels and was pursuing his PhD at the University of Kinshasa). Joseph hailed from an old colonial-era company mining town in the rainforest province of Maniema, and his research was there; among other things, he was convinced that foreign researchers had neglected Maniema, the major source of Congolese coltan and other black minerals, because they were focused on the conflict-ridden Kivus on the borders of Rwanda and Burundi. This, he held, had a major influence on how they had thought of minerals and mining as being sources of conflict and destruction rather than peace and what he referred to as "development" (a commonly deployed African term that does not mean the same thing as Western economists imagine it to mean; see Smith 2008). Like Raymond, Joseph had lived through traumatic experiences during the wars, and he wanted to know if socially and ecologically sustainable mining was possible and whether this activity might become a source of postwar peace and prosperity for people in his community. Joseph invited Raymond and me to his hometown and research site, and the three of us spent several months in various mining towns of Maniema province—especially the "company towns" of Punia, Kalima, and Kailo—together over three separate summers (we also visited Namoya and Kasongo). Joseph, Raymond, and I became close friends, and in whatever mining town we visited, our house became a perpetual seminar on artisanal mining and its impact, which drew in all kinds of people from all walks of life—men and women, young and old. However, we were the students, and our guests were the teachers. In the Kivus, Raymond and I spent several months, over multiple visits, in Walikale, North Kivu (a district with many small mines and one world-famous mine, Bisie), and took week-long visits from Goma to Nyabibwe in South Kivu over the years. We also made several visits of about a week at a time to the mining towns of Numbi, Walungu, Shabunda, and Luhwindja, in South Kivu, and Rubaya, in North Kivu, over multiple summers between 2011 and 2018.

In 2013, Raymond, Joseph, and I organized a conference on artisanal mining in Luhwindja, South Kivu, with funding from UC Davis and the NSF (Raymond chaired the conference). In Luhwindja, the Canadian company Banro was in the midst of a conflict with artisanal miners whom they were trying to expunge. Some diggers and representatives of Banro attended the conference, along with Congolese academics and a couple of academics from abroad. We invited researchers working on artisanal mining from the Catholic University of Bukavu, the official University of Bukavu, and NGOs to present papers. This was a high-octane interdisciplinary learning experience for me, as I got to learn a lot about different academic and nonacademic perspectives on mining in the setting of a contentious conflict between industrial and artisanal miners over "development," the meanings and potencies of earth, and who had rights to the fruits of extraction. Some of the ideas that were generated during this conference also inform this book.

Slowly, as I grew to know more about eastern Congo and all that east-ern Congolese had gone through, and as I came to more fully understand the nature and scope of the work that those involved in artisanal mining did and all of the powerful forces that were pitted against them, this project took on a different kind of urgency and weight for me. I came to see those in the trade as engaged in a recuperative project, as grappling—physically and conceptually—with some of the most profound issues and changes of our time and as pioneering new ways of engaging with and thinking through global capitalism while moving across ontological orders or di-mensions (something discussed fully in chapter 3).

Probably the most intractable evidence of anthropology's colonial legacy today is the assumption that anthropologists define their field sites, es-tablishing what they're going to do and who they're going to do it with, as if they were completely in control of the social situation they just walked into ("Why didn't you talk to more X, Y, or Z?" people ask). Non-anthropologists, or people who have not yet done research, are especially guilty of assuming that we have this kind of control, but even the most humble among us find ourselves having to play into this mythology, writ-ing our grant proposals as though the world will just open up to us rather than our bending to it as we make do with the limitations that are imposed on us by others in "the field." If no anthropologists really define their field sites or what their research is going to consist of, it was especially the case with this research. In particular, I would have liked to have spent more time working and digging in holes, but that kind of work was just not possible because my very presence in a hole would have meant that I was prospect-ing and so "in need" of prospecting papers from the agent representing the Office of Mines. This would have quickly halted research, as funds would have evaporated as soon as I got started. So, for the most part, I stayed in towns that were proximal to mines, occasionally venturing out to min-ing settlements that were on-site but mostly avoiding the holes themselves. Also, although I did talk to many women, the majority of people involved in digging and trading are men, and so there is definitely a preponderance of male, and masculine, voices in this work, which was certainly not inten-tional and which I hope is not overwhelming.

The main impediment to carrying out the research were the barriers to movement that all Congolese people face from various state officials; foreigners, especially those immediately identifiable as such, have an es-pecially hard time moving through Congo when they are not part of an organization that has been granted unrestricted movement by the government (such as an international NGO). This has in part to do with the long history of state and state-like formation in Congo, focused as it has been on taxing the movement of people and things rather than controlling territory (Schouten 2019). Still, my approach was multisited, moving between different sites and comparing them with each other, but spending enough time in each one to develop a good sense of the relationships and histories that made them unique (each mine is its own world, some people say). I soon realized that, without meaning to, I was mimicking the practice of many (not all) artisanal miners, who move from site to site following news about which places have the most movement and are therefore the most full of life and the most interesting as well as the most likely to allow one to earn. Very commonly, I would run into the same people in these far-flung places, and I eventually realized that I was tapping into a mobile community made up of various kinds of workers.

Otherwise, this research proceeded much like other ethnographic research; after showing up in a town, usually accompanied by Raymond or Joseph or both, I would introduce myself to state authorities and settle in. I would explain my research to people, and over time, people would come to where I was staying and talk to me, or I'd go to where they lived and talk to them. In the beginning, I wanted to know everything about people's lives and experiences, their thoughts about local and regional histories and the world, especially but not only as they pertained to mining. I started off broadly, listening to people openly and deeply, and eventually, when I understood more, I started to define and narrow my questions, which became more pointed and focused. In general, I found that people were very eager to talk to me about their lives and experiences and to share their speculations about how the world was put together. They knew that their work—the work of digging and moving the black minerals—was somehow important for the world as a whole, though they rarely knew how much or in what ways. But they also felt that their circumstances, troubles, and perspectives—as well as their positive contribution to their places and the ways in which they innovated new techniques and social arrangements—were not appreciated by others, particularly those in positions of authority and outsider NGOs who saw them as vortexes of violence and chaos. Generally, they seemed happy and excited to have people listen to them about issues related to mining, people who didn't see them as pariahs.

The overwhelming sense from the people I spoke with was that they wanted people in positions of power to know the "truth" about their places and their work—this despite the "ethics of invisibility" (discussed in this book) that emerged around the mining trade, mostly in the context of debt accrual and avoidance. Nonetheless, unless otherwise indicated, names have been changed in this book despite the fact that very few of the people here were illegal actors (artisanal mining is legal in Congo, and there are also set-aside areas called "zones of artisanal extraction," or ZEA). Where work names or pseudonyms were being used anyway, these have been maintained. Some people preferred their names be used, and they have been. The names of companies have been retained, as have most place names. In some cases, where the people involved were publicly known or had already been written about by others—such as some of the main actors involved in the drama of Bisie—names have been retained (the section on Bisie concerning the Bagandula has been read to many of the key members of the family named in the chapter, who then translated and related the content to other members of the family via cell phone; they have agreed with that representation of events).

Seeing the World at Major Bravo's House:
An Introduction to the "Eyes of the World"

It is 2009 in Minova, a market town on the shores of Lake Kivu on the border between the provinces of North Kivu and South Kivu, only about an hour's drive from the North Kivu capital city of Goma. This is also the first town one hits when coming down from the hilltop village of Numbi, a major coltan mining town a hairy two-hour motorbike ride from here. I am with my Kenyan friend and then assistant Ngeti Mwadime, and we are beginning the process of trying to understand the total significance of coltan mining for the eastern Congolese artisanal miners who dig the substance for the tech companies that make digital devices such as iPhones and laptops. (Typically, diggers sell to a range of middlepersons who in turn sell to buying houses, which sell to smelters outside Congo; some basic processing is usually done at or near the mining site.)

Though geographically remote, the history of the mine at Numbi is, like all coltan mines in Congo, closely connected to the global demand for minerals used in cell phones and other digital technologies: during the 1990s (before the wars), in pace with the growing demand for this mineral for electronics devices, Rwandan buyers (or buyers locally identified as Rwandans but who may have also been Congolese) started coming to the mine at Numbi to buy coltan, and soon diggers implemented "improvements" to prevent landslides, which had formerly prevented the artisanal construction of deep pits. During and for some time after the wars, the mine was largely controlled by a Rwandan-backed proxy army battalion (the Rally for Congolese Democracy, and later the National Congress for the Defense of the People, of which see below), whose command over the site continued after they were formally integrated into the Congolese army, the Armed Forces of the Democratic Republic of Congo (FARDC). For some years after the formal cessation of armed conflict, all owners of holes at Numbi or of land where others rented out holes paid 10 percent of their yield to the battalion in addition to what they paid to the "other governments" (the so-called government of paper; see below), all of which was cast as tax (*kodi*) or payment for services provided. Hole owners were also expected to give the army fixed periods of time during which the battalion would bring in their own workers (often POWs) to dig for them or force the workers who were already there to work for them (a practice referred to in Congo as the *salongo*, which usually happened when prices were high). In 2011, the Congolese army tried to crack down on the battalion in Numbi, culminating in a shoot-out in the town of Minova over coltan that they were illegally transporting to Rwanda. For a couple of years after that, the battalion used the vehicles of humanitarian NGOs to smuggle coltan into Rwanda.

As of 2020, the war is technically (though not completely) over, the army has been removed from the site, and Numbi is a government-registered, "green," and "conflict-free" site where production and sale is overseen by an NGO working in combination with the International Tin Research Institute (ITRI), a consortium of smelters that oversees a "bag-and-tag" auditing scheme designed to ensure that Numbi's "conflict-free" minerals are sold to specific, registered comptoirs (buying houses) who will in turn sell to the smelters. Eventually, the minerals will go into the electronics manufactured by tech companies seeking to be in compliance with conflict minerals legislation, including Section 1502 of the US Dodd-Frank Act. As of this writing, those minerals that are not from registered conflict-free sites are technically bloody, even if there is no conflict there, so diggers and traders from other technically bloody sites nearby now launder their minerals through Numbi so that they will be clean according to the auditing scheme, which has been superimposed onto Congolese law.

To provide a foundation for the stories and material that emerge in this book, I will step back for a moment and briefly describe some pertinent histories related to the DR Congo (a country roughly the size of Western Europe, with just under ninety million people and about 250 different languages), finally to return to the ethnographic moment in Numbi in 2009. The goal is to quickly get the reader up to speed and prepped for what is to come, but it should go without saying that what follows is, of necessity, partial and incomplete. Today's Democratic Republic of Congo was first created ex nihilo, in 1885, to comprise King Leopold's privately owned "Congo Free State," which, once conquered and expropriated, became a colonized territory predicated on violent resource extraction (the European powers that mapped the boundaries did so in almost complete ignorance of what and who was there previously). The territory, which contains one of the largest forests in the world, turned out to also be one of the most resource-rich places in the world; to this day, new raw materials spring forth from Congo, seemingly like magic, to feed each new "stage" of capitalist growth and empire building in the Global North (from the ivory in piano keys to the rubber in tires to the uranium in the bombs dropped on Hiroshima and Nagasaki to the copper and tin in electronic goods to the coltan in mobile phones to the cobalt in "green" lithium ion batteries).

The Congo Free State (CFS) is well known for the way in which extreme violence and slavery were deployed as instruments of rule and of capitalist accumulation at the same time as the state advertised itself as a private humanitarian organization created for the specific purpose of fighting slavery (mainly, the Swahili slave trade) (Hochschild 1998; Nzongola-Ntalaja 2002; Van Reybrouck 2015). Colonial officials were monetarily incentivized to extract as much wealth (at first ivory and later rubber) as possible in as short a time as possible, and they employed the Force Publique, or army, to enforce the rubber quotas, with murder and rape being among the atrocities committed for failure to meet quotas (Force Publique soldiers were required to bring the severed hands of their murdered victims in baskets to colonial officials as proof of death; Hochschild 1998). Thus it is that the CFS has come to epitomize the centrality of resource-based "primitive accumulation," or accumulation by dispossession (meaning wealth accumulation based on violent force, theft, and slavery, typically in a way that drives industrialization elsewhere), to modern capitalism, as well as the Janus-faced hypocrisy of colonialism (Nzongola-Ntalaja 2002; Van Reybrouck 2015). King Leopold, the oft-titled "absentee landlord" who never stepped foot on his property, has come to personify the historical past of this place in the writings of many authors (Nzongola-Ntalaja 2002; for a poignant critique of this approach to Congolese history, see Hunt 2016). The CFS is also famous for giving rise to the first international human rights movement, which arose in opposition to it; indeed, humanitarianism as an ideology and a mode of governance has been present in some form from the beginning of Congo's history until now, accompanying and overlapping with violent extraction in ways that are complicated and which this book also illuminates (Hochschild 1998; for an especially nuanced reading, see Hunt 2016).

The subsequent Belgian Congo, emerging after 1908, concealed much of this cruelty under the mantle of "development" (see, e.g., Marchal 2003) and built up an enclaved industrial mining sector (especially copper, tin, and gold) that exploited and excluded Congolese people while developing infrastructure with a view to resource extraction. There is, of course, much more to be said about the Belgian Congo, and one of the most sophisticated approaches is found in the work of the historian Nancy Rose Hunt (Hunt 1999, 2016). In particular, Hunt shows how the colonial state governed not only through violent force but also through such biopolitical practices as health control, sanitation, and security—all aimed at problematizing (in the sense of making a problem of) and curtailing African movement. She places a great deal of emphasis on vision and seeing as instruments of colonial power while showing that the colonial state was a "nervous state" anxious and uncertain about its ability to rule (Hunt 2016). Hunt is especially attuned to the imaginative critiques, movements, and models that emerged in conversation with this nervous state, focusing on ordinary people and everyday practices in contrast to the focus on high-level actors and institutions that characterizes most work on Congo; in concentrating on the everyday lives and worlds of artisanal miners and traders and on the importance of visualization and (im)mobility, this book is strongly influenced by her approach.

Patrice Lumumba was elected prime minister of a newly independent Congo in 1961, with Joseph Kasavubu as president, but the former was assassinated under orders from the CIA when he asked the Soviets for help in putting down a Belgian-sponsored secession in the copper-rich Katanga province of southeastern Congo (Nzongola-Ntalaja 2002; Devlin 2008). General Mobutu Sese Seko seized power a few years later and governed for more than thirty years, renaming the country Zaire in a gesture of Africanization (though the name was actually derived from Portuguese) while stylizing himself as the paternalistic "father" of a nationwide corporate family based on what he promoted as "traditional" African values (for an excellent analysis, see Schatzberg 1988). While Mobutu initiated nation-state building projects (regardless of whether one deems that a good or bad thing), major declines in the global prices for minerals (mainly of copper and tin after 1974) in an economy that was tied to their export led to a long-term, spiraling collapse of state and economy; this was exacerbated by the interventions of the World Bank and the International Monetary Fund (IMF) (including the imposition of interest-bearing loans and extreme austerity measures beginning in the 1970s and lasting throughout the 1980s; see Young and Turner 1985; Hesselbein 2007). This culminated in the radical liberalization of the economy under the guidance of the IMF (the so-called Mobutu Plan; Hesselbein 2007). Most Congolese attribute this process of liberalization to Mobutu himself—referring to his "policy" of *debrouillez vous*, or fend for yourselves, which included the legalization and liberalization of artisanal mining after 1982.

Some analysts have referred to what happened in Congo during this period as the "jettisoning of the bureaucratic state" in favor of patron-client relationships and the development of reciprocal relations between state leaders and local and regional power brokers (what William Reno refers to as "warlord politics," in which power was centered and distributed around mineral enclaves as sources of sovereignty and mutual enrichment for high-level state and nonstate actors; Reno 1999; Ferguson 2006). Others have argued that long-term practices of horizontal governance were simply reasserting themselves after a short period of surface-level state-centered bureaucratization (Bayart 1999, 2009). In any event, formal employment largely disappeared in favor of a system in which state officials, teachers, and others imposed their services and papers (*ma karatasi*, or *ma document*) onto subjects in exchange for direct compensation and officials sometimes compelled civilians to work without pay (the latter being a practice carried over from colonial times, which Mobutu called *salongo*, referring to unpaid labor on community development projects). Mobutu continued to hold on for so long because he was supported by the United States during the Cold War (visiting the United States and meeting with different presidents over a thirty-year period) and used the Congolese army to fight against the communist regime in Angola; after the fall of the Soviet Union, this support base also evaporated, leaving Mobutu in crisis (Nzongola-Ntalaja 2002).

In eastern Congo, the gradual collapse of the state from the mid-1970s onward was compounded by ethnicized conflicts going back to the time of the Belgians, who had brought in Rwandans to work on plantations and in mines (Mamdani 2002; Autesserre 2010). When eastern Congolese land was liberalized in the early 1970s, many people of Rwandan descent bought land in North Kivu for herding cattle and making milk and cheese for sale. Then, in 1981, citizenship was officially revoked for all Congolese who could not trace their ancestry in Congo back to 1885, the inception of the Congo Free State (Mamdani 2002). This fostered a community of (often) land-rich but politically disenfranchised Kinyarwanda speakers, or speakers of the Rwandan language (whether putatively "Hutu" or "Tutsi") identified as a singular group of outsiders living within the borders of Zaire. In South Kivu, the denial of citizenship fed ongoing conflicts between agriculturalists and Congolese Banyamulenge, pastoralists who were understood to have come to the eastern Congo from Rwanda in the nineteenth century; later, in 1995, all Congolese of Rwandan and Burundian descent were implicitly categorized as refugees when parliament demanded they return to the countries from which they were said to have originated (Mamdani 2002; Autesserre 2010; Lemarchand 2009).

The primary precipitator of the First and Second Congo Wars was the overflow of the Rwandan genocide into Congo in 1994; the genocide was a three-month period in which Hutu Power groups associated with the ruling regime led a genocide against Tutsi following the murder of the Rwandan and Burundian presidents under mysterious circumstances (their plane was shot down).[4] The Hutu genocidaire (called Interahamwe, meaning those who work/fight together) fled Rwanda along with other Hutu, fearing reprisal from the Tutsi-dominated Rwandan Patriotic Front, led by the US-trained Paul Kagame, which recaptured the country following the genocide (their members had been training in the Anglophone postcolonies—former British colonies—of Uganda and Tanzania since an earlier genocide in the 1970s). The refugee camps in Goma, North Kivu, became recruitment training grounds for Hutu genocidaire and for child soldiers—ironically, unintentionally financed by the UN system and humanitarian NGOs from the north (the Forces for the Democratic Liberation of Rwanda, or FDLR, sold aid and food to buy arms and recruit displaced children as soldiers; Mamdani 2002; Prunier 2011). The Hutu in the camps in turn recruited Congolese Hutu to their side and committed acts of violence on Congolese Tutsi or historically marginalized Congolese people who were identified as Tutsi by other Congolese; these Hutu forces were supported by then President Mobutu, who in turn received aid from France (Emizet 2000; Mamdani 2002). The Rwandan Patriotic Front invaded then Zaire, launching a genocide against Hutu of Rwandan and Congolese origin living in Congo; the invasion quickly grew into an international assault and national uprising against the Mobutu regime referred to as the AFDL (Alliance of Democratic Forces for the Liberation of Congo).

Mobutu, the self-styled personification of the paternalistic Zairian state, was quickly ousted by the AFDL, a coalition of Congolese dissidents and foreign nation-states (mainly Uganda, Rwanda, and Burundi but also Eritrea and Angola) led by Laurent-Désiré Kabila, a former Lumumba supporter who had taken part in the Pro-Lumumba Simba Rebellion against Mobutu during the 1960s. The First Congo War of 1996–1997 was genocidally violent despite the fact that it lasted only a few months, with the AFDL marching easily to Kinshasa. After the war, Laurent Kabila tried to establish sovereignty over Congo's territory on multiple fronts: he revoked the existing contracts of the foreign mining company Banro in favor of a more nationalist mining policy (more on this below; but also see Geenen 2013) and insisted that the foreign African states that were still occupying the east (mainly Uganda and Rwanda) leave the country. These countries and their proxy militias resisted, citing the continued presence of the Hutu-dominated FDLR in the east. To be sure, the threat was real enough: the FDLR eventually formed a large counterstate, or state within the nation-state of Congo, and posed the risk of returning to Rwanda to take back the government (for a rich and nuanced ethnography of this group in Congo, see Hedlund 2019).

The Second Congo War, or Great War (formally 1998–2003), was initiated by the Rwandan-backed Rally for Congolese Democracy (RCD) insurrection in conjunction with the Ugandan-backed Movement for the Liberation of Congo (MLC). The RCD was a militia-cum-government that mobilized disenfranchised Congolese groups such as the pastoralist Banyamulenge in South Kivu and North Kivu Tutsi, promising them land, salaries, and political and military positions in the new RCD government (the RCD government ultimately split into RCD-Goma in North Kivu and RCD-Bukavu in South Kivu). A number of African countries came to President Kabila's aid against the Ugandan- and Rwandan-backed governments, and the "Great War" became "Africa's first World War" (the First Congo War is also sometimes called Africa's First World War), involving at least nine African countries, which in turn received funding from foreign countries and corporations (Clark 2002; Prunier 2011; Bowers 2006). In addition, various indigenous Mai Mai groups (meaning "water water," a phrase referring to their ability to turn bullets and other deadly foreign technology into life-giving water) formed to resist the Rwandan- and Ugandan-backed invasion, though they also punished civilians for "hosting" the enemy invaders and used them to extract minerals for their war effort (I have more nuanced things to say about Mai Mai and the war in general in chapter 2 of this book).

The wars, in which over five million people died, were also coterminous with the post-1980s "digital revolution" and the global demand for minerals used in digital devices, almost all of which are found in great supply in the eastern Congo; especially after 2004, the demand for tin from a rapidly industrializing China also fueled the mineral boom. During the middle of the Great War, in late 2000, online speculation and international demand fueled a tenfold, or 1,000 percent, increase in the price of coltan ore on the global market that is vividly remembered in eastern Congo to this day (in the province of Maniema it is called *bisikatike*, or "may it never end," while in the Kivus it is called the *fois deux*, or "the doubling"). The coltan price hike was also one of the most violent periods of the war, characterized by forced labor "at the barrel of a gun," to quote one NGO report (Luca et al. 2012; Mantz 2008). As many scholars have pointed out, the global demand for minerals didn't *cause* the war, but it did finance combatants and incentivize neighboring countries and international corporate players to get involved in the trade and remain in the war (Jackson 2002, 2003). One popular eastern Congolese understanding has it that minerals actually prevented the enemy (*adui*) from winning the war, because in the east they got bogged down in mining and in competing with each other to control mining and so never made it to Kinshasa.

Though these wars were not caused by the demand for minerals and were to a great extent driven by localized conflicts over access to land rooted in colonial appropriations and resettlements (Autesserre 2010), they also cannot be fully understood without taking into account global events and processes: For one, the neighboring countries involved in the war were also responding to the budgetary crisis brought about by the post-1970s structural adjustment programs that followed decades of high-interest loans from the Global North, which bankrupted African treasuries (see, e.g., Mbembe's discussion [2001] of "private indirect government"). And, in many instances, militias were financed by foreign companies that bought minerals or tried to secure mineral deals. According to a 2002 UN report, eighty-five firms from twenty countries—including Anglo American, Barclays, Bayer, Cabot, HC Starck, and Standard Chartered—collaborated with occupying armies or illegal militias during the war (UN Security Council 2002). The UN recommended economic sanctions on twenty-nine firms from Britain, the United States, Germany, Malaysia, Hong Kong, and Belgium, but these were never implemented.

While armies financed their efforts from the boom, ordinary people also broke into a business that had long been monopolized by companies and that few people knew much about or understood until that time (this is especially true of the so-called black minerals, such as coltan, which, for the most part, were not liberalized in practice until the wars). The war brought buyers, and thousands of people who had been dispossessed of their lives and things (land, livestock, wares) were able to enter into the relatively open field of artisanal mining, sometimes recovering some or all of what they had lost, which they struggled to convert into fixed and enduring forms of value (land, houses, shops, marriages and children, schooling, etc.) in a time of war and tentatively postwar precarity. Agricultural land was also unsafe during the war, and many entered into mining or conducted business at mines to benefit from the relative security at these sites, which were controlled by armed groups. The debts that miners and traders accrued during the coltan boom and after the dramatic price drop that quickly followed the boom left them "addicted to mining," as some people in the eastern Congo put it to me (it is interesting to think about this in parallel with the "technology addiction" that many Americans see as central to their lives). Because of personal debts, their dispossession from land, the dollarization of the economy following the 1994 refugee crisis and the coltan boom, hyperinflation, and the destruction of infrastructure (making transportation of food and produce difficult and expensive), many people found themselves unable to return to other pursuits such as agriculture or foraging/hunting.

After the Sun City Agreement in South Africa and the official (but not actual) end of the war in 2003, a faction of the Rwandan-backed RCD led by Laurent Nkunda refused to integrate into the Congolese army (Nkunda apparently feared being prosecuted by the ICC). They formed the CNDP (National Congress for the Defense of the People), which took off after 2004, fighting against the Congolese army, indigenous Mai Mai, and FDLR, until most of the CNDP's troops finally underwent *mixage*, or integration, entering into the FARDC, the Congolese army, in 2009. The CNDP was later followed by another smaller breakaway movement of former RCD and CNDP called M23 (a reference to the date of the Sun City Agreement). Throughout these conflicts, the RCD, CNDP, and M23 all claimed that their actions and attacks were due to the continued presence of Hutu FDLR in the eastern Congo, but these groups were also expropriating Congolese wealth (mainly minerals but also timber and other resources) and exporting it through Rwanda. To this day, there are approximately 150 armed groups operating in eastern Congo.

Back in 2009, Ngeti and I are in Minova—it is the first place you come to when you leave Goma and before going to Numbi—and we soon realize that sitting down with local state authorities and explaining this research to them is unavoidable. So we are with these state authorities in the home and headquarters of Major Bravo (not his actual name, though this is the name he went by), who is the de facto—not de jure—state head in this town. Major Bravo's wife, who has just brought us wheels of cheese from the cows of Masisi, is explaining to us that she and the kids want to leave the bullet-pocked house atop the hill because she's convinced it's haunted by the ghosts of the Belgians who used to live there during the colonial period. She and the major are, in short, not happy in this place that evokes memories of a difficult and violent past, and they want to return to their home in Masisi, North Kivu. Though he is Congolese, Major Bravo was, just a couple of years ago, an "enemy" combatant whose Rwandan-backed RCD and CNDP battalions invaded Minova on more than one occasion. In 2009, the Second Congo War has technically been over for six years, and Major Bravo has been integrated into the Congolese army along with other armed groups through mixage. This mixage was a tense and complicated process, in part because many of the new, lower-level soldiers were former Congolese combatants (including the indigenous militias known as Mai Mai), while the higher-level officers, like Bravo, were typically former Rwandan-backed RCD or CNDP. The latter were perceived as foreign invaders by the former, their subordinates, regardless of their integration into the Congolese army. To make matters worse, these commanders were often able to use their positions in the army to obtain remunerative opportunities from mining and other forms of extraction, while their lower-level soldiers, sometimes former Mai Mai, were not. These high-level soldiers also generally remained loyal to their former commanders, many of whom were not Congolese.

Major Bravo is the descendant of Rwandans who were brought to North Kivu by the Belgian colonial government in the 1930s. He still works alongside other former RCD and CNDP, who also now occupy leadership positions in the Congolese army, again because of the mixage. A couple of years after this, Major Bravo will abandon the FARDC to join his fellow former CNDP soldiers in the Rwandan proxy militia M23, which will try to recapture North Kivu. After helping M23 to take over the city of Goma for a time, Major Bravo will disappear, allegedly killed in battle at the hands of the FARDC.

For a while in 2009, Bravo and I will come to know each other a bit. Because I'm from the United States, Major Bravo sees me as a potential friend, a natural supporter of Rwanda because of US support of Rwanda after the genocide and, by extension, its indirect (most eastern Congolese insist very direct) role in the Rwandan invasion of eastern Congo during the First and Second Congo Wars. His end goal is to someday resettle with his family in the United States, and I get the strong sense that he wants me to think he has been a force for good in the region. Bravo likes to cast himself as a bringer of order to chaos, a strong sovereign leader who rises above and transforms what he portrays as the competitive opportunism of civilian government officials (the "people of documents" or "people of papers") who otherwise vie for authority and the right to collect "tax." It is common for Congolese soldiers to talk disparagingly about these so-called people of lies (*watu wa uongo*), but sometimes Major Bravo personifies the differences between what is sometimes called the "government of paper" and the "government of guns" in such a way as to make it clear that he's not just talking about different state authorities, but what he perceives to be essential racialized differences. Tutsi, he likes to remind me, are organized and can "see ahead," or plan for the future. Hutu, like other Bantu, have no sense of the future or of planning because they have no cows and so can't "see ahead." Livestock, Bravo claims, generates wealth over generations, causing those who care for them to understand the concept of incremental progress. People without cattle, and especially miners and hunters, are not capable of this, the major believes, because they live in a constant present. This is Major Bravo's particular rendering of the colonial "Hamitic hypothesis," the racist ideological foundation of colonial governance in Belgian Rwanda and Burundi (Mamdani 2002).

Bravo has presented his life history to me in heroic terms: he grew up in the Congolese North Kivu district of Masisi, where he remembered Hutu and Tutsi living together as one people. His father was an ambassador to a European country under the Mobutu administration. But when the Hutu Interahamwe launched a genocide against the Tutsi in Rwanda in 1994, Bravo went to that country with other Tutsi patriots to protect his "brothers and sisters." A couple of years later, he returned to Congo with Laurent Kabila's AFDL during the First Congo War, believing that Mobutu had helped the French back the Hutu against the Rwandan-backed Tutsi in Operation Turquoise and was enabling the Hutu genocidaire in Congo (see, e.g., Prunier 2011). As Bravo sees it, Laurent Kabila later "turned his back" on the Rwandans who had put him in power when he ordered them to leave the country, so Bravo then took a leading role in the RCD insurrection, which initiated the Second Congo War in 1998. I will later learn that the people of Minova remember and judge Bravo for his actions during the war, when his subordinates committed many acts of violence and theft against the civilian population. But I did not discern that from their behavior toward him in 2009, which seemed genuinely aimed at making friends with Bravo. Some later explained to me that they did this out of a combination of fear and the hope that they might someday transform Bravo's approach to the community for the better; this is one of the first intimations I had of a bottom-up practice of peacebuilding aimed at incorporating and collaborating with potentially dangerous others.

While the civilian state authorities are supposed to operate independently of the army, in practice they are under Major Bravo's authority. Back in 2009, this was the rule once one left Goma and Bukavu—outside the city, the "government of paper" gave way to the "government of guns" (whether a particular location is in the space-time of the government of paper or the government of guns remains context-dependent to this day). At Bravo's house, we are joined by several other representatives of the government, all of them Hunde or Havu from the area. Three of them had been Mai Mai insurgents who fought against the Rwandan-backed RCD invasion during the war and so had technically been Bravo's enemies. Sharing the sitting room with us is a DGM (Direction General de Migration) representative, an ANR (Agence Nationale de Renseignements, or intelligence) representative, a *chef de territoire* (a state-recognized administrative chief), and another army major named Choma Choma (Burn Burn), who headed his own Mai Mai battalion until 2008 (also not his "real" name but the name he used). Choma Choma has been integrated into the Congolese army and given the rank of major, and now he reports to Bravo, his enemy just a year ago.

With time, I would learn that one of the main customs when a stranger comes to a town that has been "hit hard" by war is to regale the person with stories about the war in that place, demonstrating that the people there have a special kind of knowledge that other people don't share (as in, "Here in Shabunda, we really know what war is!"). This is probably why Sylvester, the ANR representative, stands up and begins to perform the RCD invasion of Minova for us. His eyes open wide as he simulates the experience of being shocked and awed, of seeing something for the first time: He dances around the room while reproducing the sounds of grenades, bazookas, and AK-47s. He points to the bullet and mortar holes in the walls around us—Bravo's headquarters. He stretches out his arms as if to show the unbelievable size of the tanks: "They had so much technology! The machines from America!" he intones. Major Bravo, who was in that invading battalion, smiles in bemused consent while observing the performance. "Where did they get those weapons?!" Sylvester asks the room, indicating that this was power far beyond Rwanda's capabilities. Then he casts his eyes at me in quiet reprimand as if to say that we both know the answer to his question. The machines, he knows, were the outcome of a friendship that Rwanda had forged with the United States at Congo's expense (more on this in chapter 2).

At some point during this performance, a Danish woman working with the Red Cross abruptly pops in, escorted by Bravo's immediate subordinate soldiers (they are also Tutsi), to interview Major Bravo as part of a surprise visit. She knows that here there is "still war," because there are Mai Mai and Hutu FDLR battalions nearby, attacking communities and mines in this area. The Dane wants to ascertain whether Bravo is illegally holding any POWs and, if so, how he is treating them. When the Dane sees me sitting with Bravo on the couch, she looks surprised, even alarmed—to me, she seems curious about the nature of the relationship, but we greet each other as if it's always normal for people to be wherever they are. Bravo is charming and inviting to the Dane, offering her cheese and coffee. The Dane spends some time showing us a Swahili graphic book entitled *How to Treat a Prisoner of War*; it uses illustrations to teach the rules of international law as they pertain to the treatment of prisoners. Bravo admits, to my surprise, that he indeed has a prisoner in the basement—a man he claims to be Mai Mai and who he assures the Dane is being held only because there is currently no one to take him to the court in Goma. He offers, nonchalantly, that he has no plans to interrogate the prisoner. When the Dane leaves, Bravo dismisses her work as "useless" and "ignorant" of the subtle and complex realities of Congo—it's fine that she has her business, he says, but she has to understand that he also has his. Although he doesn't say this explicitly, he seems to be suggesting that, after all, her business wouldn't exist without his.

With the Danish interloper gone and the performance of the battle ended, Choma Choma, the former Mai Mai major, begins the work he came here for—giving Bravo a report on the ongoing fighting between the Congolese army and the Hutu FDLR and Mai Mai in this area. The FDLR are working with a certain Mai Mai group, and together they have been attacking small, less-protected coltan mines near Numbi. "Animals of the forest travel together" (*Wanyama wa msituni hutembea pamoja*) is the common phrase these state actors use to explain this alliance, which also consists in a sharing of indigenous Mai Mai knowledge with those who may be Rwandan. Major Bravo listens and asks questions, but his contempt for his subordinate is obvious to me. Choma Choma is dirty and unkempt, speaks no French and only very broken Swahili, and has other habits that people in the towns associate with the forest. At one point, Choma Choma tells me he has a father in America and wonders if I might know him. Bravo finds this hilarious. He has a hard time keeping a straight face and begins to not-so-indirectly ridicule Choma Choma. "Mai Mai are not straight in the head," the major says. "They eat cat brains and smoke pot so they won't feel fear. But they have no knowledge [meaning secret or mystical knowledge]. They can't even tell you what the army is or who's in it." Looking away from Choma Choma, he adds in a whisper, "They don't even know who their own parents are." Major Bravo proceeds to tell us all in front of Choma Choma that, during the war, his battalion killed 1,500 Mai Mai in a single day. He says the Mai Mai just kept coming at the guns because they were high and hypnotized. "They have no magic," Bravo intones, staring at Choma Choma in a manner that seems to me to say, "I am doing you a favor, man. I am trying to teach you something important." Choma Choma smiles back at him in inscrutable silence.

This single moment in a single place more than ten years ago certainly doesn't represent this region, but it does capture something about the multiple scales and the competing, whirling forces that converged tensely in tentatively postwar Congo, especially around mines. (Minerals and mines are somewhere in the background thanks to the mine at Numbi and the neighboring mines subject to attack, but they are certainly not the only, or even the main, cause of what is happening). Enemies—perhaps reconciled, perhaps not so much—are making plans in a house that is also a base camp while someone who may or may not be a Mai Mai insurgent is hidden in the basement waiting for a visible legal bureaucracy he may or may not get to see. A representative from the international community emerges abruptly only to disappear just as quickly, as if to communicate that the "eyes of the world" (a phrase we will return to) are, if not always watching, omnipresent enough to keep everyone on their toes. No one is sure who is who and what each person is up to beneath the surface, but the Dane does make sure to leave behind a black-and-white storyboard describing what the near future is supposed to look like to a soldier who is following his own different, but nonetheless carefully laid, plans. The "friends" in the living room perform a reenactment of the past that shows the international community in a wholly different light, suggesting the existence of violent global networks, which, though mostly speculated about, are still more real than the comic-book representations of "universal law." The ghosts of dead Belgians watch this performance from the shadows. Technology that Congolese feel has long been concealed from them ("The Belgians kept us from having technology") or which their own work helped to make ("Coltan is in the bullets the Rwandans kill us with"—many diggers say and believe) is depicted as the source of death and the destruction of ordinary time.

While they remember the death machines introduced by these global networks, the men also allude to a different kind of technology, one synonymous with the forest and the leaves—this is the "magic" of Mai Mai (something discussed at length in chapter 2). While Bravo scorns this alternate power, seeing it as unreal and holding belief in it to be delusional, his angst betrays how powerful many eastern Congolese hold the "power of the leaves" (*nguvu ya majani*) to be. The power of the leaves is, like Rwanda's technological superiority, the outcome of a friendship, but one of a different kind—with ancestors who live in an alternate dimension and belong, in some sense, to a different time (they exist in a parallel present that affects the visible present, but their habits and ethics belong more or less to the past). Eastern Congolese, especially in the forest, say that the entire forest waged war on the invaders and that ancestors appeared to insurgents in dreams, instructing them in ways to render superior foreign technology impotent or to make themselves invisible to the occupying powers whose visualizing techniques and capacity for spatiotemporal expansion across geographic space were far superior.

There are, in short, a range of colliding forces at work in the situation, and different scales and dimensions come together at a moment's notice; moreover, there is a great deal of uncertainty, opacity, and flux in general. As we will see, the mining and trading of minerals participates in and magnifies that flux and uncertainty at the same time as it is the technique some people have employed for steadily rebuilding their lives, and incremental time, following the violent and chaotic time of war.

The meeting with Major Bravo also suggests that a lot of effort goes into trying to make certain things visible while also concealing certain truths and powers from others; the things that are most immediately visible are not necessarily what's actually true and certainly not what's most important. Related to this is the collective understanding that there is a global visualizing regime that is unified and nearly total that is directed against Congolese people and Congolese territory by powerful outsiders who want something from Congo and who are able to mobilize the power and the networks needed to get it. These outsiders may be war makers or humanitarians—or they may be both at the same time—and the difference between them, if there is any, often seems superficial or irrelevant with respect to the larger thing that is happening (mainly, surveillance and control of Congolese people in Congolese territory with a view to acquiring Congolese resources). While there are people who come to Congo with a plan of making certain things visible to the world and to Congolese (exposing violence or infractions of law and making certain "universal" laws visible), these actors (like the Dane) often come across as concealing their true motivation. Even if they are not, there are other agents in the background (maybe they are actually the same agents all along) who are more important than those who are immediately visible (they can be inferred through the visible tanks). Many eastern Congolese understand them to be "seeing ahead" (*kuona mbele*), gradually exercising a far-reaching plan to control Congolese territory, encompassing space-time and the essence of the earth itself while immobilizing Congolese and closing them out of the earth and forest.

This brings us to the book's title, *The Eyes of the World* (*macho ya dunia*), a phrase that I heard more than a few times in eastern Congo, especially around large artisanal mines (e.g., "The eyes of the world are on Bisie mine"). It referred to the fact that the world was watching Congo, particularly its resources, and that important things that were happening there could only be understood in light of this fact (thus, for example, rape as an instrument of war could only be understood in the context of a humanitarian industry that was already watching what was happening). The phrase "eyes of the world" was also sometimes a reference to the idea that people and things were being watched by technology (e.g., satellites, cameras that see underground, and digitized tags for "conflict-free" minerals) that Congolese were subject to. There was usually an implicit contrast to the people I was working with, in that diggers were often said to be unable to see or unable to see far (diggers said this of themselves, especially with respect to mineral price, and other people mockingly said it of them, blaming it on their temperaments). One other implication of the phrase "eyes of the world" was that mining and minerals had the potential to invert the assumed hierarchy of the world order and to bring the world to Congo, such that Congo suddenly became a center that set the pace for the rest of the world, rather than an alleged "dark corner" that lagged behind or defied civilization or modernity.

Seeing, which is what the eyes of the world did, was complicated and ambivalent because, on the one hand, seeing made "movement" (a value discussed in detail in the next chapter) possible: people understood that the only reason there was a trade in Congolese minerals was because, long ago, outsiders had seen them in Congo and knew that they were there.[5] At the same time, seeing and transparent visibility were extensions of the colonial gaze and a form of predation that threatened to curtail Congolese movement. This was brought home, for example, when the company Mining and Processing Congo (MPC) came to the mine at Bisie with maps and papers and when they used aerial GPS in an effort to make an argument for their sovereignty over the mine at the expense of artisanal miners (see part 3). In recent years, the themes of movement and visibility have become quite overt in eastern Congolese mining, as state and international actors try to use instruments and practices of transparent visibility, such as bar-coded tags, to regulate the movement of Congolese involved in the mining and trading of black minerals. In comparison to gold and diamonds, which are also prevalent in this region, these substances are especially ame-nable to being made visible in compliance with new laws operating on an international scale (for example, Section 1502 of the Dodd-Frank Act).

Regulating Mining in the Postwar Era: Conflict Minerals Narratives, Dodd-Frank, and the "Handcuffs" of Traceability (a Brief Introduction)

The context of the fieldwork on which this research is based changed dra-matically over its course, as eastern Congo went from being almost entirely a war zone to being a tentatively postwar context where artisanal miners, displaced by years of war, fought to retain their newly found livelihoods in the face of returning companies, some of which locals blamed for the wars in the first place (although those I spoke to were more likely to point out long-term continuities, hence the focus on the "eyes of the world" in this book). When I first visited Congo in 2003, the city of Goma was a field of smoking ash that had recently been incinerated by the neighboring volcano—this in the middle of the Great War. Over the years, as the Congo-lese army reasserted control over Congo's territory, the once devastated city sprung from the ashes, rebuilt largely by artisanally mined mineral wealth. (While many pundits like to draw attention to the role of international humanitarian NGOs in rebuilding Goma, those on the ground know that the incomes from that work go to a very small group of people, whereas the wealth of artisanal mining, opaque and difficult to calculate as it is, is "touched" [kuguswa] by everyone and has fed many other businesses.) Goma became a haven for a host of international NGOs, which flourished under the new conditions of security in the city, dramatically hiking real es-tate values and becoming a source of well-paid employment for a minority of educated Congolese. These foreign NGOs were always quick to seize any opportunity they could to claim responsibility for the peace from which they benefited.

Over the course of this fieldwork, many rural areas went from being almost entirely under de facto military control to, gradually, coming back under the authority of the government of paper, or bureaucratic administration. In the last few years, it became very rare to see the Congolese military near mining sites, except for remote forest areas where there were gold mines, as the state enforced the Congolese Mining Code's restrictions on the presence of soldiers at mines (which is not to say that they were not involved in the mining trade at all, as higher-level soldiers were often in the background, invisible to most). There was a larger, global set of forces feeding into this: during the same period, a group of international NGOs joined journalists and documentary filmmakers to argue that there was Congolese "blood" in computer and electronic devices.[6] Central to the discourse was the notion that the demand for minerals was actually causing the wars and that illegalizing these "conflict minerals" would therefore put an end to the war (Autesserre 2010). In 2010, these NGOs convinced some American senators to write Section 1502 of the Dodd-Frank Act, which compels US companies that think they might source minerals from Congo to exercise "due diligence" to ensure that minerals come from "conflict-free" sites and to submit documentation indicating that they have done so to the SEC. Soon after, President Joseph Kabila temporarily banned artisanal mining in the east and prohibited military and other state figures from direct involvement in mining (for an excellent overview, see Cuvelier et al. 2014). But the real reasons why artisanal mining was shut down turned out to be more complicated than this story of humanitarian motives suggests, as Kabila was also responding to the demands of foreign mining companies who were facing opposition from artisanal miners at Bisie and elsewhere in the east (see part 3, on the artisanal mine at Bisie).

An array of researchers have documented the devastating effects of the embargo on regions dependent on mining and its contribution to the proliferation of armed groups; their findings are corroborated by the material presented in this book.[7] It is now generally accepted that the embargo of Congolese conflict minerals actually exacerbated violence and war in the east by pushing miners into gold mining and militias (Stoop 2018; Diemel and Hilhorst 2018). Certainly, what came to be known in Congo as "Obama's Law" perplexed and angered most of the people with whom I worked. Everywhere I went, diggers and traders posed similar questions: Why is the "international community" suddenly so worried about purchasing "blood minerals" now that the war is over, but during the war the whites were lined up to buy them on the banks of Lake Kivu? Why is it OK for the "international community" to buy a company's minerals but not ours from the same place? Is it the minerals that are bloody, or dirty (*chafu*), or are certain kinds of people bloody, or dirty? And didn't the "international community" realize that those at the bottom were hurt the most by this and that driving people out of their only consistent way of making a living would lead to more violence as young men joined militias or simply robbed people in the forest (they consistently argued that peace came from "many hands touching money," a phrase I return to in the chapters that follow)? In particular, diggers and traders were shocked and angered by the *timing* of closure—the abruptness of it and the fact that it happened so many years after the Great War, at the same time as companies were coming in and claiming land that had, in many cases, been discovered and developed by artisanal miners. All of it smacked of conspiracy, a larger plan at work among high-level global actors, whether companies or entire nation-states (people mainly brought up the United States and China).

Faced with pressure from NGOs and civil society to reopen Congolese mining and buy Congolese minerals, high-tech companies such as Apple, Intel, and HP, operating in the context of Dodd-Frank, needed to ensure that the smelters from which they bought processed ore were doing so from blood-free sites through a reliable conflict-free or blood-free system. After 2013, a consortium of Congolese state and international nonstate actors engineered a partial reopening of artisanal mining in certain validated, or "green," artisanal mines containing what these organizations refer to as the "3Ts" (tantalum, tin, and tungsten—or what eastern Congolese call the black minerals). Most notably, the International Tin Research Institute (ITRI; a consortium of smelters) sponsored the ITRI tin supply chain initiative (ITSCI) with the help of a US NGO named Pact Global that worked alongside the Congolese government (primarily the Ministry of Mines).[8] They developed an auditing intervention based on bagging and tagging minerals at certified conflict-free green sites, using visible digitized tags to monitor and regulate who could buy and sell, to whom, and where. The implementers of the scheme tried to make it so only certain partnering comptoirs could buy minerals from these specific sites and then sell to licensed smelters outside of Congo.

The reception and consequences of this "bag-and-tag" intervention are discussed in detail in chapter 9, "Game of Tags." While it is credited for, among other things, reopening Congolese mining after the effective embargo, one of the many problems of what those in the mining trade call *ma tag* (tags) is that only a very small percentage of eastern Congolese mines is validated (in 2018, Congolese research NGOs working in this area estimated it to be about 10 percent; see chapter 9 for an explanation of this figure). The rest remain technically invisible to the monitoring apparatus. This means that the vast majority of artisanal miners, subject to a new but familiar regime of visibility, are technically dealing in blood minerals; their product is now subject to confiscation by various state authorities who are watching the tags. Even those who were able to buy and sell under the new system called the tags *craca*, or handcuffs, because they restricted to whom they could buy and at what price, burdening lower-level actors like diggers and placing them at the mercy of registered comptoirs, who set the price (their minerals could also be taken away if they didn't have the tags). The term *craca* drew a direct connection between the new internationally mandated measures and the abuses of the colonial and postcolonial state.

A conflict between incommensurate and contradictory social theories came to life in the imposition of these regulations—specifically, conflicting understandings regarding what practices and ways of being were generative of and responsible for peace, prosperity, and well-being. Moreover, conflicting views about the value of artisanal miners and artisanal mining became visible through the imposition of the tags, or *craca*. Though others tend to see miners alternately as either victims or criminals (again, they are, for the most part, not illegal actors), their tactile and conceptual work, which moves from surfaces to depths and everywhere in between, is the very source of life and rebirth in the eastern Congo despite the many dangers it entails—not only for themselves but to the order of the world and our taken-for-granted understanding of it. The next chapter begins to unpack these themes.

ONE

The Eyes of the World: Themes of Movement, Visualization, and (Dis)embodiment in Congolese Digital Minerals Extraction (an Introduction)

The main argument of this book is that artisanal mining in the eastern Congo is not only—or even mainly—a form of extraction but a kind of conflict-ridden collaboration among different actors, modes of being, temporalities, social-spatial arrangements, political orders, and understandings of peace and well-being generally conceived. In the DR Congo, a long history of predation and exclusion has fueled a transregional and transdimensional economic and ethical system aimed at producing peace and prosperity through what Congolese in the trade often refer to as "movement" (in French, *mouvement*) and "many hands touching money." The eastern Congolese theory of movement includes literal movement (mobility of things and people), collaboration across differences, making transformative connections with powerful people and forces outside of Congo, and the positive transformation of space-time and reality through the exponential concentration of all other forms of movement. Since minerals do not start as resources but rather are made into them, movement is also transdimensional (for example, according to many, ancestors have to be on board, while "clean" people speaking French can chase minerals away). In application, this theory and practice is pitted against the exclusionary practices of mining companies, as well as those of conflict minerals regulators from the Global North, whom eastern Congolese involved in the trade understand to be more or less one and the same. These foreign actors are understood to be engaged in a project of trying to purify Congolese minerals to ensure they're free of "blood" (meaning violence but also, in many eastern Congolese interpretations, the life and humanity of Congolese involved in the work of digging up minerals). Their emphasis on transparency and exclusive ownership, which comes to life in the tracking of conflict minerals among other things, is challenged by a bottom-up ethics of invisibility (see chapter 3) and practices of reciprocity and collaboration that resist exclusion.

For anthropologists, mining has long been a privileged site for accessing the totality of global capitalism and the convergence of different scales of the so-called world system. In reflecting on her fieldwork on industrial mining conducted almost a half century ago, June Nash (1993) referred to it as a synecdoche for modern capitalism because it condensed all of its different dimensions in what were, on the surface, geographic "peripheries" (Jacka 2018).[1] More recent work has continued to show how extracted minerals not only feed the development of "centers" in the world system but also collapse and upend core-periphery relations as mines and mining towns become global centers in themselves (Lane 2019). In recent years, anthropologists have been more attuned than ever to mining, largely because we live in an age of rapidly expanding and intensifying mineral extraction, an industry in which diverse actors compete over ever-scarcer resources on a global scale as capitalism outstrips the earth's capacity to sustain life (Jalbert et al. 2017).[2] *The Eyes of the World* echoes some of these earlier works in that it connects local and regional worlds in the eastern DR Congo to global processes through a close-up focus on artisanal miners and traders of digital minerals. This ethnography complicates more dualistic stories of mining (e.g., mining companies versus indigenous people), focusing instead on what emerges when violently dispossessed people make their way in a tentatively postwar situation by entering into liminal places (e.g., holes in the ground and the ruins of industrial mining) and collaboratively try to convert the potentials that cohere in earth into futures for themselves.[3]

I'd like to delve directly into these issues now by introducing the world of Congolese artisanal mining through a single individual, the Machine, a man I met in a bar in Walikale, once home to the largest artisanal mine in this region.

A Machine in a Hole

Between fast drags of his cheap cigarette, the fifty-something-year-old "Machine" informs me with a big smile that he is probably the luckiest person in the world because he has intelligence (*akili*), always ends up on his feet regardless of what happens to him, and knows how to cooperate (*kufanya kope*, or do/make cooperation, in Swahili-French; a related phrase is *kuwa souple*, also Swahili-French, meaning being supple or willing and able to bend or compromise). Those who do business with him trust him, and trust (*uamini*) is essential to this business that is built on the forwarding of money and things (food, tools, etc.) from patrons to diggers, which are in turn circulated among the diggers as an expression of what they sometimes refer to as *upendo*, or love (as in the phrase, "no one has love for another like a digger"). This forwarding of money makes possible the movement of those who work in or under the ground, workers who often push themselves beyond the limits of what many Congolese would consider humanly possible or desirable (for example, wriggling "like a snake" or mole under the ground or in a cramped and cavernous space). I find myself focused on how much smoke comes out of the Machine, which is a lot, but he assures me that, because of how he moves while working, the smoke from the cigarette doesn't have time to linger in his body and cause harm. He advises me not to follow his example because the warnings on cigarette boxes were made for the likes of me, not him.

The Machine dramatically enacts movement by breathing fast and performing the actions and rhythm entailed in digging, blowing in and out with great force while passionately miming the work of artisanal mineral extraction with his whole body, as if to prove his point. He then goes on to performatively embody the movement that artisanal mining enables by hopping around the nearly empty bar (it is still early in the day), pouring imaginary people imaginary drinks, eating fictive meat, paying for fictive medicines for the fictive sick, and pouring out invisible beer on the ground for the ancestors. He is demonstrating some of the multiple dimensions of movement and what it looked like during the "time of movement," before the artisanal mine at Bisie, where he worked, was closed by the government (something discussed in more detail in part 3). The point of the exercise is that the physical movement of his body, enabled in some sense by others, also flowed out to others, even passing beyond the visible world into the invisible dimension of ancestors. This movement allowed his life and the lives of others—including the lives of the dead—to continue incrementally.

Worn out and laughing, the Machine explains that the smoke from the cigarette fills the hole in which he works and makes it feel like a home, almost like a kitchen, so he forgets his worries and is comfortable in the darkness (one interpretation of this is that the hole, a place of interdimensional movement, also has become a home, challenging older understandings of home). All diggers smoke, he says, because they're always in the dark, deep in a hole. This phrasing carries certain connotations, and it comes up often among eastern Congolese artisanal miners, in different ways and contexts: diggers (*creuseur* in French, *wachimbaji* or *ma creuseur* in Swahili and Swahili-French) are often said to be "in a hole" and unable to "see ahead" (*kuona mbele*) into the future, in part because they lack knowledge about forces that affect them, such as the prices for the substances they dig up from the ground. Sometimes this quality of nonchalance about the future is also deemed to be part of diggers' very natures, perhaps induced by the demands of the work they engage in. "Being in a hole" (*kuwa ndani ya shimo*) also references the fact that diggers engage, out of necessity, with concealed, subterranean forces and entities—ancestors and, according to some interpretations, demons—that others would prefer to avoid. This sacrifice is necessary because, according to those who are closely involved in this work, minerals do not merely exist in the ground—rather, these substances are gifts provided by ancestors or other spirits, and they materialize and disappear in relation to how their living descendants are being treated (although it is also possible to temporarily chase ancestors away with the noise and violence of war, or even the semblance of war, which certainly complicates matters). On another level, being in the ground involves being in and with dirt (*chafu*, meaning dirt and disorder; in contrast, *udongo* means something more like generative soil), a substance and a way of being that others associate with disorder, lack of education, and flight from cosmopolitan civilization, or (post)colonial modernity (the lifestyle of what, in DRC, is sometimes called the *evolue*, or evolved, a modernist concept that colonialism institutionalized). The irony of this association is that were it not for diggers, "you wouldn't be able to buy a bar of soap" (a loaded, racialized icon of colonial modernity)[4] because money and things wouldn't circulate—a fact that diggers are quick to point out.

While they enable the smallness of digital devices, the materials the Machine digs up are socially dense, in that their heaviness and relatively low weight-to-value ratio bring into being vast divisions of labor (many different kinds of diggers, cleaners, and work managers; porters; different levels and types of middlepersons; hole owners; on-site creditors; and all kinds of ancillary businesses, from toolmakers to purveyors of food and other necessities). This division of labor emerges in part from what Michael Nest refers to as the "physical amenability of coltan ore to artisanal production" (2011, 37); for example, it is easy to separate out tantalum from dirt using only water, whereas gold requires chemicals such as mercury (water—including the use of generator-powered hoses to break up rocks and the washing of ore in sluices—is also essential to the coltan-mining process). In this vein, the Machine at one point calls coltan the "mother of all minerals," not because of its ubiquity in digital devices (he does not know about that), but because of how it conjures into being multitudes of people and vast divisions of labor, thus generating lots of regular, rhythmic movement.

Many eastern Congolese argue that, because it puts so many people to work and feeds so many other businesses, coltan and the other black minerals are more likely to generate real, enduring peace than other forms of work—certainly more so than the mining of gold or diamonds. These latter minerals produce unequal wealth (some get rich, while others remain poor or may even starve), are more easily controlled by armed groups, and are easily smuggled to foreign countries. Unlike coltan, gold incites people to selfishness, which limits its capacity to regenerate social relations in the wake of war. As one digger put it to me, "[With gold] what's yours is yours, and what's mine is mine—if there's washing that needs to be done, I'll do it myself, because every part of the process is valuable." Others described gold deals as inherently violent affairs, where everyone is advised to bring a gun to the table (although in my experience this was not generally the case). Despite its association with collaboration and bottom-up peacebuilding, outside of Congo, coltan and the other black minerals are widely understood to be the worst of all the artisanally mined "conflict minerals," mostly because of the humanitarian apparatus's publicizing of coltan during and after the war (HCSS 2013; Autesserre 2012).

In other words, these different minerals—gold and diamonds existing at one end of the spectrum and the black minerals/3Ts at the other—materialize different potentials of the qualisign of movement (*qualisign* is a semiotic term that refers to a quality, like movement, that becomes a meaningful sign when it is embodied or materialized).[5] Largely because of their material qualities, the black minerals tend to index the more positive dimensions of movement for those involved in the mining and trade of minerals—incremental progress, social network formation, collaboration, circulation of things and money, and transparent accumulation. In contrast, gold and diamonds often (but not always) index some of the more negative potentials of movement—fast wealth acquisition and expenditure, accumulation by some at the expense of others, and the secret expropriation of Congolese wealth/earth by armed, often foreign, groups. At the same time, specific black minerals each have different value potentials in relation to one another; for example, the opaqueness of coltan's purity (the wide range in percentages of tantalum it contains) makes it difficult to depend on and plan around from an investment perspective, encouraging coltan's association with uncertainty and deception in relation to cassiterite.[6]

Diggers know these minerals in an intimate, tactile way because they mine them artisanally (mainly with small tools). That term, *artisanal*, technically means that these minerals are extracted by groups of workers without any direct control or supervision from a company (*small-scale mining* is another common term, also disseminated by institutions such as the World Bank). Rather, extraction is done by people who have organized themselves into complex systems which, in Congo, interact stressfully with modes of regulation, including state and nonstate systems and actors (more on how this works, and the meanings of it, in chapter 3). The word artisanal comes to diggers through their interactions with state officials, who sometimes invoke the World Bank–drafted Congolese Mining Code in an effort to tax them. Mining differs from other nominally artisanal work (e.g., artisanal cheese making in the United States) in many ways, two of which are especially notable. For one, while many artisanal miners have some basic tools (they often refer to their shovels as "diplomas"), they usually do not "own" their product (a group of workers typically keeps half of what they've dug, minus expenses incurred at the hole, handing over the rest to the renter of the hole). Perhaps, more importantly, in other forms of artisanal labor (such as cheese making), the value of the product comes in large part from the transparency of the processes and techniques that go into making whatever is being made, which are made explicit and visible to consumers as marketable values (something that is desirable and also has exchange value; Paxson 2012). In contrast, artisanal mining is more or less invisible—or rendered invisible, mainly by the global supply chain—such that the end product (an iPhone, for example) appears to simply materialize, with no history. Its value seems to be a direct expression of its intrinsic qualities rather than the different forms of work and the different places, materials, and histories that went into it.

The Machine's name suggests that he and others understand their work and its value in relation to their understanding of the history of industrial mining, which is part of the collective memory of people in this region, even though comparatively few people ever worked for these companies, even when they were operating (for example, *Siku ya posho*, or "day of the portion," is the word for Saturday through this region and references the day workers received their pay and rations; a week is also a *posho*, or portion/wage). Though the name Machine seems to speak to a kind of alienation, it is also a praise name implying that he is at least as productive, perhaps more so, than a machine (although there is something in this name about the transformation of people into things that we should keep in the back of our minds even if it's not the main intention of the name). Another meaning of the name Machine is that he works consistently and without interruption despite—and really even because of—the inconsistencies and dramatic fluctuations of the mineral trade. Price and demand (not always selfsame) rise and fall unpredictably, but he is capable of working regardless of these shifts; this capacity is also part of, and connected to, trust (*uamini*) and friendship (*urafiki*). Other people can depend on the Machine and so are likely to forward him money, a materialization of friendship and trust that in turn allows his movement to continue consistently. The name thus suggests that his self-conception and very existence are indissociable from the overall situation of active movement.

My new friend isn't the first "Machine" I've met while studying the artisanal mining trade of the eastern DR Congo; it's a fairly common name—again, suggesting the capacity of the person to work vigorously without stopping, as well as his generative relations with others (thus referencing the value of consistent temporality over temporal rupture). Sometimes the term *machine* is also used to talk about the highly organized and productive nature of artisanal mining itself—the complex division of labor among miners and the multiple organic committees and subcommittees that work to mitigate conflict. The word *machine* can also be used to refer to diggers' collective rate and rhythm of extraction ("We are better than a machine!"), as well as the sound their work makes (for example, workers who pound rocks as part of the process of "cleaning" the ore are sometimes called *drumeurs*, or drummers; the rhythmic, almost piston-like sound of their work sets a consistent tempo for other workers at the hole).

Many of the other miners at this bar use a sobriquet as well, usually one that speaks to and also helps to make a global social imaginary. Just at our table at the bar, we have a Chuck Norris, a Rambo, a Snoop Dogg, and a P. Diddy. "Have you ever met an Obama?" Raymond asks the Machine with a laugh. "Obama?!" The Machine points excitedly to a small house. "He's right over there! He works real hard, but he smokes too much pot!" On the one hand, these names can suggest rebirth—becoming a wholly new person through the movement that mining manifests and enables—and index the fact that miners are connected to a social field that is global in scale; diggers help to produce this interconnected world, and to some extent, they draw their identities from this interconnection. But in many cases, these names are also designed to render these workers less visible to people and to state actors, typically so they can avoid creditors. There is simply no such thing as an artisanal miner who doesn't have a debt to someone, because this work cannot take place without a regular flow of value (in the form of food and materials), usually from middlepersons to diggers. But an array of unpredictable and uncontrollable factors can make it so that these debts don't get repaid in a timely fashion, and work comes to a stop (see chapter 3). In some cases, indebtedness can become indentured servitude, because state and nonstate authorities can compel the indebted to work off their debt in the hole; debt can also be acquired through the accumulation of fines from violations of law or custom, accrued at the hole or nearby. And so, diggers like the Machine have to come up with strategies for dealing with people who could bring them harm, especially when there's little or no movement: "These days, we are like snakes in the ground, always looking for a hole to crawl into," the Machine says, in reference to the indebtedness caused by the closure of the mine at Bisie.

The Machine didn't start out as a digger. Like many, he got into this work during the Second Congo War (or Great War; officially, 1998–2003), when all of his things were taken by the Rwandan-backed Rally for Congolese Democracy (RCD). Before that, the Machine sold miscellaneous wares. After those were gone, he heard about the price that coltan ore was fetching and so went into the forest to make a living for his family. But because of the various crises he has endured (including the collapse of coltan prices and the Congolese government's temporary closure of artisanal mining in 2010, followed by the de facto international embargo), he has never been able to return home. Now he talks as if home is no longer a place but rather the time when there is movement around mining. For example, he wistfully remembers a French priest and soccer coach who befriended him as a youth and insists that there is a house waiting for him in rural France, property all his own, should he ever manage to get there. But even if he were to get there, he jokes, it wouldn't be a real home, because he wouldn't know any of the foods there and would eat in the way a *creuseur* digs, always working to discern what a thing is and separating out the good from the bad—carrots from peas, peas from potatoes. "Better for Bisie [the mine] to open again," he laughs. "I'll be at home there." In other words, the Machine is not rooted to any specific geographic location in the world—and, tellingly, neither are his ancestors whom, he claims, follow him from mine to mine. He depends, rather, on the situation of movement, which can potentially transform any place into a home—and even into something greater than or beyond the concept of home. And so, for the Machine, as for other diggers, mobility—and also the more expansive concept of movement—is at once normative and desirable, while immobility is synonymous with abjection and crisis (see also Chu 2010).

Unlike a chemist or comptoir, the Machine does not know these minerals in their final form but through the process of their becoming, a process that also "makes" him as a person. Although he doesn't have access to expensive "testers"—he doesn't even have a magnet for separating out the iron from the other metals, as most négociants, or buyer/sellers, have—

the Machine has learned to distinguish different minerals from each other. Coltan, he explains, is distinguishable from other minerals because it looks oily, like goat shit with a lot of fat in it, and is easily distinguished from cassiterite (the ore from which tin is derived), which looks more like black pants that have been washed with too much cheap detergent and are now faded. While he does not know exactly what these minerals are used for, he suspects that the Europeans and Chinese use coltan to make roads that last a long time. And, like others, he has heard on the radio that coltan was used to make the bullets with which the Rwandan enemy killed Congolese people during the war. He is, in a word, projecting his own desires and anxieties, as well as collectively held ones, onto this substance, and these imaginings reflect his desires for movement and freedom from containment or enclosure, as well as his dread of war and displacement (a negative and violent form of movement, which results in dispossession, a form of immobilization).

The Machine also doesn't have any information about the reasons for the fluctuating global demand for these "information-age" minerals, although they have deeply shaped his personal life. He couldn't have known, for example, that, in late 2000, Sony had run out of the tantalum it needed to meet the Christmas demand for Sony PlayStation 2. This shortage fueled online speculation, culminating in a dramatic price bubble: coltan prices rapidly rose tenfold, only to drop down to close to their original price in 2001. For a while, the Machine was able to ride the boom (*bisikatike*, or "may it never end") and make enough money from his own sweat to become a middleman himself. But when the price for coltan fell from $100 a kilo to $10 a kilo, it left thousands of people in the forest holding "worthless sand." The resulting indebtedness on the part of those, like the Machine, who financed their enterprises using various sorts of loans (usually interest-bearing ones) affected the entire region. It impacted everyone, from urban entrepreneurs to low-level diggers, many of whom continued to stay in the mining business to pay their debts or escape debts while living in the forest far from home. Many claim that the end of *bisikatike* was the moment when eastern Congolese became "addicted" to mining because their debts made it so they had to look for other minerals in the forest in order to return home.

After the coltan bust, the Machine had debts that he needed to pay off to various financial backers, so he couldn't return to the city of Goma, his original home. Instead, he hung around in the forest, digging to make ends meet while waiting for something to happen. When he heard that Kumu trappers in the remote rainforest district of Walikale had discovered cassiterite and bauxite on a hill that came to be called Bisie (Kumu for "it will continue"), he went there to work, along with what soon turned out to be nearly twenty thousand others. During that time (roughly 2003–2010), the Machine says he ate and drank like a king. But, these days, a Canadian company called Alphamin Resources has taken advantage of the hard work of these artisanal miners and secured a contract from the government to drill industrially (see part 3). The Machine and his friends have been expelled from the site they discovered and developed, and a very small fraction of them now work for Alphamin, building a road for a pittance.

"Before [the closure], when my boot touched the ground, it left behind the footprint of a giant!" the Machine exclaims. "Now it's like I'm not even here." The end of movement is something he experiences viscerally as synonymous with loss of self, and he is not alone. When the government shut down artisanal mining at Bisie, it impoverished the entire area and led to widespread starvation because the region had long been depending on food flown in on cargo planes from other areas (the cargo planes stopped running when the mineral business was shut down). Closing minerals (*kufunga madini*) also froze the assets of the hole owners (really hole renters), many of whom were financed by foreigners and running deeper into debt with every passing day. The Machine jokes that, because of this immobilization, there are more bandits in the forest now than there are mosquitoes. "Bisie was our security," the Machine laments, because the movement of mining kept people from more violent ways of making a living, as the value from it flowed out to ordinary people and state figures.

Movement: A Theory from the South

On one level, the Machine is a middle-aged man in a bar in the eastern Congolese rainforest at 10 a.m., ranting, albeit with a sense of humor and irony, about his loss of relatively autonomous work and his devolution to a very low-level itinerant laborer for a foreign mining company that he despises. But, on another level, he is articulating, in bits and pieces, a kind of grounded philosophy, or theory. It is a practice and worldview that valorizes the concept of movement, which is related to the concept of touching (as in the common eastern Congolese phrase, "touching money"), collaborating across differences and dimensions, and developing relationships of reciprocity based on trust. As briefly mentioned above, movement also refers to the transformative dynamism that can occur when people and things are moving at the right rate or speed, creating a condition that allows people to break out of their superimposed situation of inequality or debasement. Movement can allow for reversals or what could also be called miracles. In practice, movement is oriented against closure or the exclusion of people from places and situations where there is life (*uhai*). Everywhere I went, Congolese narrated how the process of making people transparently visible had helped the colonial and postcolonial states to exclude them from movement; moreover, exclusion drew on entrenched colonial notions about cleanliness and the implied dirtiness (*uchafu*) of African practices and bodies. In this regard, Filip De Boeck, in his work on the Congolese capital of Kinshasa, has emphasized how the "undulating" movement of the city and the Kinois' desire to be "flexible" is threatened by the contemporary state's "ocular" regime, which wants to exclude Kinois from the possibility of being-with-others-in-the-city in the name of "cleaning" that city through gentrification and highly visible (if also spectral) urban "development" (De Boeck 2011).

The local valorization of movement in the context of the eastern Congolese artisanal mining trade certainly raises some conceptual questions. For one, it would be easy to see the Machine's emphasis on movement as an effect of high-speed global capitalism, which he is describing and even celebrating when he talks about the way his body and the bodies of others move in response to the global demand for minerals. There is certainly something to this: As mentioned, when the price for coltan spiked as the result of the temporary decline in supply matched by online speculation, it precipitated massive movements of people—people moved in sync with the global demand for substances—and that time of "may it never end" is still remembered for that (and some of those people were forced into labor by soldiers).[7] That energy, embodied in mineral resources, could potentially be harnessed and redirected into enduring and relatively stable forms of value (the ideal for many was to own a house in the city of Goma). But if one were to have a chance at harnessing that energy, one had to flow along with the movement of things that were in demand, participating in a form of mobility that was also immobility as people became things, and "the country," or rural areas, became like the city in many respects (for example, many mining areas largely ceased to be places for growing food and instead became places of fast volatility dependent on other regions for sustenance).[8]

Consider, for example, the cargo planes, the main way of getting around over long distances in the mining regions of Congo during the period of my fieldwork (they were mostly old Russian Cessna planes). Cargo planes emerged as a mode of transportation during the coltan boom, which was contemporaneous with the Second Congo War (the main transport companies were Rwandan-owned). Traveling on a cargo plane means moving as cargo, a commodity, which means that you must seek out a transport company—a business that moves things back and forth from a city or large town to a mining region. These companies transport food, palm oil, gas, and miscellaneous goods like plastic chairs and mattresses to the mining zone, and they transport minerals back to the city. You do not look for an airline or plane company, as they would have you rent out the entire plane, and you would spend $3,000 for a thirty-minute flight going perhaps one hundred miles. So you go to one of these transportation companies, usually on a street with a lot of wholesalers.

Using American dollars, you then buy a receipt for goods rather than a ticket. This is actually a receipt saying that you have paid for the transportation of a quantity of things, mainly yourself. The owner of the company, or an employee, will then take your cell phone number and call you when a plane is ready. His job now is to be in contact with the one or more plane companies that sometimes make runs to this area or that have agreed to do so. He rents space on the plane to transport things. The wholesaler does not actually know when a plane will take off—certainly not what time and probably not what day—but he has a general sense of what has been happening lately. No one will know for sure until enough produce or material has been assembled to make the flight profitable. At that point you will get a call, and you have to be ready to move from wherever you are onto that plane, as the call might come only two hours before takeoff. Depending on the rate of movement—which, if it's a mining zone, is determined by the demand for minerals from the place at any given time—you may be able to leave soon or not so soon. Perhaps you will be promised that you can leave on a Friday, only to find that that plane is going to a different town than the one you wanted to visit because there is more business right now moving in that direction, so they will call you when a plane is ready. When everything finally comes together, you pass through the multiple state officials at the airport (the ones lower down on the hierarchy, less directly relevant to the matter at hand, hang out next to the plane itself in an effort to extract money by physically blocking your movement before you get on the plane). Finally, you climb aboard and sit, a commodity on top of other commodities.

Clearly, this aspect of movement—in which movement is determined by the demand for commodities, and people's desires seem to become a function of this demand—is real enough. It was also brought home by the prices of food in mining areas—in general, where there was artisanal mining, food was expensive because people weren't growing it but were digging and selling minerals instead (although this was not true in every place where people mined artisanally).[9] Even if they were to grow food, they couldn't grow it or move it quickly enough to pay for their continual, never-ending expenses. When this was the case, people usually said they didn't have time to grow food or that it took too long—unlike mining, which could quickly earn people some of the money they needed to pay for things. They also complained about the roads, which was partly a complaint about time (that it would take too long and be too expensive to transport food over long distances). In some places, mining had at least temporarily ruined the land and made agriculture impossible for many (it also diverted water needed for growing food), so they lived in a time that was vulnerable to the changing prices of things because they could no longer depend on growing food. Rather than consistent growing and harvesting seasons, they experienced periods of fast movement around mining, followed by abrupt stops that could be excruciatingly long and dangerous. These spatiotemporal fluctuations were part of what was meant when people said they were addicted to mining—that they couldn't go back to the slower pace of other activities (this moralized and made personal what was really a much more complicated issue they generally didn't have immediate control over). The mining town of Kailo in the province of Maniema showcased an extreme example of this dynamic: there, market day lasted for about an hour in the very early morning once a week; a very small chicken cost $40, and many adults claimed they had never so much as seen an egg or a can of sardines. (Joseph, Raymond, and I ended up sharing our cans of sardines with as many people as possible; these sardines, a common item in other rural and forest villages, were the talk of the town for weeks. Later, we paid for eggs to be brought to some people in Kailo so that they could at least have them once, to accompany the beer that cost $5 but was only $1.50 in Goma.)[10]

Movement, then, was often experienced as something that happened to people, or that impressed itself upon them, as they were forced to move at certain times, at a certain rate, or in a certain direction. While movement was almost always superior to closure or immobilization, it could also lead to closure, as too much uncontrolled movement could manifest in what was referred to as "disorder"; for example, when there was so much movement at a mine that too many competing state authorities got involved, making it difficult for people to move or sell their things for a profit, this hectic immobilization was called disorder. Generally, the key to making the most of movement was to try to embody and convert the energy of movement into material forms (coltan, cows, houses) that would allow that energy to be stored and expended in a beneficial and more or less stable and controlled way; however, many of these material substances (such as coltan) were also volatile and unpredictable, because their prices were subject to rapid fluctuation and their value was almost never fully transparent.

However, when people in the mining trade talked about movement as a desirable quality, they weren't mainly speaking about space-time compression, or even speed, but about a collaborative, transformative practice in which people came together around certain tangible materials (such as coltan), converting them so they could experience this personal and collective transformation. This collaborative practice did not necessarily involve fast movement, and it was sometimes aimed at slowing movement down through the development of regular relationships (agreeing to buy at a certain price so the transactors wouldn't be affected by rapid fluctuations) and maintaining a certain rhythm or tempo that culminated in predictable or positive outcomes. While these transactions could sometimes involve deception (see chapter 3), they also involved establishing convivial relationships and being flexible, a way of being that was contrasted to the ways of state officials, especially soldiers during wartime, who presented obstacles to movement (again, the phrases used were "being supple" and "making cooperation"). On a broader level, movement was associated with peace because wherever there is movement, there is circulation of *makuta* (money) so that "many hands can touch money," nourishing communities and bodies; in contrast, war, and the violent contraction associated with it, came when people were excluded from movement (there was a common axiom to this effect: hunger comes before weapons, and hunger follows weapons; *njala mbele ya silaha, na njala kisha silaha*). People also contrasted the consistent tempo of movement in the mining trade (the digging and the exchange of minerals) with the violent abruptness and rupture of war, when anything could happen unexpectedly at a moment's notice.

More than simply a synonym for mobility, movement entailed bending the limits of what was ordinarily possible to become something different from what one was generally defined as or the role one assumed (e.g., a representative of the state or an ordinary person). It involved not only moving through space but across all kinds of norms or barriers and converting certain forms of value into different forms of value (relations with ancestors into the commodity coltan into houses). Participating in movement could mean getting literally dirty as well as engaging in the kind of "dirty work" that colonialism built a whole racialized system around—at once depending on it, vilifying it, and relegating it to the colonized (Mbembe

2019). Movement can also uplift dirty work into a kind of calling, elevating it and the places in which it takes place so that those involved in it can break out of the degraded positions to which they have been consigned. Through movement, the sylvan worlds in which many Congolese live—and which urbanites look down upon and associate with death—can become cosmopolitan places (places of great mobility and movement across barriers), sharing in some of the qualities of other highly valued or desired and desirable places, like the United States or Kinshasa (and large holes are typically conferred names reflecting this fact). Above all, the practical philosophy of movement and the related idea of "many hands touching money" have enabled and supported a technological world and a form of capitalism whose practitioners see the likes of the Machine as backward and criminal subjects who need to be regulated so that their technological instruments can remain pure or conflict-free.

Movement was thus irreducible to market logics, nor was it simply an expression of neoliberalism run amok ("the market" unfettered), in part because it often involved reciprocal relationships that offset whatever was happening to price at the level of the "world market" (the *soko ya dunia*). Most importantly, the collaboration across differences that movement entails has little in common with, and is often dramatically opposed to, the idea of private property that is central to contemporary global capitalism. Mining companies that closed people out of the forest because they had purchased papers from the government and now have sole access to land as property were thwarting movement as eastern Congolese understood it. This was especially true when these companies invoked exclusive rights or set about enclosing space that belonged to indigenous people and ancestors. The practice and concept of movement also goes beyond political economy, in that the transactions cut across dimensions, entering into spheres that a Euro-American would consider religious (for example, transactions and communication with ancestors).

While I would hesitate to try to lay out a genealogy of this theory of movement, one can point to a few important factors that are irreducible to high-speed capitalism and war. One is the long history, in African and Congolese life, of actively and agentively forging connections with others across boundaries. Scholars of Africa have employed different terms for this, including the idea of a *longue durée* "politics of extraversion" (Bayart 2000), or the practice of developing long-distance relationships across differences with powerful others. The related idea of "wealth in people," a very well-worn concept in the anthropology of Africa, is also relevant here: it refers to the fact that, in much of Africa, social relationships are the end of wealth acquisition, and material wealth is a means of building social relationships and clients as opposed to simply more material wealth (Guyer 1993). Achille Mbembe (2019) has framed the concept of "transactionalism" in more expansive terms that are close to what I have in mind when thinking about the eastern Congolese concept of movement. As he puts it, "The [precolonial African] world itself was a transactional world. One was always transacting with some other force or some other entity just as one was always trying to capture some of the power invested in those entities in an effort to add the latter to one's own originary powers" (Mbembe 2019, 107–8).

In Congo, it is impossible to understand the value of movement outside of the long history of exclusion and immobilization—especially during the time of King Leopold and the Belgian Congo. Scholars of the history of the state in Central Africa have drawn attention to the centrality of immobilization dating back even to precolonial times, showing how the colonial and postcolonial states built upon and expanded an enduring political system that had been predicated on temporary immobilization. In particular, they have pointed out the existence of a long-term strategy of state and state-like control based not on sovereignty over territory but on the control of circulation, of people and things, mainly at geographical choke points (Schouten 2019). This "political geography" was, and continues to be, "built up around control over the circulation of people and goods rather than territory or population" and only came into existence in places and times of movement (Schouten 2019; see also Cowen 2014, Fairhead 1994).[11] In the present day, this political system based on extracting value from movement continues, accounting for why one finds most or all of the state actors at airports, roadblocks, and well-trafficked forest paths on the way to mines and often nowhere else (Fairhead [1994] refers to this state presence on roads as "paths of authority"). As Peer Schouten has put it in his work on "roadblock politics": "From [precolonial] long-distance trade networks to colonial roads to today's supply chains, there has never been a moment when exchange across Central African space has not been imbricated with organized violence. On the aggregate level of roadblock geographies, a general politics of circulation seems at work; one which depends on efforts to put things into motion but, paradoxically, is only activated when these aspirations encounter obstacles" (2019, 937).[12]

If immobilization was long an aspect of governance in the region, the system of violent extraction and exclusion set up by King Leopold's Congo Free State intensified the assault on bodily mobility; it included prohibitions on the touching of valuable things. Under King Leopold's Force Publique, the extraction of rubber entailed the establishment of quotas for rubber brought in, with the severing of hands—the precondition for spatio-temporal expansion through touch—being one of the most remembered forms of punishment for not meeting quota (Hochschild 1998). Violent immobilization and forced mobilization (a kind of extreme form of immobilization) were central to extraction.[13] After King Leopold's Free State, under the Belgian Congo, violence was combined with systematic exclusion and infrastructure-as-surveillance; forest dwellers were forcibly relocated to the road so they could be made visible and taxed (my interlocutors referred to this as being "closed out" of the forest, a phrase they also used for contemporary conservation interventions by NGOs and the enclosure of land by mining companies).[14] And it almost goes without saying that Congolese were excluded from the fruits of mining, as mines—mapped and owned by whites—became separate enclaves bound by strictly enforced racial hierarchies and violence (mining companies selectively brought in Africans to work for them, typically not from the area in which the mine was located). Artisanal mining was illegal until 1982, which meant that, until then, artisanal digging and trading (particularly of gold) had happened mainly under the cover of darkness. A widely told story has it that whites told Africans that gold was poisonous and that if they touched it, they'd become sick; these warnings were, in part, a mechanism of enclosure, an early (if also ridiculous) effort to use concepts of toxicity, public safety, and humanitarianism to keep Africans physically removed from, and unwilling to touch, the materialized capacity for movement and transformation that lay buried in the earth.

Colonial land law laid the foundation for and continues to provide an operational framework for exclusion and immobilization in Congo. Current land law has its origins in the Congo Free State, which in practice recognized only European-style proprietorship as ownership; the closure of land was thus both physical and conceptual.[15] After the Congo Free State, this practice continued, with Belgian colonial law recognizing only land under cultivation as "land occupied by natives" and refusing to recognize mobile African populations (such as so-called foragers) as having rights to land (Long 2011). In the 1920s and 1930s, large swaths of land were given over to public parks in which no human activity save for "research" was allowed. In 1966, under Mobutu, a new law made everything emerging in and from the ground (from forests to land to minerals) property of the state; a 1973 revision to the law stipulates that land occupied by communities can be held under customary arrangements, but the specifics are not spelled out and have been perpetually "forthcoming" since that time (Long 2011; Ushudi Ona and Ansoms 2011; Geenen 2013). Up until the outbreak of war in 1996, land lay in the hands of a small group of state actors who were in a position to give it over to private interests; communities had almost no legal rights, and large concessions were allocated to national and foreign-owned businesses. In 2006, a new constitution was put in place, reiterating state control over land, though now in terms of sovereignty rather than ownership (Long 2011). At the same time, a "customary" system, shaped by colonial and postcolonial interventions, policy, and understandings, allocated land for the use of local populations in the name of local "custom" (*coutume*), or reinvented tradition.[16]

Immobilization was compounded by years of war (see chapter 2), which ended up becoming a de facto instrument of accumulation through dispossession by displacing thousands of people—alienating them from their land and things, as well as from incremental social time, while making them available for supply chain capitalism as artisanal miners (Smith 2011). In recent times, the assault on mobility has continued in new forms under the umbrella of postwar peacebuilding. Diggers often pointed out to me that they were at once physically endangered and immobilized because they were prevented from moving where they wanted in order to dig and sell; in contrast, minerals and gorillas—both of which were protected by international regulations and a regulatory apparatus—were safe and free. Diggers blamed what they saw as a degrading, even racist, situation on the international apparatus of INGOs (henceforth simply NGOs) working in combination with Congolese state officials (see chapter 9). They plaintively complained that minerals and animals (mainly gorillas) were "sacred" to the "international community," while their lives were dispensable (people joked—but they were also serious—about how no one cared what happened to you in the mines, but if you touched a gorilla, a satellite would see you, and the police would come to get you later). In some places (not everywhere), they were certain that, even if no one seemed to be watching, if you so much as touched a bag of minerals that had been tagged "conflict-free," a punitive apparatus would kick in that was identical to what would happen if you poached a gorilla (see chapter 9). Again, this apparatus was often said to be triggered by satellites that were watching in the sky. According to them, diggers were deemed less important to the world—in a sense, less "human"—than either gorillas or rocks, which is why the satellites were watching these valuable things, like cameras in the world's largest convenience store (my metaphor, not theirs). There is a focus, in all these Congolese speculations, on exclusion and the foreclosure of movement, which includes being prevented from touching valuable sources and conditions for spatiotemporal expansion (in the extreme, by having your hands or limbs cut off; or in more prosaic versions, through stories about the dangers of touching).

Congolese understandings and practices of movement also reflect a larger set of processes that are happening on a global scale, though they are experienced differently in different places. A now copious literature on accumulation by dispossession, in the past and in the present, has drawn attention to the ongoing importance, for capitalism, of the dispossession of people from preexisting social systems and modes of circulation, a process that frees up land and people for capitalist exploitation, at once mobilizing and immobilizing them. Accumulation by dispossession also entails visualization, as people become transparent to the state in ways they were not before (e.g., by being removed from land in the countryside and coming to the city, where they are compelled to work for cash and can be subject to various forms of policing, or by being subject to regimes of regulation, as in the case of "conflict-free" minerals regulations and interventions).[17] In this vein, a number of scholars have argued that the struggle over mobility is the primary issue of our time; they show how climate change, urbanization, and global migration are "each part of a common phenomenon of unequal and uneven mobilities that impact everyday life at all scales" (Sheller 2018, 14; see also Chu 2010). As Achille Mbembe has put it,

> Indeed, wherever we look, the drive is decisively toward contraction, containment, and enclosure. By enclosure, contraction, and containment, I do not simply mean the erection of all kinds of walls and fortifications, gates and enclaves, or various practices of partitioning space, of offshoring and fencing off wealth. I am also referring to a matrix of rules mostly designed for those human bodies deemed either in excess, unwanted, illegal, dispensable, or superfluous. (2019, 96)

Similarly, Saskia Sassen (2014) has used the concept of *expulsion* to refer to the profit-driven evacuation of people across the world from their social support systems, ways of making a living, and the "biosphere," a process that directly enriches global corporations and political elites. Sassen's concept of expulsion is particularly apt for the eastern Congo, where more people have been displaced by war than nearly any country in the world and where many of those displaced ended up involved in the extraction of the minerals used for digital devices (Norwegian Refugee Council 2020). For Sassen, expulsion results from the global inequalities produced by the intersection of global finance capitalism with the imposition of austerity measures on the former liberal democratic states of the Global North and postcolonial nation-states alike. Through the concept of expulsion, Sassen is able to link seemingly unconnected events happening in different parts of the world, including mass incarceration in for-profit jails; mass unemployment and underemployment; the expulsion of people from their homes due to new, more complicated forms of algorithm-generated debt (especially subprime mortgages); the expulsion of people from agricultural land in the Global South; and the expulsion of people from forests for the extraction of minerals. Of particular relevance for understanding Congo, and all postcolonial African states, is the fact that the debt-driven austerity measures that were imposed on the Global South through World Bank and IMF structural adjustment programs in the 1980s and 1990s primed these states for a new wave of recent land expropriations, in which bankrupted postcolonial political elites now sell off land to foreign investors, creating new waves of displacements (this is certainly at the core of what happened at Bisie, described in part 3).[18] Computerization, though it is experienced differently in different parts of the world, is one of the conditions of possibility for contemporary expulsion, and it is crucial for understanding the mutual imbrication of mobility and immobility on a global scale. As Sassen puts it,

> Some of the major processes feeding the increased inequality in profit-making and earning capacities are an integral part of the advanced information economy. . . . One such process is the ascendance and transformation of finance, particularly through securitization, globalization, and the development of new telecommunications and computer-networking technologies. . . . Among the most significant [new trends] over the past twenty years are technologies that make possible the hypermobility of capital at a global scale; market deregulations, which maximizes the implementation of that hypermobility; and financial inventions such as securitization, which liquefy hitherto illiquid capital and allow it to circulate faster, hence generating additional profits (or losses). (2014: 24–25)

And, of course, computerization requires minerals that need to be extracted from the ground and (along with mobile phones) makes possible the decentralized (mineral) supply chains that define the new economy and constitute the market for digital minerals.

In short, the eastern Congolese value of movement emerged from an array of experiences that were at once historical and contemporary, geographically local and globally shared. In general, the eastern Congolese idea of movement stressed the importance of connecting with others, of collapsing and moving through and beyond imposed geographic and ontological boundaries, and of engaging directly with, and creating the world anew, through earth while converting forms of value into seemingly incommensurate forms of value (e.g., relations with ancestors into money). Through the work of digging and trading, Congolese movement helped to make digital technologies, and so the digital age, possible, but Congolese understandings of movement differed markedly from the dominant Euro-American understandings of the digital age and the future, which entailed disembodied, frictionless transcendence from earth rather than collaborating with and through earth (see also Tsing 2004). In particular, if Congolese understandings of movement entail being connected to the world—with others, with dirt, and with ancestors—in a manner at once tactile and imaginative, Euro-American understandings of the digital have more often entailed escape from the embodied, physical world (certainly, from those aspects of the physical world that Congo epitomizes for Euro-Americans). Lately, the Euro-American mythology of the digital age—consisting in

Figure 1. "One nation, one network: wherever you are, we have you covered":
Vodacom billboard advertisement showing the mutual imbrication of the network
of diggers and the cell-phone network (Bukavu, South Kivu)

Figure 2. Moving people and things (Maniema)

Figure 3. House art showing different dimensions in which "movement" can take place: a water spirit locally referred to as Mamba Mtu (crocodile person), Vodacom cell towers, and a truck with an "X-ray" view of the engine (Walikale, North Kivu)

transcendence of materiality and history, as well as regulation and control of bodies through vision—has also come to haunt Congo in the form of "blood minerals" regulations and interventions employing digitized tags intended to purify technologies while curtailing and regulating Congolese mobility/movement (see chapter 9).

Before moving on to a more detailed analysis of the Congolese material, I want to flesh out this understanding of the digital age and its relationship to Congo, or rather the idea of Congo, a little more clearly.

Taking the Red Pill: "The Congo" and "the Digital Age"

Silicon Valley believes in the religion of transparency.

—Franklin Foer, quoted in Spencer (2018, 145)

When I first heard about the Congolese coltan boom during the war (circa 2000) and that this ore was essential to digital devices, I felt that all the disparate, disconnected fragments that made up the world were brought together in a moment of clarity. It was as if I had taken the proverbial red pill, suddenly realizing that, underneath the digitized matrix, which promised personal freedom and transcendence from historical structural limitations, there was a grimmer reality in which people were being consumed to make this brave new world possible (Smith and Mantz 2006). And I was certainly not alone: for example, in an article entitled "The Dirt in the Machine," the *New York Times* informed readers of the shocking fact that "a black mud in Africa helps power the new economy" (Harden 2001).

At least for a while after the boom, coltan fed the dystopian imagination in the West, playing a major role in more than one novel.[19] In John le Carré's post–Cold War spy novel *The Mission Song*, the struggle over Congolese coltan is at the center of the plot, symbolizing the ambiguous shadow economies and resource wars that define global politics after the end of Cold War politics and ideology—the author's erstwhile bread and butter (le Carré 2008). And in the television show *Terminator: The Sarah Connor Chronicles*, we are informed that the endoskeletons of the relentless AIs from the future are made out of coltan. As the title of the *New York Times* article suggested, the substance came to signal that these technologies weren't as clean or benign as we thought, that there was dirt in the seemingly pristine machines that we owned but which also increasingly seemed to own us.

The coltan price hike made the temporal simultaneity and politico-economic interdependence of seemingly disconnected worlds suddenly visible, and this was part of what was compelling to so many observers. It wasn't just that people discovered that Congolese coltan is used to power digital technologies, but that digital speculation and new technologies were *converted into* violence in Congo, especially given that the internet was helping to drive the investment that fueled some of the violence there. But the shock of the narrative, and its subsequent power to produce legislation and peacebuilding interventions aimed at making coltan mining conflict-free, owes at least something to the fact that digital devices were found to be composed of *Congolese* earth and blood (although it is still not clear what percentage of coltan in use actually comes from Congo)—that the devices contained something of the "heart of darkness" within them. They were no longer able to unambiguously stand for a disembodied future and were instead shown to be associated with places that Euro-Americans see as incommensurate with modernity and mind. This was certainly the message that many eastern Congolese have received over the years: mainly, that suddenly white people (*Wazungu*) didn't want to buy their minerals anymore because they were "bloody" or "dirty," concepts that were intermingled with one another, and that these minerals therefore dirtied their otherwise clean technologies (*chafu*, or dirty, implies disorderly and polluted; in Congo, uses of the term often reflect a colonial past that saw Africans and their practices as dirty, a past that postcolonial cultural elites and officials perpetuate in various ways).

This requires a bit of unpacking. It is a commonplace that, in the grand metanarrative that "the West" tells itself about itself, Africa is the absolute Other, the negation of Mind and Reason—as Georg Hegel famously put it, "no historical part of the world" (Hegel 2004; Taiwo 1998). For Hegel, who theorized what was to become "common sense" about Africa's nonrelationality to the rest of the world most explicitly and cogently, the continent was more of a state of being than it was a place, a condition in which "completely wild and untamed" humans had yet to use their reason to distinguish and separate themselves from "mere nature" and begin to "develop" their humanity through practices of mastery and objectification like enclosure and private property. And so, Africa existed for Western philosophy as a nullity, a benchmark against which one could observe and comment on the "universal" development of mind and spirit (Geist) over historical time in the form of ever-expanding reason and freedom. Africa was also an illustrative example of the violence and horror of the State of Nature for, as Hegel put it, "there is nothing harmonious with humanity to be found in this [African] type of character." Later in the nineteenth century, King Leopold's Congo became (and still remains) the part that stood for the whole of African savagery in Western thought. If Africa is known, as Achille Mbembe (2001) put it, "under the sign of absence," Congo is its master trope, long represented in maps as a big white void, as if waiting to be filled in by an exogenous consciousness.

The Euro-American story about "the Congo" became complex early on, in that the place came to represent a twofold backwardness, or darkness, that has always been impossible to fully disentangle. On the one hand, this was the primordial violence of the native whose identity was seen as part and parcel with nature (irrational, untamed, violent, and Other) and, on the other hand, this was also the "modern" violence of an unfettered capitalism that was morally willing to equate people with things, thus to enslave them. This view was first articulated in Joseph Conrad's *Heart of Darkness*, which continues to be such a controversial novel because it is both a racist portrayal of Congolese "natives" living in a putative state of nature as a backdrop to the actions and torments of whites *and* a critical indictment of extractive colonial capitalism under Leopold's reign (Conrad, *Heart of Darkness* 1899; see, e.g., Achebe 1977 and Ngugi 2017 for critiques and countercritiques). This twofold narrative tradition continues to this day; it relies on ambivalence (in the sense of pivoting back and forth between horror and romanticism) about Africans living outside of History *alongside* a critique of rapacious capital taking advantage of them.[20]

It is not coincidental that the anthropologist who wrote the most consequential critique of how anthropology employs temporality to "construct its object" conducted his fieldwork in what is now Congo. In his *Time and the Other: How Anthropology Constructs Its Object* (1986), Johannes Fabian argued that the "denial of coevalness," or existing in the same time, is central to a "politics of time" in which the anthropologist portrays interlocutors as existing eternally in a bounded, frozen space-time that is outside of the flow of other people's experience of time. While anthropologists have for the most part learned something from such critiques, when it comes to Congo, the rest of the world barrels forth, in denial of the actual "extraverted" and dynamic history of the region (Bayart 2001; Comaroff and Comaroff 2014). In general, this narrative reproduces the West as the agent of progressive and positive transformation, making it more or less impossible to see the actions, tactics, and knowledge of Congolese as anything more than examples of lack or abjection. For example, Euro-American representations of the allegorical journey down the Congo River continue to represent a movement backward in time that also entails a corruption of soul because it implies journeying to an earlier *stage* of human development that lacks the variety and volume of freedom and reason that is imagined to exist in subsequent stages.

In the Western imagination, the antithesis of Congo (typically associated with water, darkness, and mud) is probably not best represented by a geographical place (New York, Paris, etc.), but by the technologically mediated surpassing of geography itself: the "information age," in which the contents of interconnected minds finally annihilate the limitations of place and history and the violence inflicted on bodies. Especially in earlier (circa 1990s) formulations of the potential of the digital, the mind was understood to have finally secured its divorce papers from the body, thus annulling the troubled marriage of mind and body in Cartesian and post-Cartesian thought. And the mind got to keep the house. Unleashed from the confines of our physical bodies and our putatively singular identities, we could now go traveling in our virtual *Second Life*, doggedly waging *Warcraft* in brave new worlds that were more real than our "real" ones.[21] Some people, including entrepreneurs like Elon Musk, have even envisioned a "transhumanist" future in which our minds, finally uploaded onto clean computers, will be able to leave the messiness of our diseased bodies, and their blood and affects, behind for good (Lewis 2013). A journey back into the not-so-distant past of the 1990s gives a better sense of the unilinear cultural evolutionism entailed in early conceptualizations of the information age, in which mind would finally be unleashed from the body and matter.[22] For example, in the early days of the internet, the scholar W. J. Mitchell described the internet as an inchoate, deterritorialized city without matter and bodies in words that dug into my mind and stayed there to this day:

> This will be a city unrooted to any definite spot on the surface of the earth, shaped by connectivity and bandwidth constraints rather than by accessibility and land values, largely asynchronous in its operation, and inhabited by disembodied and fragmented subjects who exist as collections of aliases and agents. Its places will be constructed virtually by software instead of physically from stones and timbers, and they will be connected by logical linkages, rather than by doors, passageways, and streets. How shall we shape it? Who will be our Hippodamus? (1996)[23]

The television show *Star Trek*, especially the very high-modernist "original series" of the mid/late 1960s, offers an excellent entry point for understanding the knee-jerk Euro-American vision of a potentially disembodied future that emerged at the dawn of the digital age; it also clues us into the fact that this vision had less to do with the technologies themselves, which mostly did not yet exist, and more to do with an already-in-place set of cultural concepts about the mind's "natural" evolution away from the body through progress. This vision included transporter beams that routinely annihilate and recreate the body for long-distance travel, mind/consciousness uploading onto computers and into androids, "evolved" aliens with big telepathic brains and withered bodies, and people using disembodied mind/brains to regulate their cities (similar to today's so-called smart cities). To be sure, even in this modernist Euro-American understanding, disembodiment is not understood to be unambiguously positive, and there is a lot of nostalgia and anxiety surrounding the threat of losing the pleasures and powers associated with embodiment.[24] But, whether positively or negatively valued, the apparent inevitability of technologically induced disembodiment is one reason why, outside of dystopian science fiction, many people are disinclined to think of highly embodied forms of labor in and underground as growing alongside technology and as contemporary with it (despite the fact that artisanal mining is a rapidly growing activity globally)[25]—even less that diggers would be able to offer some insights into the contemporary world and tactics on how to live together in this emerging moment.

At the turn of the twenty-first century, the desire for virtual transcendence and disembodiment took shape in the context of the deregulation taking place in US society, which many now refer to as "neoliberalism" (Harvey 2007). In particular, computerization and the growth of the internet was one response to a crisis of accumulation that emerged during the 1970s and 1980s, a time of wage compression and declining corporate profits in the Global North (Streeter 2010; Neubauer 2011; Kergel 2020; Zuboff 2019). For corporations, personal computing and the internet promised a new way to sell goods to consumers and to increase consumption overall. On the part of ordinary people in the United States, personal computing and the internet seemed to allow for entrepreneurial activities and more "flexible" work schedules in this period of long-term wage deflation (Streeter 2010; Besley 2007). Also, as government-subsidized services such as education and health care declined in the 1980s and 1990s, computerization seemed to enable market-based solutions to these problems (think of the rise of online medical sites during this period; Kergel 2020). And computerization managed to make this transformation seem not like a loss for people but a gain. Now that we all had computers, we might not need the state, older bureaucracies, or even our employers in the same way anymore; we could all be entrepreneurs (President Ronald Reagan talked about the age of the entrepreneur, specifically in the context of computerization; Streeter 2010). For a time, it even seemed as if we could all be governments—part of the lure of personal computing in the '80s and '90s was that each individual might take on powers formerly reserved for large corporations and states. (The 1983 film *WarGames*, in which a teen starts and has to stop World War III by playing a video game on an early version of the internet, made this fantasy comically transparent.) Moreover, a host of thinkers and humanitarians would go on to imagine that this brave new future would also become the future of the entire world—through a process of market-driven cultural evolution toward the digital age—and that therefore the way to "develop" Africa was to "leapfrog" past the Industrial Revolution and into the digital age through, for example, gifts of computers to rural schoolchildren (Ginsburg 2005).

Mitchell's augury of a future in which humans finally evolve beyond the material was paralleled by the voices of many others who claimed that capitalism had changed fundamentally and was no longer dependent on physical labor. A variety of thinkers propounded the notion that capitalism was becoming "immaterial" and no longer worked mainly by accumulating surplus value from physical labor (Hardt and Negri 2001; Castells 1996). While this genre of thought includes a lot of interesting ideas that I certainly wouldn't want to discount outright, it's important to draw attention to the general framing of it—mainly that the temporal movement is toward immateriality and away from materiality. In this literature on "virtual" capitalism, hard minerals—raw earth itself—often emerge as symbols of a historic past in the process of being transcended by the mind. For example, in his ethnography of "virtual migration," Aneesh Aneesh writes that "although the global economy still produces steel and aluminum, the majority of labor, especially in the United States, has shifted to the manipulation of symbols" (2006, 9). One of the problems with the literature on disembodied labor and information capital is its teleological futurism and its technological fetishism—its tendency to privilege the most recent manifestation of technological "progress" in the Global North as the driver of history. (There are many instantiations of this. I think, for example, of other anthropologists who have informed me that African studies is backward because of its focus on land, unintentionally implying that Africans who are also focused on land are backward—a strange thing to think during this time of massive and unprecedented corporate land grabs in Africa!; see, e.g., Franz 2012)

During the 1990s, one of the reasons why this story about technologically induced transcendence was so compelling was because many of us seemed to be living the dematerializing promise of digital technology in our daily lives. It was not just that many of us suddenly had access to personal computers and, a bit later, technologies of interconnectivity, but that investors had seemingly solved one of capitalism's periodic crises—the stock market crash of the late 1980s and early 1990s—by investing in new tech start-ups. For a moment, it was as if technology had solved the historical problem of capitalism's perpetual crises, rather than just displacing crisis onto another geographic space (Harvey 2001). In the end, we all know how that played out—ordinary people thought that the futuristic "new economy," based on network technologies, defied all the old rules of political economy and investment (Spencer 2018). In the late 1990s, most of us who were in our twenties or thirties knew some optimistic young person who left her job or grad school, lured by mysterious "futures" and "options," to join a tech start-up in the hopes of making bank (they almost always lost their jobs after the overvalued company went under). Some may have hoped that the unfettered Mind of the internet had changed the rules of the game forever, but, when the tech bubble burst, all that capital with nowhere to go went into real estate, which soon became a new, old site of speculation and the formation of a new bubble. After the loans, debts, and defaults, thousands of people were out on the streets, face to face with how important Mitchell's bygone stones and timber really were.

There are, of course, still plenty of people who spend most of their time in virtual worlds, and others who look forward to uploading their minds onto computers (Hansell and Grassie, eds., 2011; More and Vita-More 2013). But, in general, Euro-American attitudes toward the digital seem less Panglossian these days, and for the moment, the dream of disembodiment no longer seems quite as believable or desirable as it was in the '90s. It is, after all, hard to believe in unfettered information as an intrinsic good in an era in which whatever can be thought or said, by anyone, can be deemed true *because* it is thought or said (or tweeted), and in which communication leads increasingly to walled-off worlds of sequestered storylines—infinite pizzagates of the mind—rather than some kind of utopian rapprochement. But, more than this, there is a general awareness that technology did not fundamentally change global relations of inequality and even the way things are made; rather, much older and enduring forms of labor like artisanal mining are gaining momentum in the ruins of more centralized, "Fordist" capitalism (Tsing 2017).

The growth of literature on ecological collapse in the Anthropocene is one among many indicators of a shift in thought, reminding us of ecological limits to technological innovation. In this and related literature, there has been a renewed focus on matter, epitomized by the Latourian literature on the agency of objects and the irreducibility of networks. This is echoed in literature on the digital age, as authors draw attention to materiality over virtuality: that "the cloud" is in fact made up of warehouses of wires cooled by Appalachian streams and underground networks of cable enabling immaterial data to zip along the trawled ocean floor (Blum 2013; Starosielski 2015; Parks 2015). Much of this literature maintains the idea of a mind/matter split, inverting their significance without challenging the polarity, and mostly without complicating the uninteresting banality of matter for that matter (although see, e.g., Massumi 2015 and Povinelli 2016 for ways to move beyond the materiality/immateriality divide).

It's critical to keep in mind that the Idea of Congo, and of Africa, and the Idea of the Digital have emerged together, in unacknowledged tandem, if not always in direct conversation with one another.[26] For example, the idea of an immaterial future based on transparent vision is not easily dissociated from the colonial-era unilineal cultural evolutionism of the nineteenth century, which, in Africa, became an instrument of colonial administration and exclusion (later morphing into the seemingly, but not really, value-free economic development narrative of the twentieth century). It was long taken for granted, in many quarters still is, that the West was leading the historical march of mind over matter and nature and that Africa was the place where the body and nature ruled supreme and were epitomized—the so-called state of nature, bereft of mind and reason. The Silicon Valley "religion of transparency," as Foer puts it, emerges from the Enlightenment project of shining the light of reason onto the irrational darkness that was imagined to lie in the hearts of people and in the heart of the emerging world system that Conrad began to unpack. How well this contemporary ideology of transparency, which these days comes to Congo in the form of auditing initiatives and other interventions, resonates with an older, mission-driven narrative about the need to bring light to a dark continent.

If Africa served as an unchanging benchmark for Hegel's dynamic story of progressive unfoldment—one that he at least thought enough about to feel the need to mention, if in passing—Africa is rarely even mentioned in Euro-American understandings of the contemporary, emerging only as a point of extreme contrast or juxtaposition, often in the form of a quick montage. Yet Africa remains an unspoken point of reference against which the progress of the technological is measured. Change continues to be depicted as coming to Africa from outside in the form of technology, or materialized mind, even though, if anything, and to paraphrase Jean and John

Comaroff, Europe is "evolving toward Africa" in multiple ways (Comaroff and Comaroff 2014). This imaginary division of the world into the material and immaterial—in which Africa is seen as epitomizing nature and the body, with mind coming to it from outside in the form of technology—is one of the reasons why so many technological interventions directed at Africa seem so explicitly colonial: whether it's about getting Africans to leapfrog into the digital age (unilineal cultural evolutionism), finding ways to make slums visually transparent to outsiders (bringing light to a "dark continent"), or protecting Africans from supply chains (generally cast as more evidence of Africa's primordial violence) through enclosure and monitoring that ends up benefiting foreign mining companies at the expense of excluded Congolese.

Euro-Americans may have imagined Africa to exist outside of the digital age (or only as a supplier of "natural" resources for the minds of others to use), only entering into it in the form of "contamination" (e.g., in the form of blood minerals). But a geographic division between "production" and "consumption," in which Africa is imagined to produce materials that others use, doesn't hold well here in part because eastern Congolese are very much "consumers" of digital products, with the fastest cell-phone uptake of just about anywhere in the world; they are also users of the most current technologies and platforms (Facebook, WhatsApp, Twitter). Moreover, mobile phones and smartphones are absolutely essential to the organization of the trade in digital minerals: knowing which routes and places are safe, knowing what the price is at any given location, negotiating various matters with buyers, and going over the heads of abusive state authorities all require cell phones, and these days internet connections (available through cell towers) can also help. Indeed, as my research progressed over time, miners and traders were more likely than me to be on their smartphones using Facebook, even in very geographically remote places unreachable by road. But to say that Congo was part of a digital age, as well as a mining age, does not mean that Congolese shared the same experience of digital technologies or that the technologies determined their experience of them in ways that were predictable (for example, eastern Congolese were less enamored of the Euro-modernist utopian telos of disembodiment and more attuned to the potential for mobility and immobilization-through-transparency/visualization).[27]

In sum, the ideal of digital disembodiment is deeply entrenched in Euro-American thought, rooted in the Christian and Enlightenment humanist

Figure 4. "Play Station 2 here" (home of the clean/upright people, or *chez wasafi*) (Kindu, Maniema)

dichotomy of mind and body; by the same token, Africa, "the very figure of the strange" for the West (Mbembe 2001) has been equated with the opposite of mind, hence the body and nature. This orientation comes to life in practical interventions that see Africa as synonymous with the body, the antithesis of mind, and which try to control the movement of African bodies to maintain the purity of mind, materialized in technology. There's therefore a tendency for Euro-Americans to double down on colonialist dualisms when they hear about coltan, sometimes inverting their value—demarcating Congo as the nature that is outside of culture or the source of the dirt in clean devices. Rather than reiterating the idea of a digital divide, or insisting on the importance of materiality over immateriality, following the practices and substances that comprise the so-called digital or information age shows how far-flung lives and places, and the virtual and the nonvirtual, are interconnected—highlighting the relevance of places and practices, like artisanal mining, that seem beyond or outside the digital and the contemporary. Following the embodied and conceptual practice of diggers and traders also drives home the fallacy of progressive transcendence from the earth through the digital, a bill of goods that we've been sold at grave risk to the world—mainly, that through computerization we might finally free ourselves from all limits, including ecological ones. Paying attention to the lives and arrangements of Congolese diggers may also help us to break out of the false dichotomy of materiality and immateriality, in part (but not only) because the grounded "material" work of miners also involves engaging with the disembodied spirits that make minerals available (see also Coyle 2020; Adunbe 2015; High 2017). The point is not mainly to draw our attention to embodied suffering or abjection in the Congo (though these are real enough and irreducible to the digital age), but to see how certain forms of work that emerge in the wake of destruction and collapse challenge taken-for-granted understandings of the world while offering alternative models for being together in the world (Tsing 2014).

In contemporary eastern Congo, the long-term Congolese practice and ethics of movement—emerging partly in response to the historic experience of violent enclosure—once again confronts the Euro-American Enlightenment vision of transparency, practices of exclusion and extraction, and the idea of Africa/Congo as the "heart of darkness." As we'll see, some Congolese who experienced the collapsed market for Congolese minerals understood that Wazungu (whites) felt that the potential presence of Congolese "blood" in technologies such as the iPhone made these technologies impure for Europeans. According to this understanding, these consumers of particularly expensive digital devices like iPhones were paying for something that they wanted to own in a clean form, "untainted" by their bloody earth and bodies (hence disembodied), so some of those involved in the trade and its regulation helped to create a system (or at least the appearance of a system) that allowed Congolese to experience movement while at least creating the illusion of cleanliness (*usafi*) for foreign others (this system is discussed at length in chapter 9).

"Are You Here to Take Pictures?" Visibility, Immobilization, and a Congolese Digital Vernacular

"Are you here to take pictures?" In 2013, the head of the local diggers' cooperative at a coltan mine not far from Goma, in North Kivu, is concerned—not so much because of the pictures I might take, but because of the fake ones I might use. Humanitarian NGOs have made a business out of taking photographs from other mines where there are soldiers and child workers, uploading them to the internet, and using them in their online publications to depict this particular mine or mines like it—places where there are in fact no child laborers, pregnant women, or soldiers (for a mine to be considered "free of blood" according to the Congolese Mining Code, it needs to not have any of these). His apparent empathy about this practice surprises me a bit. "NGOs are a business," he explains, "If they can't show that there's blood [at the mine], they won't have jobs." He is also a businessman, so this is not a stretch for him to understand. Like everyone involved in the mining business, he was hit hard by "Obama's Law" (meaning Section 1502 of the Dodd-Frank Act) and Joseph Kabila's 2010 closure of artisanal mining, and he has witnessed the real consequences of what people in the mining business refer to as the "government's whip." The phrase connects the postcolonial regime's deployment of sudden, violent rupture (in this case in the form of a sudden, temporary ban on mining) to the brutal colonial past.

Although the main referent is the humanitarian industry centered in Goma, the digger's comment does reflect something about his experience of digital technologies—mainly, that they have rendered his work "visible" in a way that is false and which actually conceals the truth while, in the process, also impeding his movement, or his ability to move minerals and his life forward. No one has bothered to check up on the facts—in this case, which mine is which—because the people who do the work of creating reports are rarely willing to take the time or the risk to actually go to the places where minerals come from or to follow up on those who have. Moreover, a business has emerged around a culturally resonant Euro-American narrative about the bleeding heart of darkness, which has a past that is difficult and (for some) not profitable to disrupt.

The immediate background to the digger's question is that, since the coltan boom of 2000, journalists and Congolese and international NGOs had been writing reports and taking photos and videos to show that the demand for Congolese minerals like coltan was contributing to a war economy while enriching the nations of neighboring countries like Rwanda and Uganda as well as foreign companies. As one Congolese NGO director in Bukavu said to me in 2016, "We [the NGOs] told [President Joseph Kabila] that war is continuing because [his] soldiers are out there forcing people to dig and killing people. People like Colonel Sammy and Tango 4. And when Kabila got this pressure, he decided to close minerals (kufunga madini)." But Kabila's directive was ineffective in and of itself: while the goal was to remove soldiers from mining, the mines continued to operate and the soldiers continued to protect the "closed" sites, even if they were not physically on-site. Moreover, soldiers still brought in their own people to dig, even if they were not physically at the mines. The embargo was felt in earnest only when the awareness of consumers in the Global North combined with the strictures imposed by Section 1502 of the Dodd-

Frank Act: "When more people started to realize that their phones had the blood of Congolese, they said, 'We don't want to continue to touch the blood of people,' and companies started refusing to buy." As the director explained, "People really started paying attention when that movie *Blood in the Mobile* came out." Ironically, this was seven years after the Second Congo War had ended. "What worked about that movie," the Congolese NGO director continued, was its visual and virtual immediacy. "It showed ordinary people how minerals have blood in them. Now you see if you hold a mobile phone, you're touching the blood of people who have died in Congo." In the understanding of the Congolese NGO director, then, seeing (through film and reports) was connected to an awareness of touching, which ultimately led to "closure" and a cessation of Congolese movement in the mining trade, creating a new set of problems.

In posing the question "Are you here to take pictures?" this digger was implicitly identifying characteristics of digital technology that are now familiar to just about everyone in the world. He did not use the now ubiquitous phrase "fake news" (this was, after all, a few years before that became a thing), but if he had known of it in 2013, he very well might have. At the same time as he is attuned to an aspect of the digital that is globally experienced, it's clear that his understanding of technology emerges in the way it does for reasons that are geographically and historically specific. The context that grounds his experience of technology is at once contemporary (having to do with the ubiquity of humanitarian NGOs in tentatively postwar Congo) and historic (having to do with the *longue durée* of postcolonial exclusion and surveillance). His understanding of how the technological "eyes of the world" are targeting Congo was also shaped by an even deeper history having to do with the idea of Congo as the heart of darkness going back to the time of King Leopold and his knowledge of the fact that the whites who are watching Congo, and him, see his place through this prism (for example, it is common to overhear European expats in Congo referring to the "law of the jungle").

This concern about being watched by the "eyes of the world"—both the extractors of minerals and the humanitarian industry, which were really the same—was a theme informing the way eastern Congolese thought about their relationship to the world and the impact of technology on their lives. As already alluded to, many Congolese (not just miners) were rather fixated on the idea that they were being watched by satellites—either those of the government in Kinshasa or ones belonging to the "international community." When they gathered together for rituals of state, as when President Joseph Kabila visited provincial capitals, they'd look up to the sky for these satellites that were watching their every move. When there were conflicts at the mines, the heads of diggers' cooperatives often expressed concern that the "international community" was watching their conflicts through satellites in real time (they worried that if the world saw conflict, whites would stop buying their minerals because blood minerals were "taboo" to whites). People believed that every act of past violence (bodies thrown into a mass grave during the war) and every future promise (resources buried in the ground) were being seen by these satellites and super-powered cameras (some interlocutors told me that, no matter what I said, they would always be convinced that there were cameras—not diamonds—on the soles of my shoes). And, of course, they knew that the true purpose of UN operatives was to recover the minerals that their colonial-era grandparents had stashed (a mythology that drew a direct connection between colonialism and the UN system); they needed digital technology and surveillance equipment to do this effectively. All of this was understood to be a high-tech extension of colonial vision: colonial authorities had come with maps and other devices that allowed them to identify Congolese wealth that Congolese hadn't known to exist, and it was through these instruments that Congolese came to know about this wealth, to have land taken away from them, and to be forced to work for others. In turn, many humanitarian interventions—especially the bag-and-tag schemes—were understood to be efforts to make Congo transparent to others in order to control these places and the people who lived in them while remaining invisible themselves.

I'm not arguing that eastern Congolese were opposed to all forms of technological transparency—indeed, they often sought to appropriate and redeploy visualizing technologies for their own projects (Poggiali 2016).[28] Rather, they were aware of its colonial, and colonizing, potentials when wedded to powerful institutions that oppressed them. In particular, transparency threatened movement as this condition came to life in artisanal mining—a powerful multidimensional practice with the capacity to transform space and time so that the world emerged anew through openings, or "holes," in the ground that resisted transparent visibility.

Plan of the Book

The remainder of this book is divided into two parts. Part 2 focuses on what minerals and digging make and mean to people in Congo—the worlds that are produced through these substances and the practices surrounding them, focusing on the eye-opening experience of war (chapter 2), the practices of multidimensional movement in and through the supply chain (chapter 3), and geographic locations (ruins) that become vortexes, or focal points for action and imagination (chapter 4). Part 3 (chapters 5–9) consists of shorter and quicker, more staccato and less sprawling, narratives, as the themes and histories that comprise part 2 of the book come together in an event—the eruption of the vortex (or "wormhole") of Bisie in the rainforest of North Kivu. Chapter 9, "Game of Tags," examines the impacts of a "bag-and-tag" conflict minerals tracking project from the points of view of those involved in the trade.

In developing the theme of artisanal mining as generative transdimensional collaboration, I have tried to employ a form that mirrors the movement that I'm trying to convey. I begin with a wide lens, employing an undulating narrative style that seeks to capture the multiplicity and movement of mining in eastern Congo and of this multisited, longitudinal research. The structure is intended to capture the tumult of events and history in which this region has been caught up (there is no "history section," for example, but histories—or rather, various pasts—emerge in relevant places throughout the text). Ultimately, the narrative zooms in closer, culminating in the section on Bisie—an actual vortex, and hole, that connected the Congolese rainforest to the rest of the world in a direct and immediate way. The forces unleashed by Bisie eventually brought attempts to control, exclude, and regulate, the consequences of which are still ramifying throughout eastern Congo in the form of digitized tagging schemes.

PART TWO

Mining Worlds

TWO

War Stories: Seeing the World through War

You thought you could take this country, but look at us now
We know you are here, but some day you'll be sent away
And that day is very soon
Our eyes are on you. We see everything you do.

—Song performed by a band of women musicians in an open-air Kindu bar
in 2013 (italics mine)

This chapter concerns the productivity of war—specifically, the ways of seeing and social infrastructure that emerged amidst the displacement, dispossession, and violent destruction of social relations and ordinary time,
especially but not only during the Great War.[1] The main goal is to harness
the insights and experiences of my interlocutors to show some of the ways
in which artisanal mining became important and meaningful to people beyond economic need and the absence of other alternatives. Of all the data
I gathered during fieldwork, stories about war and the violence of war were
among the things that my interlocutors most wanted me to write down
and tell the world about. On a basic level, they had suffered at the hands
of others, and they wanted people to know about it. In general, they felt
that neighboring countries, in particular Rwanda, had benefited from a discourse about the Rwandan genocide, but that some of these victims went
on to wage a genocide in eastern Congo that the world remained ignorant or silent about. Most people felt that this genocide, and the world's silence about it, had something to do with high-level (i.e., global) politics—
including the American demand for Congolese minerals.

The war also provides a deeper context for understanding the meanings of minerals and artisanal mining for those who engage in this work,

and this is also one of the topics that some people wanted to bring up. Specifically, war and stories about war drew attention to the regional and global demand for Congolese earth, and the power of that earth, which took a concrete, specific form in substances with unique and different qualities and spatiotemporal capacities (e.g., gold, diamonds, and coltan). The violence of war served as an extreme point of contrast to the positive aspects of movement and embodiment that digging enabled and to the relationship-building and rhythmic tempo that artisanal mining produces and draws upon. Finally, war helped produce the current social and spatio-temporal arrangements that made mining possible while also setting the stage for the growth of artisanal mining and the emerging resistance of artisanal miners to foreign mining companies. War did this by threatening and displacing families, making people available for mining while rendering other activities less remunerative or possible, introducing a new system for moving minerals, and generating a shared awareness—spread through observations and storytelling—regarding the forces, people, and companies that threatened Congo. These conflicts would come full circle later on in my research, as the displaced army of diggers found themselves pitted against foreign mining companies, some of which were protected by former soldiers.

In any event, the subject of war was inescapable in and around the mining enclaves of the eastern Congo during my fieldwork. Memories of war and portents of war to come haunted many conversations and at times seemed to come into view, erupting again in the present, such as when the Rwandan-backed militia M23 took the city of Goma in 2011 or when Raia Mtomboki, "the infuriated citizens," invaded Walikale in 2015. Like an earth tremor along a fault line, war was understood to be an ever-present potential behind the veil of daily reality and also a process that continued to take place beneath apparent peace. Because talking about war almost always entailed some attunement to surfaces and depths and the disjunctions between them, war was also generative of moral and philosophical debate (see also Ferme 2001, 2018). The nature and causes of war were not immediately visible, and it was widely understood that the apparent combatants (the ones engaged in the fighting) were in some fundamental way not the real combatants, who, in reality, hid behind the scenes and worked, invisible to most, through their proxies. The real warlords—powerful entities manipulating militias and less powerful nations like "marionettes," as some people put it—had to be discerned and interpreted. After all, "war comes from above," many repeated—though, as many also pointed out, war was also facilitated by people acting from "below" and was fed by low-level instigators (including villagers or family members who were in conflict with one another) pursuing their own particular projects and agendas, taking advantage of the existence of militias to exact violence in ways they wouldn't have been able to otherwise.

Most academic discourse about the Congo wars has been preoccupied with the determination of "causes" of conflict ("Minerals didn't cause the war! It was ethnicized land-tenure practices dating back to the colonial period!"). These discussions tend to shift back and forth between local and global determinants—mainly, between the global demand for resources, on the one hand, and local ethnicized conflicts with a colonial history, on the other. But one of the main points eastern Congolese who actually experienced the wars firsthand make about war is that it opened a vortex in which different places, dimensions, and times became visible and concrete and that it collapsed the multiple scales at work in the world, bringing them all into existence together at once, even in the most geographically remote places. If we put aside the debates concerning the causes of the Congo Wars—and the shifting back and forth between local and global explanations—and instead focus on the lived experiences and insights of people whose eyes were opened by war, we get a better sense of not only how war is understood but also how eastern Congolese today are directing their actions as well as the futures they are imagining and producing. Again, this leads to a deeper understanding of the significance that artisanal mining and minerals hold for many.

One of the main things people dwelled on was the way war forever upended ordinary life and experience, polluting the memory of it and making it difficult (to this day) to return to ordinary routines. For example, in his efforts to convey an extreme but exemplary case of the violence of the Rwandan-backed invaders, one man in Kalima, Maniema, narrated the story of a woman whom, he alleged, Rwandan-backed Rally for Congolese Democracy (RCD) soldiers compelled to grind her baby's head and body in a large mortar using a giant pestle ordinarily used for cassava. How, he asked me rhetorically, would it be possible for that woman, or any other person seeing the event, to return to this ordinary and necessary task in the present after the war? How could one prepare food, and if one couldn't prepare food, how could one eat, raise children, and generally go on living? Sociality, and the possible futures emerging from sociality, had been rendered impossible because of a preponderance of memories of violence connected to otherwise ordinary activities. In a related vein, people lost the ability to build relations through things or even to touch the things that were theirs (note the contrast with the touching of money that digging enables): your things could be taken away from you by soldiers, and you could be made to carry them away for these same soldiers (an extremely common story that nearly everyone had some version of). If you tried to touch those things again, to get them back, you might be killed (a number of people also told stories around this theme, involving innocent but ill-begotten efforts to get back one's radio or car battery or what have you from soldiers). Moreover, the ideal of stable domesticity was violently imploded and often seemed to be purposefully mocked by assailants during violent invasions (e.g., the Rwandan-backed invaders, in particular, were commonly described as "guests who raped," a phrase that drew attention to the implosion of the family as well as to the fact that war brought together elements whose combination was inconceivable, such as the concepts of guest and of rape). In general, war put bodies together and tore bodies apart while instrumentalizing bodies and body parts in ways that were at once new and old (redolent of earlier colonial atrocities, for example); war also tore people away from embodied social situations and their textured, incrementally made lives and social worlds. In doing so, war dispossessed people of their capacity to produce incremental time through social relations and social relations through incremental time (Smith 2011).

In disrupting people's ordinary lives, war had also brought an altered state of consciousness, including an awareness of violence, of aspects of the Congolese past (for example, the cruelties committed during the "time of the Belgians"), of minerals, of technology, and of the way the world was put together to constitute a system. People touched and were touched by these suddenly visible interpenetrating spatial and temporal scales, often in ways that were very painful. For many, minerals were one of the main media through which these forces and scales came together and through which they could be touched. Many spoke of the knowledge brought about by war as a movement from a state of childish ignorance to one of adulthood. As one man who grew up in a mining zone remembered, "When we were kids, during the time of Mobutu, we were ignorant of war. We didn't know what it was when the AFDL [Laurent Kabila's army, which invaded during the First Congo War] came. We thought it was a game (mchezo). But we saw with our eyes, and we will never allow that to happen again, to be invaded again." He placed his fingers around his eyes to widen them in a demonstration of what open eyes look like. In addition to "opening the eyes" (kunfunguka macho), many people said that war helped them "to see clearly" (kuona claire), implying discovery of hidden secrets. One person, who had worked for the UN during the Second Congo War and had direct experience with many terrible instances of violence, compared the wars to

the Hundred Years' War in Europe—a time he understood to be one of total destruction but also of rebirth. Like many others, he also used the motif
of "seeing" (*kuona*) to describe war as at once terrible and productive:

> War was terrible, but it was so important because it opened our eyes to so
> many things. Before, we had heard stories about the *Wasimba* [Referring to
> the Simba Rebellion, a widespread 1960s resistance movement initiated by
> Lumumba supporters following his assassination] and the power of forests,
> but now we were seeing these things for real in our daily lives [in the indig-
> enous militias known as Mai Mai]! And we were being recruited [by ances-
> tors] in dreams to get this power. And we learned about the value of minerals
> and the prices minerals are getting. About how the nations are connected to
> one another, their plan for Congo. And we learned what a body can do, the
> uses of a body.

I will return to the idea of "the uses of a body" below, but here I would
just point out my friend's focus on visibility and, more specifically, on
learning to see through the murkiness of war and to observe the multiple
forces and scales operating during war while also being exposed to the gaze
of other people (mainly of "armed actors" but also the buyers of minerals,
who were looking for Congo's resources). I am not suggesting that one way,
or a single regime, of seeing came to dominate during war, and I'm also not
interested here in trying to identify the different regimes of visibility. Suffice it to say that the way in which Congo and its minerals became visible
to foreign others was not exactly the same as the way in which ancestors
became visible to Mai Mai insurgents or the way in which the cargo planes,
going back and forth carrying minerals, became visible to people for the
first time. Or how the prices of 3T minerals became visible to people who
had not known what they were or what their value was before, or how
the secret mineral business of the UN became visible, or how old colonial
maps suddenly appeared in the forest, as people went on treasure hunts
looking for coltan. Or how suddenly there were state officials with their papers in the forest who had not been seen there before. Suffice it to say that
war brought hidden things out into the open to be seen at the same time as
it produced dramatic volatility and uncertainty.

This was a theme with many different permutations: for example, gold
and coltan were sold out in the open during the war and for years afterward. Unscrupulous people also came out into the open during war and
later disappeared—mostly soldier types but also civilians who, during or
immediately after the time of war, used fake papers obtained from the pro-

visional government to take away the property of their deceased friends' wives, or who set up their friends in fake gold deals that ended up with the friends being mugged at gunpoint by soldiers in the less visible outskirts of town. Question: "Where is that guy who did such and such? Where are those types of people these days?" Answer: "Oh, they're still around; they're just hiding right now." What was happening was that the secret potential of ordinary social relations (of friendships, for example) was becoming visible to people, as was the ability to "crack open" friendship, to accumulate by using friendship as a resource, dispossessing your friend of her things and yourself of a friend. War removed certain limitations, such that nearly anything, or anything invested with power, could become a useful resource, and this fact was also becoming visible because it was taking concrete forms.

During war, being visible to armed actors often meant losing control of your body and your things, such as when the soldiers follow the visible trail of beer bottles to where the diggers are and then force them to work or carry their own stuff away for the soldiers. Ordinary activities like cooking, eating, and drinking became dangerous because people with guns were watching and making mental connections after focusing in on material objects. If a new group of soldiers came to your village and saw you with matches or salt (even worse if there were canned sardines or cookies), it meant that you had been in contact with or "hosted" another armed group (their enemies) because during that time all the roads were closed off and only armed groups had these commodities. If a village was deemed to have hosted the enemy (*adui*), the whole village could be punished for something they had no control over. But, at the same time, seeing the ways in which things were connected to one another and how Congolese earth was connected to global networks became an important way for ordinary people to survive and sometimes even prosper. Or at least they might recoup some of what they had lost. During the wars, new things and technologies that had never been seen before (cell phones, for example) suddenly became visible ("What is that thing that soldier has?"), and their importance and significance became clear to people. The plans of powerful foreign others (the Americans, the Chinese and, at a more everyday scale, certain mining companies) also became visible and concrete. A major part of what was becoming visible was the fact of Congo's dispossession by others who were accumulating at the expense of *Wa Kongomani* ("Congo man" or Congolese) and, in turn, *Wa Kongomani*'s dispossession from "resources" that connected them to the forces that governed mining, like ancestors.

I want to develop this theme through a speculative story told in South

Kivu, an example of the larger regional analytics that have emerged to think about war. In this province hit hard by war, the themes of visibility and immobilization come to life in rumors about an international trade in the ropes people use to hang themselves or to murder others.

Some Ropes Are Resources That Carry the Trace of War's Power[2]

According to a widespread rumor, there is an international trade in suffering that only people in South Kivu really know about—you might say they discovered it. This market takes the concrete form of a trade in murder and suicide ropes (usually raffia ropes) and knives—instruments that have been used to murder people, or to commit suicide, and that still retain the trace of those deaths and of the victims' suffering. In this trade, powerful whites who reside outside of the country are the end-of-chain buyers, while Congolese middlepersons are the suppliers. Although most people claim not to know much about the trade, they are clear about the fact that what is being bought are the spirits (*roho*) and affects of the deceased, registered in the ropes and knives just prior to death. The spirits of the dead reside in and can be sensed on the ropes; indeed, through the ropes, the dead are made visible and sensible, as well as effective for others, their end-market "consumers." According to those South Kivutiens who are attuned to this topic, ropes are especially valuable (more so than knives) because of the amount of time that deceased persons are likely to have lingered on them prior to passing and because of the material qualities of the ropes. The deceased's life force and affects (including but not limited to fear) have seeped into the fabric of the rope, which retains and registers this force and these emotions. The rope is in a sense alive with this force, and what is more, that life energy can be measured—people use the word tested (*kupimwa*), the same word they use for minerals—with the right technology.

Many said that this trade in ropes went back to the time of the Belgians and so the ropes and the stories about ropes also made visible the connection between the colonial past and events happening in the present. While no one specifically mentioned the colonial-era use of rope to immobilize people in forced labor (when ropes were used to tie people together in a line) or as punishment (people being painfully tied down or tied together in public), when I began putting together all the material in this book, I interpreted a connection (see, for example, the story of Diamond in chapter 4). Similarly, because the ropes represent the death of a person and are supposed to be proof of that death, they bear a resemblance to the violence of the CFS-era rubber boom, during which body parts (hands) stood for the death of the victim while ensuring soldiers' compensation. (Though no one ever mentioned this past to me or suggested this connection, authors have pointed out the existence of a kind of economy in severed hands, which was detached from the rubber economy that it was supposed to have indexed and encouraged.)[3]

As with other high-value resources desired by outsiders, South Kivu-tiens do not know the exact purpose of owning these ropes, and this lack of knowledge regarding markets and end uses is part of the meaning of these rumors. But most assume that this force gives power to its possessors and enables them to hurt others—to spread violence on a global scale and to benefit from that violence (perhaps, tangibly, in the form of a promo-tion). In this way, the dead become sacrifices for the war makers, enabling them to continue spreading more death and becoming powerful through death. The ropes are thus instruments and symbols of necropolitics, or the organization of power around the production of death (Mbembe 2019), par excellence, and they in turn make necropolitics concrete and visible. As one South Kivu woman put it to me, "This is not [the kind of] power that can bring development to a place. It's power that enables [them] to do bad things, to bring war to Congo and to ruin the world."

The agents who are involved in the rope trade are said to be the same state-connected ones that are involved in the mining trade, and the ropes and minerals follow the same channels and end up in the same hands. As with minerals, the ability to use technology to see the power poten-tial of the object is intimately connected to the ability to control the "re-source," to extract value from it, and to spread more violence. Technology unknown to Congolese enables buyers to know which ropes are authentic, meaning they have actually come into contact with a person's life force for extended periods and are not fake (there is also said to be a trade in counterfeit ropes). Like mineral comptoirs, the end-of-chain buyers are said to have cameras that can "see" the spirit/force on the rope or a com-puter into which they can feed the ropes in the same way they do with minerals. Some speak of a hypodermic needle that is used to give the rope an injection; if the rope jumps, it is an authentic murder rope because it is animated by the life of the person who died, whose life force and emotions are still attached to the rope.

While the eyes of the world are said to be on these ropes, the market is so inscrutable and elusive that it is not always clear that it still exists—or ever existed. Some people say that the international buyers were in Congo for a time and are no longer around. That would mean that now people are responding to a market demand that no longer exists when they kill others on spec and try to sell the ropes in the hopes of a payoff. This underscores the more broadly relevant point that, while the whites who buy the ropes "see ahead" (*kuona mbele*) and plan for the future on a global scale, the middlepersons are able to see only the immediate demand in front of them and have little understanding of the larger picture. This lack of knowledge is paralleled in the mineral trade: diggers know there is a world market price for coltan (*bei ya soko ya dunia*), but they don't know what it is or what's behind it (they typically imagine the "world market" to be an actual physical place where minerals are traded out in the open by people from all over the world).

Are people trading these ropes, and are they real "murder ropes"? Or are people actually murdering others in the hopes of accessing this market? These matters are difficult to know. On several occasions, police in South Kivu have arrested people they believed to be involved in the trade; in a couple of cases, the alleged perpetrator is said to have confessed, but in each case the rope was said to have already been passed on to the buyers and was never found. Some people insist that there are those who have been persuaded into murdering others through the incentive of a return. The rumor is that the perpetrators were instructed by an intermediary or via a cell-phone call to murder so-and-so and, when the deed was complete, to call a certain number so that the pickup of the rope or knife would be arranged and the money could be transferred via cell phone. In these stories, invariably, the "buyer" never comes to pick up the rope or knife, thus revealing the stupidity of the middleperson, now an unpaid assassin duped by his greed and lack of knowledge of the "true" nature of the market.

In some versions of the rope rumor, there's an interesting twist involving humanitarian organizations: it turns out there was an international humanitarian NGO—some say it was German—that wanted to know the causes of suffering in Congo. They developed a technology that entailed inserting the ropes of suicide victims into a computer that was able to read and thus to see the thoughts of the person and everything that was bothering them when they died. The idea is that the NGO originally wanted to help Congolese but, because they were offering money to people who would bring them ropes, they ended up unwittingly incentivizing murder because they didn't appreciate how bereft and desperate the war had made people and how some had become inured to violence. When the NGO left, people didn't realize that it took the market with it, and so today they are still trying to sell to a market that no longer exists. According to this perspective, the naivety and temporal inconsistency of NGOs—here one minute and gone the next—precipitated violence and death.

This idea made sense in the larger context of a tentatively postwar economy dependent on mobile, foreign NGOs.[4] It also sat alongside a larger vernacular critique of the international humanitarian community as a business with its own agenda, in which humanitarianism was a fake veneer concealing direct involvement in war and extractive economies (as mentioned, it was commonly held that UN officials were there to recoup the wealth of their grandparents who had worked for mining companies). Or, in a more generous reading, genuine humanitarian intentions had been hijacked by unscrupulous actors from outside the humanitarian field. One extreme example of this theme was the rumor, in the east, that UN representatives murdered Congolese and cannibalized their flesh. On a more prosaic (and empirically verifiable) level, it is common knowledge in Congo that certain UN actors and NGOs have long been trading illegally in minerals (including what could be called "conflict minerals") under the guise of humanitarianism (for example, during my fieldwork, armed groups regularly used the vehicles of humanitarian NGOs to transport coltan out of Congo and into Rwanda). And most everyone asserts that these organizations (particularly Monusco, the military branch of the UN) have interceded on behalf of occupying militias, at times seeming to hold the Congolese army at bay when it has tried to reassert national sovereignty against the Forces for the Democratic Liberation of Rwanda (FDLR) or the National Congress for the Defense of the People (CNDP) (which led some people to believe that the UN was on the side of these illegal armed groups, perhaps with a view to acquiring Congolese resources). A more quotidian version of the corruption of humanitarianism (which I once myself witnessed) was fake Congolese "NGOs" coming to rural areas from the city, promising to assist women who had been raped during the war, poor families, and women entrepreneurs; they collected fees for registration in advance from women "members" in anticipation of a future time in which the white "owners" of the NGO would come from abroad to distribute aid (money for school fees, small businesses, etc.). In the case I observed in the rainforest town of Punia, Maniema, the NGO disappeared with the money its members had collected over the course of months, without even paying for their food and lodging at the Catholic church.

There is, in short, what Michel Foucault referred to as an "analytics of power" at work in these rumors, a grappling with how power is exercised in specific places and times (1990; see also Ferme 2001), here specifically regarding the economization of war and the deeper implications of extraction (see also Jackson 2002). The ropes, which are rarely if ever actually seen, and, more importantly, *the stories about ropes* make visible the hidden secrets that are crucial for understanding how the world works. They reveal the dead, the history of violence in Congo before and during the war, and the powerful social networks that govern the world. Stories about murder/suicide ropes also reveal something about the dangers entailed in being compelled to sell whatever one can get hold of—of living in thrall to sudden shifts in global market demand with nothing else to fall back on. Finally, these stories reflect local thought concerning the uses of (post) colonial technologies related to visualization—that there are technologies and techniques for seeing which, by making Congolese visible, may also endanger their lives. I believe the stories also suggest that Congolese have to develop their own ways of seeing, as well as ways of remaining invisible to others, if they are to avoid having their bodies converted into sources of profit and power for these powerful others.[5] Storytelling is, of course, one of those ways of seeing, and part of what was important and transformative about war was that it made stories that had been told in the past become "real" in the present, later to be narrated as stories.

The Uses of a Body: Imploding Spatiotemporal Orders

Most people associated the time before the war with the political and cultural order surrounding President Mobutu Sese Seko. They looked back on this time as static and repetitive: "Things went on and on without change," as several people put it. While the vast majority of people were not nostalgic about that period (a few older people certainly were, and they sometimes wore the dress associated with that time to communicate something like this), they did see it as a time of relative innocence. For example, people pointed out how different the behavior of Congolese soldiers was before the war, in comparison to today; their memories tended not to extend to the longer history of violence under the Force Publique, going back to the time of King Leopold. Some noted how it was possible to engage in a tactile, convivially conflictual way with armed state representatives without having an experience that threatened to culminate in death (for more on embodied politics and resistance in Congo, see Covington-Ward 2016). Several explained that, during the time of Mobutu, soldiers would often stop you and demand, in Lingala, that you "Raise your hands!" (*moboko likolo!* in Lingala, or *haut les mains!* in French). After that, they would take everything you had. When they were finished, they would pose the question "Have we stolen from you?" (in Lingala, *tobotoli yo?*). You were supposed to answer "No." Then they'd repeat the question, perhaps with a slight variation: "Have we cheated you?" Again, the answer was supposed to be "No" or "No, I gave it to you." But if you did fight back or engage the soldiers in argument, the soldiers would put their weapons to the side and fight with their fists rather than with guns. As one middle-aged man remembered,

> They knew amazing fighting techniques, and they were big guys. Real soldiers! Not like soldiers today. They'd say, "Oh, you think you're a soldier? Come on, let's go," and they'd beat you with their bare hands [laughs with apparent fondness]. But they could never kill you, no! Just steal from you. These days a soldier will just pull out an AK-47 and shoot you. This habit of killing came from the Rwandan invaders.

There was a meaningful system at work here, even if some might call it paternalistic and kleptocratic, based on a concept of reciprocity (protection in exchange for money), and the fist-fighting, however unfair, was one tangible way in which this conflict-ridden collaboration was enacted. That system was also understood to have constituted a long-enduring, spatially bounded national family with Mobutu as father, and in these narrations, the beatings from soldiers are made to seem like fatherly punishments in retrospective comparison to the present (Schatzberg 1988; Covington-Ward 2016). After the war, that old family system died, and it came to be associated with death at the same time as people found themselves enveloped within a larger spatiotemporal regime that was changing the scale and pace of life and belonging ("This habit of killing came from Rwandan invaders"). People joked about the Congolese government being "dead," killed by the war, and they meant it literally. As one man put it, "During that time [of war], when you were dealing with a state official, it was like talking to a memory or a corpse. Or someone dressed up in the costume of the dead. Even now [2018], when people talk about the law, it's like they're talking about the wishes of ancestors, rather than a system that actually exists."

By far the most common way that people talked about how war opened their eyes was through stories that highlighted extreme violence, the ruthlessness of the enemy, and how war turned otherwise ordinary people into monsters. These narratives also focused on how people's capacity for mobility and incremental progress had been destroyed and how home had been turned into a place that was no longer safe and could no longer be lived in—it was not home anymore, in short. This implosion made people mobile in ways that were new and often indistinguishable from being immobilized (being forced to flee to the forest, for example). People usually focused on how violence upended family life, invading the assumed security of the household and imploding it while making family intimacy impossible. Many thought the invaders committed acts of violence to make their victims docile so they would dig minerals for the armies without compensation or resistance; others felt there was no politico-economic justification, that the invaders did it because it made them "happy" (people felt this about the "invading" armies—rarely if ever the Mai Mai, of which see below). Perhaps they did these things, some felt, because they came from Rwanda or were formerly disrespected people in Congo (e.g., the Banyamulenge in South Kivu). According to this view, they felt resentment about their own situation with respect to the comparative wealth of other Congolese (there was often an element of empathy or pity for the perpetrators in these accounts). At the same time, many recognized that these atrocities had historical precedents, that current violence was an iteration of past violence, and that the past was reemerging before people's eyes in the present. (Leopold's Congo Free State—especially the violence the Force Publique unleashed on villages during the rubber boom—and the Belgian Congo were main referents, but in Maniema the Zanzibari slave traders also came up.) A few people narrated stories they had heard regarding these historic pasts, showing the resemblance to the violence and forced labor they had seen during war.

Many narrated instances of spectacular rape, in which armies (almost always the RCD or CNDP) publicly humiliated families, implanting images that made it impossible for people to go back to their ordinary lives.[6] They claimed that these armies (particularly the RCD and CNDP) created a boisterous atmosphere, signaling their arrival with singing that could be heard long before they showed up, heralding the end of the world. People noted how these attacks uncannily resembled wedding processions, or at least sounded like them at first ("they sounded like they were on their way to a wedding"), and how this familiar-sounding thing turned out to be its opposite, something that destroyed sociality and made being together impossible for a long time afterward. In expressing what seemed like joy or exuberance, it sometimes seemed as if the perpetrators sapped the collective effervescence they felt from the energy of the social relations they destroyed. Many people described scenes in which soldiers raped people—men and women—in front of others, often in front of the spouses. They described scenarios in which family members were compelled to have sex with one another (fathers with daughters, mothers with sons), often with the laughing encouragement of soldiers. Some people talked about how the RCD soldiers prevented people from turning away and acted as if they wanted to get them to see the fun in it—they wanted the victims' eyes to be opened, so they would remember. When it was over, sometimes they'd perform conviviality by patting the victims on the back and telling them what a good job they did. One of the main takeaways of these stories was always that the victims couldn't go on living together, and so society dissolved from the bottom up; the victims of rape weren't only isolated individuals but the entire social fabric. The memory of violence seeped into family life in the same way that the memory of death seeped into the fabric some people used to kill themselves or others.

The destruction of social worlds came to life in and was remembered through materials and substances that could be seen and touched; some of these substances, such as minerals, also had the potential to enable movement and the development of relations with others, which is one of the reasons they were so fraught and so struggled over. They could thus be used to train others to submit. One man from Maniema, emphasizing the relationship among violence, seeing, and memory, talked about the time a Mai Mai group took his village to a river in the forest to dig for gold. After a while, the soldiers caught one man bent over in the river, surreptitiously pocketing a nugget of gold. After grabbing him and seizing the gold nugget from him, the soldier held it up for all to see and asked the villagers, loudly and ominously, what should be done to the man, their fellow villager. Silence. So one soldier commanded, "OK, stop your work and watch!" The soldiers murdered the "thief" in front of his fellow villagers and, instead of allowing him to be carried home to be buried, forced the villagers to bury the victim on-site so that his spirit wouldn't be able to find its way to the village to continue "living."

While they were training their victims' eyes on the violence, perpetrators were also concerned about being seen. They realized that they were committing acts of violence in a temporary moment of war and that eventually things would go back to normal, so they usually acted quickly, often making efforts to conceal their identities by wearing big jackets and baseball caps pulled down low. They were trying to quickly get away with something before the "eyes of the world" caught up with them. But when the humanitarian industry did get involved in Congo, it gave a new impetus to violence, becoming another instrument and object of necropolitics.

For example, one common explanation of rape, especially mass rape, was that it was a strategy, for male soldiers, for attracting the eyes of the world in order to get things they wanted—in this view, violence against women and all that women made possible became a path to success for some precisely because the eyes of the world were on Congo. Many former RCD and CNDP soldiers insisted that rape was primarily a form of revolt by conscripted junior soldiers (almost all of whom had been kid-napped and forced to serve) against officers, who were allowed to have women companions or what were sometimes described as concubines. This was understood to be a way of shaming the officers while also po-tentially putting them in trouble with their superiors or the "international community," since the blame for rape would ultimately fall on the officers' shoulders. In this explanation, rape was politics by other means, carried out between men over the bodies of women, with a view to high-level mili-tary commanders and the international community. A related explanation, reserved for specific instances of mass rape, was that perpetrators commit-ted horrendous acts to garner the attention of the international community (e.g., the Red Cross and Amnesty International) with a view to negotiating a salaried position in the Congolese army (FARDC). This was said to have happened especially during and just prior to the period known as *mixage*, when diverse militias were being incorporated into the Congolese army. To get the attention of the international community, armed actors had to be well known and highly ranked—and the more notorious they were the better. According to this narrative, calculating rapists with an eye for publicity converted the suffering of others and the destruction of people's lives, networks, and social time into linear incremental progress for them-selves.[7] This theory mocked the dominant view of rape articulated by Euro-peans and Euro-Americans in the press and in the humanitarian industry in Goma and Bukavu, which had it that rape happened because the only law in Congo was the "law of the jungle," as one humanitarian worker de-scribed it to me back in 2006. Rather than these practices occurring because Congo was the proverbial heart of darkness, outside of the rule of law and the "eyes of the world," they happened *because of* Congolese people's inser-tion into a larger legal-bureaucratic and humanitarian apparatus, a system predicated on (selective) seeing.

The commonness of rape drove home the fact that war entailed the implosion of social relations emerging through families and especially through women. In those areas where Mai Mai were prevalent and popu-lar, many people felt that Ugandan and Rwandan invaders raped as part of a concerted effort to destroy Mai Mai's indigenous power so that it wouldn't hurt them later. Some said that this power was indistinguishable from the knowledge and power of both women and "the leaves" (*majani*) of the forest since women gathered the forest medicines that Mai Mai (men and women) used to defend Congo from invasion, and they also possessed the knowledge of these medicines. In this vein, some Mai Mai soldiers told me that at least some of their magic was literally the instrumentalized power of women, in that some Mai Mai medicines came from women's bodily fluids and contained an augmented version of women's power. Thus by "ruining" women, the invaders hoped to ruin indigenous Mai Mai technology, the only viable defense against the enemy's superior wealth and foreign technology (see below).

The implosion produced by war was powerful in itself, especially when it was materialized in bodies and body parts, which could also become resources for those wanting to perpetrate violence and extend their own power through violence (note the similarity to the rope rumors); bodies were said to have been disassembled and used as technologies or weapons against the enemy (note the similarity to the mass collection of Congolese body parts, especially hands, under the Congo Free State). In this regard, one former Mai Mai insurgent, a woman, discussed how her Mai Mai group first came to use the flesh of the RCD soldiers they had killed in the war. Her alleged discovery that the Rwandans were cannibalizing Congolese people "opened her eyes" and helped incite the ancestors to teach them how to adopt this technique and use it against the enemy:

> I was in a group of Mai Mai, and we were on a mission to sabotage the airport in Kindu [the capital of Maniema], but we were captured by the RCD. And they took me into a back room, and I saw all of the bodies of the dead Congolese they had killed in a giant pile, like a hill. And that's when I realized for the first time that the Rwandans were eating us, that they were gaining in power that way. We were a sacrifice for them. And when the ancestors realized what the Rwandans were doing, their plan, they gave us the knowledge to do to them the same, to use the Rwandans' bodies against them.

She explained that the flesh of Rwandan corpses, once mixed with medicines from the forest and cut into the body, became a technology that would transmit a charge to the initiate, warning of the enemy's approach (the dead flesh was recognizing the approach of its friends). This speaker was reiterating a common theme surrounding Mai Mai, in that their acts of violence were often figured as an appropriation and redirection of the victims' life force. Many people described seeing Mai Mai insurgents wearing necklaces from which hung the severed penises of "the enemy" or carrying impaled heads on sticks in a public demonstration that death and dismemberment were "normal" (*kawaida*) to them. Some theorized that these people had captured the generative powers of their victims and were redirecting that power. They held that they had been instructed to do these things by ancestors, some of whom had lived through the violence of the "time of the Belgians" and who understood something about the power of violence and about power in general.

The implosion of family often resulted in physical displacement, which usually took the form of being torn away from home (which had already ceased to be home), another eye-opening situation that often involved being introduced to dangerous new places and scales that the person had never seen before (and perhaps no one had seen before). People described walking around an alien world that was unrecognizable and filled with the potential for danger and death. Several people—men and women—told stories that involved being kidnapped as children from their homes and forced to serve in armed units, sometimes moving between groups that were enemies to each other or escaping from one group only to find themselves in another (such as the RCD and the Mai Mai; see Smith 2011). Many people described being caught by an armed group who believed them to be the enemy (*adui*) and near-fatal scenarios in which they had to convince the armed group they were not the enemy (typically someone else gets murdered in front of the near victim for the same offense, but the narrator narrowly escapes by forging a connection with the perpetrator around a shared memory—a word or a name that the victim utters or some familiar physical characteristic that the victim has). One young man spent three months walking from village to village across almost all of South Kivu. At each point, he had to convince a different group that he was not their enemy and that he had passed through a different route than the one through which he had actually passed (if they had known the actual route through which he has passed, they would have seen him as consorting with the enemy or assumed that he had been accepted by the enemy as a "guest"). Needless to say, these odysseys were always frightening experiences, often involving some shocking revelation (e.g., being made to go to a town's Mai Mai office and having to pass under a doorway with decapitated heads posted on sticks in order to enter).

The emergence of the child witch (*mtoto mlozi*), a figure that was understood to be historically new, was one major consequence of the violence meted out on families, as well as forced mobilization/immobilization. People were very specific about the fact that this new figure was brought about by the collapse of the family in the wake of war.[8] These children, held by some to bring sickness and death to neighbors and family members, were also felt to carry the trace of the violence and the severing of social ties that they had experienced and been witness to. The phenomenon of child witches was complicated and opaque: some people felt that these children were not actually witches but were simply being accused by their extended families, who could not or did not want to take care of them for various reasons, usually financial (they were often children who had lost their parents in the war and found themselves living with in-laws or other extended kin). Others believed that these were children who had come to witchcraft (understood here as a way of improving one's situation by blocking others from improving theirs) as a means of survival after they had been separated from their families or that they had been unwittingly bewitched by other witches while separated from their families. Some children who were accused of witchcraft indeed believed that they had been unknowingly "recruited" into it when they were separated from their families during the war. Several told me that they found themselves networked to groups of people who gave them food and played with them but who later turned out to be witches (the food and the toys turned out to be cursed instruments used to bewitch them, and the games they played were acts of bewitchment). As Filip De Boeck has pointed out with respect to tales of child witches in Kinshasa, these rumors spoke to the disintegration of the family and the emergence of the violent, dangerous child—most dramatically in the extremely common phenomenon of child soldiers (2000).

In general, the idea of child witchcraft brought home the feeling that people had become so lost, so torn away from their families and home connections—simultaneously mobilized and immobilized—that anything was conceivable. In a story that brought home all of the aforementioned elements of these war narratives, one coltan trader told me of how he "escaped the war" when the RCD came to the town of Shabunda by hiding in the forest with eleven other coltan traders. Already in a state of terror at night, the group of twelve, asleep in a circle, grew more afraid when they awoke to discover a young girl among them. One man yelled out to the others to wake them out of their fitful and uncomfortable slumber, claiming that he had captured the girl while she was touching his head, purportedly in an attempt to bewitch him. They took the girl to her family, who lived in the forest (the sense this urban narrator had was that everyone who lived in the forest was a witch, and so the family was probably also a "witch family"), only to discover that she had also lost her parents and was displaced by war. Apparently, her new family of related kin took sides with the coltan traders, agreeing that the girl was a witch. The narrator said that the girl, surely a witch, "disappeared" by morning to avoid being killed (the implication was that she disappeared through mystical means, but some of those who heard the story thought it may have also been that the coltan traders murdered her).

The Social Infrastructure of War and the Global Conspiracy to Grab Congolese Wealth

At the same time as people's eyes were opened to the depth of violent implosion, a new social and economic system, operating on a larger regional scale, with different spatiotemporal consequences and capacities, was becoming visible through the movement of minerals. The introduction of the cell phone was one of the first things people saw signaling this new set of long-distance relationships and rhythms. Awareness of this technology's value was simultaneous with many people's recognition of the value of coltan and the coltan business (people didn't realize that coltan was also hidden in the phones). As one man, not directly involved in the coltan trade, remembered,

> The Second [Congo] War brought the cell phone. Before that, soldiers had giant Motorola walkie-talkies for short distances. Nokia came first, the 1110 model. Rwandans came with them during the war, and we all said to ourselves, "What kind of thing is this? What is he using?" We didn't understand its value or what it was for. It wasn't until 2001, with coltan, that we realized the value of cell phones. Then we bought the phones from MTN through the Rwandans and used the MTN network for one dollar per minute so we could conduct business with them, especially coltan. They sold us the SIM cards, which were each five dollars. The [SIM] cards were so expensive. We said, "These people are killing us twice—first with their bullets and now with these cards!" And this is why, as soon as people could, they left MTN for Vodacom and Airtel, which was one dollar for ten minutes of airtime and the SIM card was free. They were remembering the pillage of MTN.

Along with the phones, the war came with knowledge of the minerals used in the phones and of a secret (to some) geography (of valuable mining concessions, for example), whose existence became more widely known during the war. As already mentioned, up until the war, artisanal mining rarely included the black minerals, which belonged to the industrial mining concessions, many of which continued to mine industrially until the outbreak of the first war in 1996. Ordinary people lacked knowledge about these industrial minerals—they did not know what they were and what to do with them. Where would they sell them, how would they wash them, how would they carry them, and where would they carry them to? Even more importantly, what was their value? As one trader in Maniema remembered,

The [Second Congo] war brought this knowledge [of 3T mining] because the war came with buyers. Artisanal mining of the 3Ts started with the RCD and the war—they started a network, because they brought the first comptoirs [buying houses]. They brought the system of flying in planes from Goma and Bukavu, going back and forth. They did this because, when they were at a certain point in their war, they realized they needed money to continue, so they decided to take the concessions of [the company] Sominki. They began in Bukavu [the South Kivu provincial capital], moving first to Shabunda [South Kivu], and then to Maniema—Namoya, Kalima, and Punia [old company mining towns further west, away from the border with Rwanda and Burundi]. Maniema wasn't hit hard by the war like the Kivus, but the minerals from [the old concessions in] Maniema financed conflict elsewhere.

In short, there was little to no artisanal mining of the black minerals prior to the war because there was no "society," as one person put it, no network around mining. In lieu of industrialized mining, the creation of some kind of social infrastructure was necessary because of the material properties of these minerals—mainly their heaviness and per-kilo low value relative to gold and diamonds. The war brought that network into being by producing an army of people who had been displaced and stolen from, their families often destroyed or disconnected in the process, and by introducing a larger social infrastructure that enabled the business of mining to take place and for minerals to flow out in a known direction. In doing so, the new business also created the conditions of possibility for postwar peace by allowing "many hands to touch money" rather than guns. As the leader of a mining cooperative put it, "There have to be people who know how to wash it, and sort it and grade it and distinguish the different types. You have to carry it, and it has to be processed somewhere else. This means there has to be a chain of people at different places, and there has to be movement between places. That's what the war brought." Moreover, the war introduced a new tempo involving regular, long-distance travel ("The planes landed twenty times a day!") that was more or less consistent, contrasting with the sudden rupture associated with acts of violence. The wartime economy of extraction sped up processes in a way that for many was new; combined with rapidly increasing prices, it helped make a return to a slower temporality (say one based on farming) less possible or desirable.

Mining thus helped create the conditions for a kind of renewal, but during the war the vortexes that mining opened up were inherently volatile, and in the already existing context of war, the collapsing of scales could end up quickly putting an end to the social relations that mining also engendered. Take, for example, the widely known case of the descent groups of Sungura and Panya (not their actual names) in Maniema, a series of actual events that people secondarily involved narrated to me in an effort to show that marginal actors, in trying to improve their situations under precarious conditions, often unleash forces that they can't control and which can end up destroying them. These two Kumu families had intermarried in the 1970s: The descent group of Sungura had a daughter who married into the family of Panya, and the Panya husband's new father-in-law from Sungura gave him and his daughter a part of the forest occupied by the father-in-law's descent group, Sungura. A river cut through the territory of Sungura, and the husband from Panya was given land on the other side of the river, away from where the people of Sungura were living and closer to the territory occupied by his own natal family, Panya, but still within the area that was understood to be the territory of his new affinal kin, Sungura. The husband's natal family of Panya lived closer to an old colonial coltan mining town, and they had experience with mining. In 2006, after the Second Congo War was officially over (there were still armed groups in the area—mainly Mai Mai and FDLR), the descendants of this new third family, which had settled in the territory of Sungura, found coltan in the riverbed and brought in their patrilineal relations from Panya to help them mine and sell it.

Over time, the people of Panya, the agnates and natal kin of the husband, came to see the river as the border between the families, though the family of Sungura held that the river was in their territory. The family of Sungura insisted that Panya were the "wives," or dependents, of the family of Sungura, though it was the husband from Panya who had married into and relocated to the land occupied by the family of Sungura. The original husband from Panya, the founder of the new family, stood up for his father-in-law from Sungura and declared that the land indeed belonged to his affinal family of Sungura and not his own natal descent group of Panya. However, he couldn't stop the war that was to follow. The conflict began with spears and other handmade weapons, but it acquired a new dimension when better-connected people related to these lineages got involved: the people of Sungura "had a general" in Kinshasa, while the people of Panya "had a politician." The politician and the general now fought against each other through the families. Each side got weapons from these connections, and in the escalating melee, the father of the Sungura descent group who had originally given occupancy rights to the newlyweds in the 1970s was killed, allegedly at the hands of some of his in-laws. Things got even worse when the FARDC, taking orders from the general who was related to the family of Sungura, came in to try to put an end to the conflict: when they showed up, they met Mai Mai, who had been in this part of the forest for years, buying minerals from all three families. When these two parties, Mai Mai and FARDC, found one another, they began fighting, and both families were caught in the middle. The conflict wiped out much of the adult male population of both descent groups, Sungura and Panya.

Seeing the Truth of the World System

War also opened people's eyes to the long-term plans of international corporations and foreign governments. These machinations were not new: they went back at least to the 1961 murder of Patrice Lumumba, whose assassination nearly everyone knew to have been "ordered" by "the Americans" because he threatened to exclude them from Congolese wealth (that is, except for those who believed that Lumumba was actually still alive). To this day, powerful nations "felt" scorned if they found one of their rivals to be getting a better price for Congolese minerals. And so, the story went, the Americans would get upset if the Congolese president gave a special deal to the Chinese or whoever; they would use their "friends," the Rwandans, to start a war in Congo, something the Rwandans were not believed to be able to do on their own.

In areas where there was a history of company mining concessions, people understood many events of the Second Congo War to have been engineered by corporate actors working to regain control of those concessions after they had been lost during the First Congo War. In this understanding, the source of violence was not artisanal mining but the existence of contested high-value industrial concessions with rich deposits, or "reserves." In Maniema province, where there has been a long history of industrial mining of the black minerals going back to the 1930s (see chapter 4), these speculations directly involved the struggle among foreign companies for Congolese wealth. One of the most compelling arguments accounted for the 2000 assassination of President Laurent Kabila (the former leader of the AFDL who led his armies to Kinshasa with the backing of Rwanda and Uganda) in terms of a profit motive, the gist of which echoed local understandings of the assassination of Prime Minister Patrice Lumumba in 1961.

The story was based on actual events that I was able to verify to the best of my ability, which were as follows. The first war had interrupted and reversed a land sale between a private Canadian mining company, Banro Corporation, and the bankrupt parastatal, Sominki (a privately owned company in which the state had a 28 percent share), which, many averred, had become something like the personal bank account of President Mobutu under his reign. In the mid-1990s, Banro entered into an arrangement with the Congolese government to buy out the increasingly embattled and bankrupt Sominki. The contract between Banro and the government was to be 72 percent to 28 percent, meaning Banro would own a 72 percent share in all of Sominki's mines, which were mainly gold and 3T mines, although some of these mines contain other minerals (see, e.g., Honke and Geenen 2016).[9] Banro wanted only the gold mines though and, according to those who were involved at the time, didn't want to inherit the pensions of the workers in the 3T mines, so Banro agreed to take only those gold mines and to leave the government with the remaining mines. But, after the First Congo War, Laurent Kabila came to power and resisted Banro's claims, knowing that Sominki's workers would not be compensated for their jobs or lost pensions. Kabila instead planned to nationalize the concessions under the name Sominco, with King (*Mwami*) Philemon of Luhwindja, South Kivu, serving as director of the company. Sominco was formed on paper only, and Banro went to the "international court" and "got the law" (*walipata sheria*), because they had purchased papers of their own from the Mobutu regime before Laurent Kabila came to power (actually, it was a federal court in New York City). But President Laurent Kabila continued to resist, rejecting Banro's claims to 72 percent and to the gold concessions because he thought the contract was unfair to Congolese workers and Congolese (*Wa Kongomani*) in general.

This is where the speculation began: According to some, because of Laurent Kabila's legitimate resistance, Banro started a war, using the RCD as its proxy. The evidence for this view lay, in part, in the route the RCD followed at the beginning of the war, which targeted the concessions. As one man explained, "When the [Second Congo] war started [in 1998], the RCD left Goma, went to Bukavu, went straight to Shabunda, and on to Kalima, all places where Sominki had had mines. Kabila had closed Banro out of mining, but these were their spaces, which they had purchased! They needed a war!"

To make matters worse, President Laurent Kabila, the founder of Sominco, and his appointed director, King Philemon, were assassinated within three weeks of each other, two years into the Great War (King Philemon was found dead, his body charred beyond recognition after his car was set on fire, with his shot-up body in it, in Lyon). For many of my interlocuters, the synchronicity was too much of a coincidence, as it meant the entire prospective Sominco leadership, and the Sominco concept, had been shut down in one violent blow. A related but alternative theory was

that Sominki, and not Banro, had used armies to try to recover the concessions they lost during the first war, since it was actually Sominki that had
held onto the black minerals mines, while Banro had taken the gold mines.
This theory came with very specific local evidence: for example, a former
Sominki manager in Kalima joined the RCD army and led the battalion
that took over Kalima, Maniema, a major mining center. The battalion occupied the company concession, stealing its reserve stock. Later, the former
company-manager-turned-RCD-officer became the owner of a large hotel in
the provincial capital of Kindu; many speculated that he was able to build
the hotel because of the mineral value he had extracted from the concession during the war (people speculated that this was his fee from Sominki).

Some people gave more credibility to these stories by drawing attention
to larger geopolitics, specifically President Laurent Kabila's relations with
the United States and foreign companies just prior to the outbreak of the
Second Congo War in 1998. Before that war, President Laurent Kabila publicly discussed the fact that he was in danger on the radio, predicting that he
would be murdered by Europeans and Americans because he had refused
to enter into a contract with them in which they secured 70 percent of Congolese wealth, leaving Congo with 30 percent; these percentages were very
similar to Banro's arrangement, which many people familiar with and connected to the state-owned Sakima knew about. Kabila also explained to the
listening public via the radio that he had created Sominco and, in doing so,
had rejected Banro's claims to Sominki's concessions, and he claimed that
Banro had gone to "the Americans" to ask for help. Some also claimed that
President Bill Clinton had insisted that Laurent Kabila "must go" before
his presidential term was finished, and many interpreted this as a death
threat, especially since Kabila was murdered soon after. Some people went
even further and suggested that, since Laurent Kabila was trained by the
US government, he had a debt to them that he hadn't paid off, and this
was one reason for his having been murdered. The fact that Laurent Kabila
and his close entourage had visited Cuba and China after the first war was
also cited as evidence; there was a general feeling that Rwanda's invasion
in 1998 (the Second Congo War) had something to do with "the Chinese"
since, under Laurent Kabila's presidency, China obtained land and mining
rights in exchange for road construction, while "the Americans" lost out.
According to this theory, the US government was angry because its client
had made new friends—it didn't like China's new influence in Congo, so it
used Rwanda to create war and chaos so that the Chinese would leave.

In general, well-connected global actors were held to be able to enact their plans incrementally with a view to exercising total control over

Congo's wealth. They did this by destroying the conditions of possibility for social production in the areas whose resources they wanted to control; war was one element, but alone it was insufficient to complete this process. One story that I heard versions of many times in the Kivus concerned an alleged global conspiracy involving all of the African "Hamites," the Belgian colonial term for people of Nilotic descent who were held to be backed by foreign powers (see also Malkki 1995). The plan, said to have been "discovered" in the form of stamped documents that Laurent Kabila found while he was president, allegedly involved Rwanda, Sudan, Egypt, Eritrea, Chad, and even Kenya—all of the putatively Nilotic peoples who wanted to together control the African continent. In the most elaborate version, the Hamites, sometimes also called *Hima* after a Nilotic aristocracy in Uganda, had first tried to take over the country with guns. Their technology was insufficient because of the autochthonous power of Congolese Mai Mai "leaves," or magic (of which, see below). Defeated in war, the Hamites, networked together as a bloc that also penetrated the United Nations, instead decided to rely on "economics" and "disease," waging a war on the Congolese countryside and the cities simultaneously. In the cities of Goma and Bukavu, they built large real estate projects that only expatriates—the Hamites and their "friends"—could afford. At the same time, they destroyed the countryside, devastating families, stealing livestock, and taking away land in a story of accumulation by dispossession that I've already detailed.

Unable to live and to grow food, rural people were forced to move to the unaffordable cities, where they became rootless and destitute, effectively excluded from the newly gentrified urban spheres in which the NGOs and the international humanitarian community now live. Their abjection was compounded by AIDS, a disease many alleged to have been purposely spread by the Hamites, who brought sex workers to Congo to infect them. According to this theory, the Hamites, because of their sheer numbers and their connections with international backers, would ultimately be victorious in their struggle to control all of Africa by occupying Congo—its key and most "resource-rich" nation. One of the points of this narrative was that all of the different, seemingly accidental outcomes of war (hunger, disease, inflation, displaced families, the presence of UN troops, etc.) were actually part of a premeditated plan to control Congolese mineral wealth. This total and totalizing theory also dwelled on the temporality of dispossession, or temporal dispossession—the appropriation of other people's capacity to produce incremental temporality through social relations. Chiefly, while the victims of war have lost the capacity to plan, manage outcomes, or produce the future, the dominating powers carry out a step-by-step plan that succeeds because of its coherent and incremental temporal logic.

The Forest as War Machine (or How the Web of Life Waged a War against Global Capitalism and Its Technology, with Mai Mai Caught in the Middle)[10]

In Congolese reflections on war, the global and regional conspiracy against Congo is so powerful, far-reaching, and total that it demands a total, unified response. Moreover, the dispossessed cannot compete with the dispossessors on their own terms, so they require recourse to an entirely different ontological domain. This is what many eastern Congolese say happened during the onset of the First and Second Congo Wars: the forest, angry at being invaded by technologically advanced and networked others, the guests who raped, waged a war against the invaders ("The whole forest was at war!"). This idea of the forest being at war had a poetic dimension, in part because the forest is where the minerals desired by the invaders were located, and minerals were tied up in the power of the forest (they belong to ancestors and, in many people's understanding, were put in the ground by ancestors).

The topic of Mai Mai—the local armed groups that emerged during the invasion and continue to exist to this day, at times aligning themselves with the Hutu-dominated FDLR—is a huge one, which I touch on only briefly here to give a fuller sense of how the war is widely understood in eastern Congo. Importantly, Mai Mai is not understood to be simply another militia competing for sovereignty or wealth with other armed groups, as most descriptions of Mai Mai by non-Congolese authors would have it. Rather, it is widely held to have at least begun as an affect-driven reaction from an alternate spatiotemporal dimension against invasion by a technologically advanced and interdependent network of foreigners—the world order that emerges in Congolese theories of war and which also surely exists. Eastern Congolese insist that not all groups that have been given or taken on the name Mai Mai are *true* Mai Mai—some are just gangs that have taken on the brand name for its notoriety. This means that many people don't consider Mai Mai groups that have acted unethically (e.g., who have "stolen" or committed acts of rape) to be Mai Mai by definition. True Mai Mai, the supporters of Mai Mai claim (and most people I talked to, at least outside the cities, were supportive of Mai Mai, at least in principle), are those that have vowed to resist invasion by foreigners and who follow the rules, or moral prohibitions, that are required of Mai Mai initiates (being Mai Mai requires discipline!).

Again, eastern Congolese said that "Mai Mai is the leaves," meaning that these groups and their power were indistinguishable from the power of the forest, which was the source of all life ("There is no life without the forest" was a truism, regardless of whether you were Mai Mai or not). When a Mai Mai group tried to invade Walikale (North Kivu) at one point during one of my research stints in 2015, the FARDC commanders there (some of whom were former Mai Mai or had worked alongside Mai Mai) held them back by strategically planting whisks (used for sweeping households) and *sombe* (cooked cassava leaves, the most commonly eaten leafy green in the region) under the ground. The idea was that coming into contact with domesticity destroyed Mai Mai magic, which was of the leaves, or undomesticated forest.

The Mai Mai "rebellion," then, was not created or waged mainly by living humans but by the forest and all its parts (again, the "whole forest was angry" with the invaders, and the "whole forest" and all of its elements waged a war against them). Mai Mai initiates' stories dwell on the various animals and plants that were weaponized during the war—the swarms of bees that networked to defend not only their queen but the very life and soul of the forest; the fruits that became grenades; the snails that became cell phones; the branches tied around the arm that became homing devices, transmitting the location of the enemy to the wearer. Always there is some comparison being made (in Mai Mai practice and in Mai Mai narration about their practice) with technology, where the harnessed power of the forest is analogous to but also trumps foreign technology.

It's vital to point out, though, that when people talk about this enraged warrior forest, or the forest acting as war machine, they are not talking about "nature" in the Euro-American sense of a realm distinct from culture and the human; nor are they referring to an ontology completely removed from the social or the human. Rather, the power of the forest is channeled or deployed by ancestors and sometimes other spirit entities (in some areas, gods who in turn give power to ancestors). These entities approach living people in dreams and instruct them in how to wage war against the enemy (actually, there are different points of view regarding whether living humans reach out to the ancestors or if it is the ancestors who are in fact angry and reach out to living humans). The activated forest, in this view, is therefore also a social assemblage, as well as a portal into another dimension, which is also another time. For example, the knowledge that the ancestors give to initiates is understood to come from the past—it is a legacy that has been overridden or ignored because of years of colonial and postcolonial education and, for lack of a better word, "modernity" (including the attitudes and actions of the school-educated Congolese *evolue*, or evolved, a term that goes back to colonial times). Because it is a gift from ancestors, it resists commercialization: "You can't sell it, this knowledge." Many Mai Mai technologies are understood to have been common in the past, like the chain of slit drums through the forest that could warn the colonial administration in then Leopoldville about an emergency in faraway Stanleyville in under fifteen minutes. More importantly, the rage that animated Mai Mai also came from the ancestors who had lived through the violence of Belgian colonialism as well as various attacks on Congolese sovereignty during the postcolonial period (for example, the assassination of Lumumba and the Simba Rebellion that followed; in Maniema, the ancestors responsible for Mai Mai were understood to have participated in the Simba Rebellion). These ancestors, dead but not gone, remained angry about a past that was also not gone (and so not completely past), and their rage and their recognition regarding what was happening (recognizing what was happening in the present in light of what happened in the past) energized and "taught" contemporary Mai Mai insurgents.

Mai Mai were understood to take the form that they did because of, and in response to, the ancestors' recognition of the technological superiority of the interconnected world order that is organized to extract Congolese wealth. Hence also the focus on parallel technologies that aren't just mimicking Western technologies but deriving their potency from an alternate source—not Western education, or "the pen," as many of my Congolese interlocutors put it, but ancestors and the forest. Mai Mai technology was, then, adapted to making the technology of the world's global powers useless; this is even implied by the reference to water in their name, meaning that Mai Mai knowledge *turns* bullets and other weapons *into* water, such that, rather than being dangerous, they become life-giving fluid for fields and forests. According to many, Mai Mai or not, this medicine could bring down airplanes or make insurgents invisible to airplanes or any other kind of tracking device. Invisibility was a major theme in Mai Mai thought and practice—they were trying to evade the eyes of the world and using the power of the forest to assist them against technology, especially visualizing technology. Part of the idea here is that, because technology has historically been withheld from Congolese, they couldn't possibly survive a war by trying to compete through this same foreign technology. Rather, they had to find a different way—or the ancestors had to point out a different way for them—as they moved further away from received technology and its source, embracing an otherwise "dirty" (*chafu*) realm that educated Congolese have long dismissed. They had a phrase for this: "The iron gets destroyed, but the wood remains," a reference to what happens to an ax over time. It was used to mean that foreign technology is ephemeral and weak in comparison with knowledge from the forest, which is indissociable from ancestors and other spirits. At the same time, several former Mai Mai insurgents told me that Mai Mai's main problem was that they never had any weapons and they were "always dirty," as if they had turned their backs on the currencies of "clean" (*safi*) life associated with colonial and postcolonial modernity.

It struck me that there was something Afrofuturistic about the way people talked about Mai Mai, in that they were reimagining science and technology through a "local" lens while also locating a version of science, technology, and a possible future in a set of relations and ways of knowing that were very different from, if also parallel to, Western science. Mai Mai were also understood to be developing and progressing "traditional" knowledge outside the limitations of customary authority structures, which were marked as traditional even if they were actually colonial. After all, Mai Mai took off during the First Congo War, after many seniors had died or fled, so ancestors were compelled to communicate directly with these youth in a context of general social implosion and evacuation. Mai Mai and former Mai Mai were often explicit about these parallels, describing themselves as "scientists" and their techniques as counter-technology that was so advanced it was indistinguishable from magic (I'm riffing on science fiction author Arthur C. Clarke here, but the sentiment also echoes Mai Mai descriptions). One Mai Mai insurgent referred to Mai Mai as "ecologists" because the entire movement was aimed, ultimately, at protecting the forest, the source of Mai Mai and Congolese power. Mai Mai insurgents, former and current, also invoked a concept of scientific method, explaining how they took the knowledge that had formerly belonged to separated ethnic groups and shared it with other groups as they came together in the forest during the war. Formerly disconnected people, displaced by war, converged and shared their knowledge, their medicine, their dreams, and their ancestors, creating a fusion of knowledge through experimentation that was decidedly more advanced than anything that had existed before.

In other words, by displacing people and putting people together in ways that were new, the war had imploded the ethnicized boundaries and modes of identification that had been put in place during colonialism and maintained under Mobutu, and Mai Mai experimentation and collaboration was one of the practice-centered ways in which new collaborations were forged. Also, since the ancestors were the actual agents behind this activity, they were moving, changing, and improvising as well. Of course, if they saw themselves as scientists, there was much that could also be labeled "religious" about Mai Mai practice: their knowledge came to them largely in dreams, they interacted directly with ancestors, they underwent numerous rituals, and they had to observe a multitude of seemingly arbitrary rules (many of them extremely difficult to keep, like avoiding rainwater while in rainforests).[11] But the religion/science binary is drawn from the Western Enlightenment tradition, and these were not understood to be mutually exclusive, or even conflicting, concepts.

Mai Mai were drawing on older traditions (for example, the techniques and the anger of the Wasimba rebels of the 1960s), but they were also breaking dramatically with the past in terms of the social organization and distribution of their knowledge. This experimental social composition, partly thrust upon them by the war (again, the seniors had either fled or were dead) fed into their understanding of Mai Mai as "science" and what I'm characterizing as a kind of Afrofuturism. For one, women, and sometimes children, assumed prominent roles in many Mai Mai groups as the makers of medicine. While many women who were in Mai Mai were captured and forced to serve, others joined up to fight against invasion. General Padiri was only about fifteen years old when he headed what was, during the Second Congo War, the largest Mai Mai battalion in Congo; Padiri was widely considered to have acquired his powers from ancestors. One friend of mine was chosen to be a Mai Mai doctor (meaning he prepared the "medicines" used in war, but did not engaging in fighting) because his prepubescence placed him beyond sexual desire, implying that he would not be able to violate the strictures on Mai Mai practice, which included not engaging in sex. Some inferred that Mai Mai were *purposefully*, and not just accidentally, breaking down established social structures and generating power from this while also fundamentally changing society, including the state-backed system of gerontocratic, chief-based authority that existed during the time of Mobutu. Many seniors hated Mai Mai for this reason, seeing their appropriation of ritual power and knowledge as witchcraft (*ulozi*) because it consisted in youth acting immorally and selfishly to accrue power for themselves outside the moral control of legitimate seniors. Some observers of Mai Mai have even suggested that this "revolt against the elders," particularly the established and corrupt system of traditional chiefs empowered by Mobutu, was the real purpose of Mai Mai—not resistance to invasion (Vlas senroot 2011). While many recognized that Mai Mai were revolting against elders in certain places and times, only the most virulent critics of Mai Mai, all of them senior men, suggested to me that this was Mai Mai's purpose.

Some of the violence that Mai Mai committed seemed at once to be reiterating colonial atrocities and generating power from the destruction of social order, normative hierarchy, and norms in general. One young man described a situation he claimed to have witnessed when a Mai Mai battalion came to a neighboring village. The general decided to punish the village for "hosting" the enemy prior to his arrival, and among the things he did was to try to force a brother and sister to have sex with one another in public. The general did not succeed (the victims refused), so he shot and killed the brother, then married the sister to "make peace." The sympathetic narrator insisted that the general did not *want* to do any of this but had been instructed to do so by the ancestors as a way of acquiring power. My interlocutor insisted that this was nothing like the acts that the RCD performed, even though it formally resembled them; while this was a terrible ordeal for those who went through it, it was good for everyone. This "sacrifice," as he called it, was a purposeful transgression aimed at forging a kind of power, for Mai Mai and for everyone else in the village, enabling the general and the villagers to be successful in the war against the RCD while also bonding the general with the villagers (here I am simply transmitting one man's somewhat distanced, though clearly considered and concretized, interpretation of this event, and not representing what the event meant for others, or for the victims, as I clearly have no idea).[12]

Finally, the power of Mai Mai was understood to be in tension with the minerals of the forest—not because of the intrinsic characteristics of minerals, but because of how minerals were already imbricated in global markets. In fact, one Mai Mai insurgent in Walikale said that, in his group, coltan was powerful medicine and a form of Mai Mai magic—"like a battery," it contained and magnified the power of the forest and could be used, alongside other forest elements, to make the technology of the invaders powerless (I found this to be a fascinating connection, because coltan does work to hold an electrical charge in electronic devices, but I never heard it repeated elsewhere or again). The problem lay in the commercial demand for these minerals because, as Mai Mai started turning toward these minerals in search of money, they became "lost." What my interlocutors cast as their newfound greed led them to torment the communities they were supposedly defending, and as they became more dependent on minerals, they became less dependent on these communities for support. They also grew less responsive to these communities and more autonomous in general (this is part of what was meant by "being lost"). As this happened, they ceased to be Mai Mai (human vessels for ancestral wishes and the angry forest) and so became vulnerable to attack and death at the hands of the enemy (the ancestors stopped protecting them). Many people told me that minerals, and especially high-value ones such as gold, put an end to Mai Mai and presaged their violent defeat, because they grew into monsters that the ancestors could no longer recognize—"Mai Mai were destroyed by money."[13]

We can now end by taking all the different bits about war presented in this chapter and fleshing out the central themes that emerge in all of them.

Concluding Thoughts

The war consisted in the violent transfer of emotional and social energy, or what could be called force or power (*nguvu*), from one group to another, generated from the violent destruction of social relations and the expropriation of the things (livestock, money, land, saleable goods) that engendered social relations. Actually, the anthropological term that comes closest to this extracted energy is *value*, but the intensity and painfulness of this process make me think that the terms *energy*, *force*, or *power* are more appropriate; most Congolese used the term *nguvu*, or power—talking, for example, about how people acquired *nguvu* from acts of violence, or how a community lost its *nguvu*. The invaders "stole" (*kuiba*, the term most often used) social relations and the conditions of possibility for creating the future, through marriage and bridewealth, for example.

Expropriated wealth was part of this emotional, social energy—its material embodiment—but the energy I'm referring to was irreducible to material wealth. This energy actually left Congo and went somewhere else, where it augmented the power of those who were dispossessing Congolese; some people believed that this power was empirical and quantifiable and could theoretically exist on ropes or knives (it certainly had to be embodied or materialized in some form to be useful). Rather than becoming ancestors who could assist or empower their descendants, these commodified spirits entered into a market system through the trade in ropes and were purchased by people who would go on to ruin Congo and other places. Some of this power was converted into wealth and technology and came back to haunt Congolese. Congolese usually didn't know that some of this wealth was used to make computers and cell phones, but they witnessed the connection between these technologies and social destruction in their daily lives (they talked instead about bullets, implying that all their work came back to them in a condensed material form intended to kill them). They also saw these technologies (especially cell phones) give birth to a new order on the ground, one that transcended the Congolese nation-state and "opened their eyes" to how the world was constituted and to the value of minerals.

At the same time, another spatial and temporal dimension, which also happened to be the dimension that minerals come from, rose up to fight the invasion of the networked world order and their incremental plan of violent ecological and economic destruction. The agents from this dimension were somewhat successful, but the push and pull between the two dimensions (the forest/ancestors and the world order) was too much for the all-too-human Mai Mai insurgents, who were caught in the middle. They also needed money to wage war, but the commodification of forest resources ended up weakening their power because it led them to participate in acts that disconnected them from ancestors.

These narratives about war highlighted the danger of excluding people, as exclusion was depicted as creating war driven by jealousy (whether that of the major world powers, the companies, or descent groups). In this way, war was, as one person put it to me, "witchcraft by other means"; some people just said that "war is witchcraft," by which they meant that, like witchcraft, its ultimate sources were greed and jealousy directed at competitors (at the level of the village, people frequently cited examples of neighbors or family members using the presence of an armed group to "bring trouble" to a rival by going to the armed group and asserting that the rival was the militia's enemy). One friend of mine, a FARDC soldier with much experience in these matters, had a favorite phrase to describe this—he called it the problem of the "child of the *baraza*," or public assembly (here with the implication of a dispute resolution process). His phrase (as far as I know it was his own, because I never heard anyone else use it in this way) refers to a person whose father is absent and was probably not from the area. Unable to assert himself in his own family or lineage, he ends up seeking out the support of others, usually members of his own ethnic group living nearby. He makes deals with them that end up undermining his natal family or village.

In an analogous way, but on a larger scale, war was felt to be brought by people who were left out of a network or who had to resort to war because they had no other options. Many felt that all global conflicts emerged from this situation of being excluded and the affects that emerged from it. So the conflict between "Arabs" and "Israelis," which many heard about in church and were taught to view as a world-historic conflict critical to the events playing out in Congo, was essentially a struggle among descent groups carried out over generations: Abraham had two sons, but Isaac was the favorite one, the one who got everything, whose descendants now live in Israel, a nation blessed by God. The children who were excluded were arrogant and jealous, and they didn't want to be left out. They wanted to take over Israel and through Israel the world. It was the anger of these "Arabs" with "demons on their backs," the descendants of less-favored children, who brought violence to the world and who, some felt, wanted to use major mines like Bisie to earn and launder money for Al Qaeda (see part 3 of this book). Part of what was interesting about this theory is that it gave these diverse conflicts a singular, shared affective source—they are caused by people who feel left out or who somehow can't touch money (where money is much more than simply a means and measure of exchange but rather the tangible condensation of the totality of social networks and relationships). This understanding fed into a local theory of peacebuilding, which had it that that peace comes from collaborative inclusion. As we will see, this idea was fundamentally in conflict with international peacebuilding strategies, which worked to exclude illegal actors from mining, preventing them from touching money.

I want to reiterate a final point that I also hope is obvious—mainly, that war directed violence onto Congolese bodies, forcing them to work, and it recapitulated a traumatic colonial history, which then became visible and "real" for people in the present. Bodies were instrumentalized, sometimes to extract value and sometimes to enact or display spectacular violence as an end in itself. For some, digging turned this situation around, not only by bringing in money, but by allowing them to repurpose their bodies to touch and move and, through movement and touching, to revitalize and rebuild what I will simply call society and leave it at that. It is to the newly visible and widespread system that has emerged out of digging—based on a tiered and tense layering of friendship, indebtedness, and acrimony— that I now turn.

THREE

The Magic Chain: Interdimensional Movement in the Supply Chain for the "Black Minerals"

The physicality of artisanal mining—mud, water, dirt, dust—should not be confused with an absence of order.

—Michael Nest, *Coltan*

Dirt is essentially disorder.

—Mary Douglas, *Purity and Danger*

I like to play in the dirt.

—Seen on a digger's T-shirt in Maniema

On Substances That Belong to Others

Raymond and I are in the city of Kisangani, the capital of Province Orientale, at a poolside restaurant in an old hotel that, despite being run-down, retains a certain prestige. Fresh from a grueling two-day drive through almost nine hundred miles of forest road and maybe twenty-four separate "borders," or roadblocks, between here and Goma, we have come to this place for the fourteen-dollar plate of spaghetti bolognaise. Its price is high for a number of reasons, including the cost of transporting goods over these roads, but one reason is that, although it is in many ways as Congolese as just about anything else, spaghetti bolognaise is symbolically connected to a particular set of distant people, places, and histories. The dish retains some of the aura of having once been withheld by powerful and capricious others who excluded and tormented the people who can now eat in this place from which they were once excluded—if they can come up with fourteen dollars. Those who run restaurants build on that distinction, that aura of racialized difference, and pricing is part of that work. If that mystique of difference and hierarchy were to completely disappear, their spaghetti bolognaise (overboiled dried pasta with ground beef and a little tomato thrown on top) would not be so expensive and probably not as desirable.

A similar kind of devaluation happened to the US dollar: In the 1980s and 1990s, the collapse of the Zairean economy, the rapid devaluation of the zaire (the former national currency), and years of imposed interest-bearing debt to the World Bank and IMF led to inflation rates of nearly 10,000 percent (Biletshi 2013). So when dollars flooded the Congolese market during the Great War—fueled by humanitarian interventions during the Rwandan refugee crisis and then by the coltan business—they were synonymous with incremental growth and security. Dollars were a stable, enduring, and universally recognized form of value one could trust, more so than banks. At first, dollars were reserved for major purchases such as mineral deals. They were treated with great respect and care, inspected for the slightest tear or fold, their age taken into account in every single transaction. Dollars are still scrutinized for almost microscopic tiny tears, as well as dates, but as they were exchanged with and for Congolese francs, and for everything else, they started to become a bit more like the francs for which they were exchanged—they changed hands frequently, were dirtied and distressed by the constant movement (being used continually), some Congolese suggest as if by Congo itself, and now you need fifty of them to get a bad hotel room. These days, because of its devaluation, the one-dollar bill is no longer acceptable currency in Congolese cities, and street dealers display US quarters and dimes alongside other useless antique collectibles in an ensemble of exotic currencies that include colonial-era coins and Mobutu-era zaires—one time I even saw a Chuck E. Cheese token on which a sovereign cartoon mouse declared that a kid can be a kid in his domain. An economist would no doubt see this devaluation of the dollar as being principally an issue of supply—there are more dollars now, so their value has diminished—but that is not, strictly speaking, how many eastern Congolese talk about the devaluation of the dollar, which they see as symptomatic of its overextension in and through Congolese daily life—becoming "dirtied," (kuchafuliwa) as many put it, through its association with the franc (more on this below; for an in-depth analysis of the materialities of Congolese francs and dollars in DRC, see Walker 2017).

If some things, like dollars and spaghetti bolognaise, seem to have value because they contain the trace of powerful forces that are beyond Congo, other things are understood to derive their value from ordinary social life and relations. I'm reminded of this while watching the restaurant's television, which is showing me ads from Kisangani. Each one is for something strikingly ordinary: wax matches, canned sardines, powdered stew flavoring, a common skin lotion, toothpaste, fishing nets. The ads quickly and concisely portray rich social dramas set to familiar Congolese music, punctuated by what seems to me to be unnecessarily histrionic yelling. A man is saved from starvation on his trek through the forest (canned sardines), a mother manages to feed her children despite an accident at home (powdered stew flavoring), a wife saves her marriage after an embarrassing incident (skin lotion), a group of young men become heroes (fishing nets), and so on. These simple things, most of which you can get almost anywhere, are shown to be critical because they can repair damaged social relationships and situations, and they keep temporality moving in the right direction. The man can make it through the forest, the woman can continue being with her husband, the fishermen and their village can be fed, the mother's child will eat. The projects that were started are successfully completed. More to the point, people and their relationships to one another are important, and the ads have to create a feeling of social connectedness and success to remind people that these things can be the difference between life and death, stunning victory and abject failure. Were it not for the social situations, these things would be pretty close to worthless.

Meanwhile, at a table near me, a large and leathery Belgian man in brown workman's overalls is showing a fat and prosperous-looking Congolese man his big map, which he has splayed out on their table. The Belgian, who strikes me as having sprung straight from another time, belongs to a small network of older Belgians who hang around the hotel, smoking cigarettes and eating spaghetti bolognaise. The hotel owner is also Belgian, and his kids jump and swim happily in the green pool water while their Congolese mother watches on. Much later, I'll wonder if the man in the overalls was in fact the infamous Yves, allegedly Laurent Kabila's former pilot, who went on to found the company MPC (Mining and Processing Congo), buy the papers for Bisie mine from Kinshasa, and try to close out the artisanal miners at Bisie—events described in part 3. All I know now is that the men with the map are doing work—preparing for business, concentrating deeply, taking notes, and consulting with each other about the places they are seeing. The map is showing them things, revealing its secrets to them. I also want to see, and at first they don't notice me standing up, peering over their shoulders, casually following their conversation. The map is clearly quite old but perfectly maintained. On top it reads, "Les Minéraux du Zaire" in big block letters. Little red squares, dispersed throughout the map, show the mineral-rich areas of what is now DR Congo, concentrated in specific sites, some rather close to where we are. Suddenly the men are aware of me and look disconcerted, annoyed. This map belongs to the Belgian and is not mine to be looking at so, closed out, I go back to the television, which is now showing me images of Kisangani (formerly Stanleyville) to the tune of upbeat, "classic" soukous music—the old post office and the precariously positioned Congo Palace nightclub, looking like it's ready to collapse anytime onto the street below. After a while, a man comes by my table to try to sell me tourist art: "traditional" masks, romantic paintings of tree houses in the forest, and model replicas of the impressive and unique system of tall wooden tripods and giant baskets used to fish at nearby Wagenia Falls on the Congo River.

The substances indicated on the map stand out as different in many ways from the other things described in this social situation. For one, the things on the map will not help you to catch or fry a fish, and their value appears to be intrinsic to themselves. They are used, certainly, but they "belong" ultimately to others—in a way that is more profound and complete than the formerly foreign but now partially domesticated (for better or for worse) US dollar and the spaghetti bolognaise, whose manufactured separation from Congo is tenuous and rather ridiculous. The black minerals, in particular, are not known to Congolese as finished products, so they are not signs of social status (like the bolognaise, the dollar, and the hotel), nor commodified representations of heritage or a romanticized "primitive" (like the masks, the paintings, and the model fishing systems), nor focal points for nostalgia or resentment (like the Congo Palace nightclub or the post office). Generally, Congolese project their own anxieties (bullets) or dreams (roads) onto the end form of these substances, which are understood to have a social use for others that may even be destructive to Congolese (as is brought home, especially, in the stories about bullets). They are, rather, substances in the process of becoming, and their heaviness and density enable those who work on and with them to come together and be transformed through them as well, but only if they can ultimately convert these substances into something else—ideally into something enduring (houses, schooling, etc.), not just expensive sustenance eaten at the hole. For any of this to happen, these substances must first be extracted from their surroundings, and from everything else that is ordinarily important, in order for their value and significance to be realized.

For me, the social situation at the hotel throws into stark relief the issue of what is valuable, what value is, and where this value ultimately comes from, a question that the trade in minerals often provokes those involved in mining to think about in various ways, not always directly or succinctly: wealth often seems to come from outside Congo (dollars and spaghetti bolognaise have more value than local currencies and foods), but the "source" of much, perhaps all, wealth—desired and fought over by the whole world—appears to be located in Congo. But is the apparent source of value (e.g., minerals) the actual source of value, or is there in fact another deeper source, and what is it? (The forest? People? Ancestors? What?) For those involved in the trade, these were not just abstract questions. Extracting mineral value was not a straightforward process of taking things out of the ground and selling them, but a complicated negotiation between different regimes, different forms of value, and different ontologies, which were formally made to disappear by the time the substance made it to the comptoir, or buying house. This multidimensional process entailed the substance's transformation from one kind of thing (usually, something that was put there by ancestors but was also not necessarily known to those ancestors when they were alive) into another: an invisible commodity (e.g., tantalum) within a commodity that almost looks like it comes from outer space (e.g., a MacBook). Those involved in the trade, at all levels, were actively involved in the work of disentangling these substances from their multiple relations so they could become commodities in the first place. One of the unintentional things their multifaceted work did was to help produce a situation where Congo's minerals were remote and belonged to others, like the men with the map in the hotel restaurant.

This collaborative work of converting substances from one domain (ancestors) to another (markets) is social in a sense that stretches the normative meaning of the word to include forces that are invisible (e.g., ancestors) and that not everyone is always certain even exist. It follows from this that the work of diggers also goes beyond extraction and the realm of political economy. In addition to being innovative economic actors who engineer new social forms, they are also accidental religious actors, mediating the conversion of substances that are enmired in spirituality and history into commodities that will ultimately be alienated from Congo (by putting it in this way, I'm not intending to reaffirm the intrinsic reality of categories like "religion," "economics," and "politics" but arguing that digger's practices show that these are not separate or distinct categories or modes of existence. At the same time, these categories do become socially real through the supply chain, as the religious dimensions of mining are formally "exorcised," or released, at the lower levels of the chain so that the mineral trade appears as a purely economic activity later on). Of all the various actors involved in this process, diggers are at the frontline of the movement from forest/ground to market, experiencing the capriciousness of angry ancestors and the disdainful brutality of paper-wielding state offi-cials head-on. These "foot soldiers of the digital age" (a phrase I heard used by the Congo scholar Peer Schouten[1]) engage directly with spiritual entities that most people would prefer not to deal with, and they do the physical work of disentangling these substances from the relations in which they are enmeshed. Because their work touches on multiple domains of exis-tence, it is also thought-provoking—inciting those involved to think about how best to manage ancestors, for example.

For many others, diggers are somewhat beyond the pale—dirty and reckless, they don't "see ahead" (*kuona mbele*), living only for the moment (this idea of dirtiness relates to a colonial history of Belgians interpolat-ing Africans as dirty; many postcolonial state actors reiterated this idea—see, for example, how the Ministry of Mines depicted their interactions with indigenous people as traveling down a hole, in part 3). People say they speak a language, an argot, known only to them (there is some truth in this) and that they consort with each other, and with spirits, in a way that some deem unclean (*si safi; kichafu*). In a sense, diggers sacrifice and "dirty" (*kuchafua*) themselves so that others can be clean and possess "clean" technologies, and they bear the moral opprobrium that others cast upon their work and the underground locations where it is performed (and many diggers do draw attention to this sacrifice and talk about it ex-plicitly, so it is not just my interpretation). They also sacrifice themselves so that often-predatory others—mainly, various state actors—can extract money from them and their work in the name of helping or protecting them. Because they're often vilified for personifying everything that is bad about artisanal mining in comparison with the collective memory (not the "reality") of industrial mining, it's ironic that they are also enacting and democratizing aspects of industrial mining, whose fruits had long been restricted to a few.

The Practices of Movement and Closure in the Supply Chain

In this chapter, I give an overview of how multidimensional movement op-erates by first describing the social organization of artisanal mining, then the forms of state and nonstate governance around mining and the con-cepts and categories that emerge from this system and also shape it. In the process, I describe the different spheres or dimensions across and through which these minerals are made to move. When discussing the organiza-tion of mining, I don't delve deeply into any one place but try to capture the breadth of possibilities in mining and the grammar, or structural logic, that informs the way in which mining and its regulation emerges in the particular. In this chapter, ancestors come into play but as agents—albeit old-fashioned ones (they are, after all, super old!)—that exist, along with diggers, traders, and state actors, in the present. In contrast, chapter 4 concerns how mining opens up alternate places and times, including the times of ancestors and colonialism, and how mines can create wormholes connecting disparate times and places.

Let us start off with another social situation that highlights the fact that digging is a morally charged endeavor that brings up deep-seated issues about who is responsible for producing value and social well-being, as well as the kinds of practices and people that are valuable and for whom.

Diggers as Healers

It is South Kivu in 2018. Tiger 1, whom I know as Claude, is the thirty-year-old *chef d'exploitation* (chief of exploitation, a title typically reserved for the head of operations at a *chantier*, or mine/construction site) and the head of one of the diggers' cooperatives at a major cassiterite mine not too far from the provincial capital of Bukavu. At this mine, which was originally owned by a French mining company in the 1970s and is now officially in the government-recognized zone of artisanal extraction (ZEA), two diggers' cooperatives share access to a "door" (*mlango*), or entrance to a large shaft that extends nearly half a kilometer into a hill. The shared tunnel branches out into separate chambers, or rooms (*ma chambre*), rented out by different groups whose PDGs, or president director generals, are members of one of the two cooperatives. Claude, who works for one of the cooperatives, is on call twenty-four hours a day and may find himself handling a conflict or problem at the mine at 2 a.m. (recently, there has been a drawn-out nonviolent conflict between the cooperatives over a water pump that they share ownership of in a complicated way). As we sit outside his house eating boiled eggs, Claude keeps in touch with his crew via his walkie-talkie; currently, he's working "virtually," as he jokes (using the French term), by talking with his second in command, The Enemy Has No Color (*Adui hana rangi*), who "does security" (*anafanya securité*, or is the *chef de securité*) for the cooperative.

> Is this The Enemy Has No Color? This is Tiger One. Tiger One . . .
> Don't leave the door. You can't leave the door.
> Where is the Pope of Rome (*Papa ya Rome*)?
> Don't abandon the door! If you have to leave, then everyone leaves.

If one of ours leaves, one of theirs has to leave as well.
The Enemy Has No Color, do you understand? This is Tiger One . . .

The Enemy Has No Color has final say regarding who will enter the mine; even if a digger comes with papers from the Office of Mines or even the governor, they can't gain access without The Enemy Has No Color's acquiescence. The cooperatives control the administration of the holes, while the PDGs, or hole owners, provide them with some tools and extend food and money to the diggers to initiate work. This food, or "ration" (in French—a term that reiterates and references the time of industrial mining), is understood as debt, and it is subtracted from each digger's percentage of the overall yield. The PDGs also gather statistics on how many people go in and out, transferring this data to the cooperatives, who have it on record for SAESSCAM, the small-scale artisanal mining authority, and the Office of Mines. In this case, because the cooperatives are rivals, they are also keeping track of each other, and their statistics are especially likely to be trustworthy.

The other rival cooperative wants Claude's people to abandon the door for a while, during which time these "statistics" will not be collected. Claude believes they want to bring in more diggers so they can boost production, perhaps to meet the needs of their buyer, or perhaps they want to move out minerals during the night to avoid formal or informal "taxation" from state authorities. This is why he won't allow The Enemy Has No Color or the Pope of Rome to leave the door. Increasing the number of workers could be risky, even if it's the other cooperative that's doing it, as all the men could run out of air. If an accident were to occur, the whole mine could be shut down, especially given that this is a highly monitored mine adjacent to the road. Moreover, the momentary deregulation would be perceived as unfair by the men in his cooperative, who are already hard enough to manage. Claude doesn't want anyone to get hurt, and he also will do anything he can to avoid a conflict with the mining police or really any situation that could result in closure of the mine and of movement.

There are many features of work in this mine that are shared across most, if not all, mines: The work is highly organized (here with a more hierarchical division of labor than some mines); there are diggers' organizations that interact with state actors and buyers (here a government-registered cooperative, but this is not always the case); there is some tension or competition between groups of diggers over space and time; and the rate of movement is affected by unpredictable events from above ("the needs of a buyer" or changes in price). Most importantly, at this mine, as in others, movement is desirable to diggers, but unfettered movement can lead to events that sabotage movement and put an end to business as usual (threats to safety, "disorder" brought by too many state officials, or, in some places these days, the possibility of shutdown by state authorities). At the same time, there are specific features of work in this mine that may not be found in others and that have emerged as the result of various factors, including physical aspects of the mine. This mine has been around for decades, and the minerals that people are looking for are deep into the hill, requiring shaft construction. As the diggers have come to rely on machines for pumping out water, they have also come to rely on financiers, which has contributed to the constitution of the cooperatives (that is, who is in them) and competition among them. Also, this mine is located close to the road, near a city that is a provincial headquarters and home to many NGOs, some of which are focused on monitoring "conflict minerals" in this post–Dodd-Frank era. This has led to a great deal of scrutiny of rules and concerns about the possibility of a shutdown, which would create major disruptions to everyone's life and threaten daily survival in this town and the surrounding region. Indeed, the miners here are convinced that the "international community" is watching this mine with a satellite at all times because in recent years UN representatives and some foreign notables (including people said to be European royalty) have come out from time to time, held out iPhones for them to see, and informed them that the minerals for their phones come from this place (the diggers have translated this to mean that they better not screw up or "dirty" the phones of these consumers through their petty conflicts over water pumps and whatnot— but more on this in chapter 9).

When Claude is satisfied that the business of "the door" is settled, we walk over to one of the main drinking spots in town, tucked away around a corner and up a creaky staircase made of cheap wood thrown together, precariously elevated above the street. It soon becomes clear that those in the bar are involved in a collective conversation, with some accusing the diggers of two interconnected failings—not seeing ahead (*kuona mbele*), or into the future, and not paying their debts. While most of the folks in the bar are diggers, there's also a "doctor" (or someone who works in the hospital and who others refer to as a doctor but at the same time recognize is not a real doctor), who has lived here for a short time. In addition, the former mayor of Bukavu, escorted by a small contingent of urbanites, is here to visit a local NGO; she seems unaware that she's offending people when she declares that she's arrived at "the end of Congo" (*Tuko mwisho wa Congo*), implying the furthest conceivable point, or frontier. After all, we're only maybe a five-hour drive from the city, a relatively short distance in Congolese terms. The mayor is invoking a common sentiment—that the city is the center of life and politics, and what is outside it is a frontier space—but it is one that rubs diggers the wrong way. From their point of view, it is their work that brings value to those cities and has rebuilt them since the war. Also, their work is forward-thinking, at times even futuristic (combining very old and very new techniques simultaneously), in comparison with the staid and status-seeking norms and values of the city. And diggers help to make the places they inhabit urban and cosmopolitan, even if these places are also remote and (often) sylvan.

The discussion takes a familiar turn, involving themes related to time and temporality—mainly, the fact that diggers extract and generate great wealth, but it doesn't endure into the future, or "develop" in an incremental, consistent, and generative way that becomes visible, taking the form, say, of roads, hospitals, well-built houses, or a bar that isn't made out of sticks. The main topic is the town's new clinic. The doctor is complaining that the diggers don't pay their bills, or they die from accidents in the shafts, leaving their bereaved families with the unpaid bills and with future bills of their own. Like many people, the doctor is blaming this on the personal characteristics of diggers—the fact that they can't "see ahead" (*kuona mbele*), or plan for the future. There is one hole owner, or PDG, to whom conversation keeps returning, no doubt because he's not here: he allegedly accumulated $6,000 worth of cassiterite in a single day, but he is in all kinds of debt and can't pay his hospital bills. This apparently isn't the first time that this PDG got so much money in a day—he "has a star" (*ana nyota*), a condition that exists in his blood and makes him lucky in mining but not in a way that is consistent or predictable ("having *nyota*" in the blood means having a spiritual and biological condition that entails experiencing sudden windfalls, especially in mining, that disappear just as quickly; see Saffitz 2019).

The doctor, shaking his head, implicitly addresses me and the mayor when he says, "I'm embarrassed to talk about this in front of guests, about guys who make $6,000 in a day and can't pay their bills," but the fact is that "diggers aren't good people; they don't want to help others. The problem is they don't remember tomorrow" (meaning they don't remember that tomorrow is an inevitability). In a broader sense, he seems to be saying that diggers forget that actual linear time exists outside their episodic and ecstatic experience of time. Because they are prone to accidents and even sudden death, other people end up bearing the burden of their incapacity to see ahead: "Diggers are the ones making money," the doctor reminds us, "but they're also the ones drinking all day and the ones with unpaid hospital bills." Some of the diggers in the bar agree with this: one explains that, if a digger is to buy a house, he has to get a lot of money and buy it suddenly, but you'll never see him laying a foundation and building it over a long time in small increments (thereby materializing incremental time). He says that no digger has ever done this in the whole history of this place, perhaps in the whole history of digging, because he'll reach a point where he runs out of money and is not getting minerals, so he'll sell whatever he has built so far: "The problem is a digger *can't* make a plan. He *can't* plan for life."

While many of the diggers in the bar are inclined to recognize that the social problem the doctor refers to is real, they see it as a consequence of the nature of their work. For example, the "owner" (really renter) of a hole has a lot more expenses than an ordinary person might realize, and it's easy for him to overextend himself and get himself into trouble because he relies on loans from others. (Note: While there is a small minority of women hole owners—at the time, I believe, there was only one at this mine—I am using the pronoun "he" here because those involved in this conversation were discussing male actors.) If he doesn't use the money wisely, assuming "the hole will pay" (*shimo italipa*, as diggers often say), he can easily get himself in debt. The guy who got $6,000 may "forget" that he has to divide this money fifty-fifty with his diggers, then pay debts, and then still pay for the other future "expenses of the hole," including food for diggers. Another may take out $100,000 in debt from an investor but, before starting production, buy a house and a car; when he reaches a point where the money's finished, but he still hasn't begun extracting in earnest, he may even ask for more money. Some never get to the point where they're producing, and if they do, it's not enough to pay off the debts. In making these statements, the diggers are also pointing out that there is an imbalance in these work-debt relationships, because the PDG is operating on interest-bearing loans and has signed paper contracts, and he has extended money to the diggers. These diggers, with no paper contracts, won't support him when the money is gone but will disappear to dig elsewhere: As one person in the bar puts it, "The problem is that this debt belongs to the PDGs, and the diggers don't care about it. Diggers won't help him pay the debt if there's a problem, because they've already gone to other mines."

Claude is quiet, listening, but eventually he addresses the doctor, surely aware that the former mayor is also listening: "You're forgetting that the diggers *built* the clinic," he says, "and also the market, when the government never had any thought of doing any of that." Claude is referring to a relatively new intervention, the "basket fund" (a certain percentage of product is gleaned from the overall production at the mine, then sold and invested in local "development" projects). This is an intervention that NGOs and Congolese "civil society" (*société civile*) had been pushing for many years and which the International Tin Research Institute's (ITRI) and the NGO Pact's new "conflict-free" tagging scheme has helped them to enact (see chapter 9). Because of the basket fund, there's now a clinic and a market, things that the government has never built, which is all more evidence that artisanal miners are more powerful than the state and that the state depends on them (a recurrent idea among diggers being that state officials wouldn't even be able to buy cigarettes without their work, because they come to the mine asking for cigarettes. Diggers take this kind of dependency—and the way state officials emerge wherever they are and disappear when they are gone—as evidence that the state, which in turn depends on personnel for its existence, wouldn't exist without them). Claude continues, "We built the hospital, so why can't you doctors cure us, even for free? You are profiting by charging people, and you don't even think to buy solar power, so you guys are there treating people in the dark! You're just waiting for us diggers to bring in money, thinking maybe our work will buy [power] for you, and you can't even reduce the price for us in the meantime!"

Claude is drawing attention to a situation that the monitoring of conflict minerals has helped to make visible, but the general principle has always been the case: mainly, while diggers are vilified by others for not seeing ahead, they are the mostly invisible, underground source of life and movement in their communities, regenerating people and places that have been destroyed by war. Moreover, they are more important than, and also the ultimate source of, state power. Claude knows the healing capacity of this work from personal experience, because this particular town experienced a great deal of violence and suffering during and immediately after the RCD occupation, and at one point he was within a hair's breadth of being murdered by occupying RCD soldiers himself. Indeed, Claude often says, and in this conversation was implying, that diggers' work is as important to the process of collective resurrection and healing as that of doctors—at the very least, diggers are not merely patients, or recipients of treatment, but active contributors to healing. Whereas the doctors' work is recognized and compensated, the diggers' contribution to the hospital is invisible and remains unrecognized, like most of the other things their work does for the communities in which they live. Moreover, if diggers don't always have money to pay their bills, it's not because they're "bad people" (*watu wabaya*) but because they've already forwarded value to many others, including but not only the clinic. In this vein, diggers often like to point out that, were it not for their work, you wouldn't be able to buy a bar of soap in the towns in which they work because there would be no other commerce. The implication is that their dirty work, their being in and touching dirt and earth, makes you capable of touching the apparent source of cleanliness and also looking clean, even if, in reality, you're not all that "clean" (*safi*, a term that can also imply "upright").

The value of diggers and of digging, which is what Claude was defending in the bar, was not just about the product that they produced ($6,000 worth of cassiterite in a day!), although that was the most tangible way this value was realized and the way most people talked about it. The value of digging was also materialized in the social relations and practices that emerged around digging, which echoed and democratized people's understandings of the past and future potential of mining. Moreover, the value of particular actors in the chain became concrete in relation to people's understanding of the value of other actors in the chain (not all of whom were human actors; as we will see, papers and ancestors were also actors). Therefore the supply chain was also a semiotic chain of meaningful values and a process of ontological transformation of earth into commodities through the substance's interaction with people and media, like money and papers.

To get a deeper sense of the social organization of movement and the competing values, forces, and histories that are packed into the supply chain, let's now take a closer look at some of the ways this work is organized.

Organized Artisanal Movement

The main 3T (or, colloquially, "coltan" or "black minerals") supply chain is formally defined by three types of workers—diggers (called *ma creuseur, wachimbaji*, or sometimes *ma artisanal*), middlepersons (*négociants*), and buying houses (*comptoirs*), which sell to smelters located outside the country. These categories oversimplify the situation, because some "diggers," like the hole owners, are often more like négociants and because there are many people acting like middlepersons who are not technically négociants (for example, women who come to wash or pound stones and transport sand in exchange for minerals and who may also exchange things like beer, food, cigarettes, or sex for minerals and later sell them to négociants). Although these are work categories that index ideal-typical stages in the supply chain, they are, more importantly, state-sanctioned legal categories derived from the Congolese Mining Code (the first version, in 2002, was written by the World Bank; see Trefon 2016, 103, 135), and so they emerge from state and international (read: neoliberal) efforts to regulate the trade. In Congolese daily life and practice, they are also semiotically distinct categories, with "diggers" (including the ones who are more like négociants) being perceived as closer to the forces of the ground than the négociants and comptoirs, who are enmeshed in a world of money and paper that is in conflict with these forces in the earth. Since the war ended, diggers have purchased diggers' cards from SAESSCAM, while négociants, who pay a higher amount, have purchased their négociant cards from the Office of Mines. Comptoirs, who sell to smelters, also have their own cards and papers, for which they pay even more. There is thus nothing natural about these divisions, which exist in part to increase sources of revenue for different government offices, giving more state authorities opportunities to "touch money" (*kugusa pesa*), ultimately through the bodily work and "touching" that diggers are directly involved in. Diggers prefer to sell directly to comptoirs if they can because they think they can get a better price (and, during the Great War, a time of unprecedented and unparalleled movement, those who could make it to a town to sell did). Indeed, in most places, both diggers and comptoirs would like to do away with the category of the négociant completely if they could (an exception would be a mine particularly hard to reach that no comptoir would ever go to).

Many diggers were involved in a communal enterprise, in that they came together as a group of friends or relations to rent out a space on which to prepare and dig a hole (or holes), from either the recognized autochthones of the place or from a company concession. Still, they might have to secure loans (*deni*, or debt), whether in the form of money or provisions, to proceed with work. In these cases, all the diggers could be *wenye shimo* (hole owners), and they would nominate one of their members to be the PDG, while others might act in other capacities (as director general, a second in command, or treasurer, for example). The PDG would recruit other diggers (if this was necessary) and provide them with the things they needed while also interacting with state authorities, representatives of the autochthones or owners of that land, négociants, and cooperatives. If there was sufficient movement, and they were able to successfully manage their debts, these original wenye shimo might transition out of digging, expand to open other holes, and become more centrally involved with management.[2] In times of movement, there was a great deal of this kind of upward mobility (among men at least), but when then there was no movement, diggers became increasingly indebted and unable to move up in this way.

In many holes, the *wenye shimo*, or PDGs, were not around at all, and were really more like négociants, even though they were legally diggers. Thus it is that, within the legal category of "digger," there are people who have never lifted a shovel nor carried water. These hole owners take a portion (a common convention is 50 percent) of the minerals dug, the rest of which is divided among the other workers. They also forward tools and "ration" (food, water, basic medicine, cigarettes) to these workers; they may get the money for this themselves (though this is rare), from a négociant, from an outside financier, from a cooperative in which they are also a member, or from some combination of these (most of these loans have to be paid back with interest from 20 percent to as high as 50 percent).

In most places I visited, diggers worked in a *carrière*, or quarry, which was sometimes divided into zones that were called *chantier*, or construction sites. Chantier was also a colloquialism for a development project: President Kabila's slogan, written on billboards throughout the country, was that he had Cinq Chantier, or five projects, that he was supposedly always in the process of completing. Taken from the history of colonial-era mining, the term also indicated that diggers saw themselves as working on a collaborative and generative project rather than simply taking stuff out of the ground. Sometimes carrière and chantier were synonyms, but in large mines a carrière would have a recognized landowner, or *chef de colline* (a descent group representative or a company or both), and it would be divided into several chantiers, each having a certain number of holes, or *mashimo* (perhaps as many as ten, each worked by a separate group of diggers who independently rented their spaces, sometimes with financial backing from someone else and sometimes not). A single person could be the owner of many mashimo or just one *shimo*.

Diggers held different job titles depending on the work they did and where they were located, but at the "lower" end there were generally those who dug holes (often deep into the ground or into a mountain), those who carried out minerals, those who washed minerals, those who brought in or carried out water, and those who transported minerals over great distances, away from the mine (these latter were porters, *ma porteur*, and were not actually diggers; theirs was often what could be thought of as an entry-level position that they entered into in the hopes of becoming a digger or merchant to diggers later on). If their jobs were particularly dangerous, such as going deep into shafts, diggers had work names reflecting that danger, as well as their masculine and even occult power. Diggers who went deep underground were called *hibou*, or owl, a term that simultaneously referenced the flashlights they wore around their heads, the fact that they were involved with mystical forces, and the tough soldiers of the Mobutu era, who were also called hibou (owls are generally associated with sorcery, the occult, and shape-shifting in this region). Above these different types of diggers and transporters were managerial diggers who organized labor within holes and managers who moved among the holes belonging to a PDG (at large mines, "manager" was also used in a different sense, to describe a middleperson operating between hole owners and négociants). In addition, there were people who worked for the PDG in accounting for production yields, demographics (the numbers of workers), and the cost of what was being consumed by diggers. Finally, there were diggers ensuring that "law" (*sheria*) was being followed—including the law of the Congolese Mining Code and the "laws" (*masheria*) of the PDG and the owners of the land (who may or may not, for example, allow drinking alcohol on-site or women to be in the vicinity of the site).

While work titles change and mean different things at different sites, one could expect to find such figures as the chef de colline (not a digger), or person representing the recognized owners of a particular area (the descent group); the aforementioned PDG and the director generale (DG), who rent holes and serve as liaisons with government authorities while managing accounts and keeping track of "demographics"; the *chef de groupe*, who collects fees and taxes from groups of workers while organizing the distribution of food and tools; the *chef de camp*, who is in charge of keeping the peace and implementing the law as laid down by the chef de colline and state authorities; and the *chef de chantier*, or supervisor of a group of workers in one or more mashimo, or holes (in some places this person was called the *drumeur*, or drummer, because he set the rhythm for work). Some of these titles have been appropriated from the history of colonial mining—they are the terms that were used by Belgian mining companies, and later, by state companies that are no longer operating. Others, like the drumeur, are local inventions (in some places, the drumeur was like a foreman, and the hole was his drum, while in other places, the drumeur was the person who pounded stones and in doing so set a tempo for work). Even the names that have a colonial history are now being used by artisanal miners who never worked for an industrial mining operation; some diggers describe their use of these terms as evidence of the democratization of mining since colonialism and especially since the wars.

While in most places the work of digging looked like an overwhelmingly masculine affair involving young men, the situation was complicated and varied from place to place. In the relatively few places where women were actually involved in digging and were not relegated to cleaning and sorting (work they were sometimes also excluded from), they and others likened their work to agriculture and often drew attention to the fact that they weren't digging deeply underground and were instead digging in small groups with friends or neighbors (they often dug in smaller mines near the villages where they lived and did not have work titles). The absence of women in deeper, larger holes was not always voluntary: at least in some places, women were prevented from digging by diggers themselves or by the representatives of the descent group that occupied the land. For example, in the early days of Bisie, women dug in groups, but eventually, when holes grew deeper, a consortium of male actors, dead and alive, prevented their access to holes. They—usually groups of women—could still be PDGs, or hole owners, but it was difficult for them to do this effectively if they couldn't be at the hole. Invariably, men justified women's exclusion by arguing that (male) ancestors did not like the presence of women (especially menstruating women) and the "conflict" that women could bring to the mine, including the potential for sex (especially, extramarital sex) and sex work. (By using the word *justified*, I do not mean to imply that there was anything disingenuous about this; they seemed to earnestly hold it to be the case that women threatened safe extraction.) As discussed below, the transmutation of these substances into resources required the consent of spirits, and men claimed that women's presence would interfere with the agreement arrived at between the living and the dead, thereby sabotaging extraction and even chasing minerals away. While some women agreed that these ideas might have some truth to them (such restrictions came from the *mababu*, or ancestors, who lived long ago and who still live in the earth, and these restrictions also showed up in other contexts, like Mai Mai practices), most claimed that men overemphasized this because they didn't want women to know the value they were getting from the ground.[3] Men also talked about the international community of humanitarian NGOs and their concerns about "child labor" as a reason for excluding women; at the same time, in more recent years, women mobilized the support of international NGOs to "get their rights" to dig in places like Kalimbi mine at Nyabibwe.

Still, women were often present at the carrière, unless they had been forbidden to be there by the PDG or the chef de colline (their children often accompanied them and sometimes helped with the work of sifting and washing, the money from which was usually used for school fees). Depending on the mine, women sorted and washed minerals and, at larger mines, also waited at "the door" (*mlango*, or entrance to the hole), where men passed off some of their material to these women, who turned around and sold it privately to licensed négociants, giving a cut to these "smuggling" diggers. This was almost always an illicit agreement between a specific digger and his woman friend(s), aimed at undercutting or outmaneuvering the hole owner. While they were less likely to be licensed négociants (I only met a couple licensed women négociants during my whole time in eastern Congo), they often operated as middlepersons at the lower level of the chain, especially at smaller mines in remote areas. Most importantly, they were often engaged in many other kinds of work at some remove from the mine, including food, water, booze, medicine, and loan provision;

Figure 5. Digging (Maniema)

Figure 6. A mobile phone sheds light on digging in a hole (Maniema)

Figure 7. Cleaning ore (Bisie, North Kivu)

Figure 8. Women sorting and cleaning with children watching (Maniema)

Figure 9. Children playing while their mothers work nearby at a small mine (Maniema)

Figure 10. Sifting ore at the comptoir's office in the ruins of the company mining town (Kalima, Maniema)

through this work, they often subsidized the work of diggers for periods of time in exchange for a later return of minerals, and so they were essential to the overall process of movement. In many mines, women dominated this commerce, especially if the mine was relatively accessible to a town and did not involve walking great distances through the forest with goods.[4]

In addition to being "ordered," then, mines are also highly variable ("each mine is its own world"), and this multiplicity responds to changing contingencies, rendering mining holes highly adaptive to the circumstances of people, places, and times.

Movement, Collaboration, and Deception in the Supply Chain

Moving "up" from the realm of diggers, the supply chain reveals itself as a fraught and ethically charged space with a diversity of actors collaborating across differences while also coming into conflict with and casting aspersions on each other. Very often, they also actively try to cheat each other. In its simplest form, diggers are at the bottom, or "low" (*chini*), and different levels of middlepersons are located "up" (*juu*) the chain; those at the bottom do indeed feel themselves to be at the bottom, with relatively less power and knowledge than those above them, and they use dramatic language to this effect (they talk about being stuck in a hole, or "living in the dark," or the fact that it's "always night" for them, and they can't see; they refer to themselves, somewhat humorously, as burrowing or nocturnal animals and pose "in the dark" type questions about the "true price" of the minerals they extract, claiming to have no knowledge of the uses of these things). As mentioned, those at the bottom would often like to move up the chain themselves (diggers would like to become middlepersons, at least temporary ones, by selling their product directly), but they are usually prevented from doing so, either because they lack the resources and connections or because state actors and state regulations block them (usually by claiming that they don't have the proper papers, or *karatasi*, especially the négociant's card).

There are several common ways in which higher-level actors try to cheat (*danganya*) lower-level actors. First, they may find ways to lie about what it is they want or are looking for or what the ore in question actually is (by saying that they are looking only for cassiterite but taking the entire ore, which may include tantalum, or by convincing those below them that there is no tantalum in the ore when there is). Second, they may find ways to deceive those below them about the value of the ore by referencing a universal standard that may or may not be "true" (the tester machine—and they may decide to use a bad one on purpose—says the percentage is X or the "world market price" has shifted to Y); in this way they seek to capitalize on the fact that those below them have very little information, may be in dire straits, or both. Third, they may stall on payment, then find a way to "dirty" the ore later on by mixing in "earth" of lesser value when they have the opportunity. If they have the means, they can create separate spaces (e.g., the comptoir house) in which they can use machines to measure the quality of the substances or "dirty" the minerals behind closed doors. Finally, groups of négociants may gang up with each other to set the price for those below them while also limiting the ability of others to enter and offer a lower price. They may do this by threatening these competitors outright or by making it legally difficult for them to enter into the trade (for example, creating cartels or using the official cooperatives in a restrictive way different from their ostensible original purpose of assisting and empowering diggers). Technology is crucial to price setting and manipulation—mainly through the use of the testers, which get more sophisticated and expensive as one moves up the chain but which can always be configured to distort reality. Generally, those "above" have information (about price and chemistry, for example) that those "below" do not have, or if they don't, they pretend that they do and use the fact that diggers are usually stuck in the place that they are in to manipulate them as much as they can (they are in a hole, as everyone is quick to point out).

Diggers do have access to information that they can use against those above them; they have, for example, a technique for scraping their stones against a harder stone (quartz) and observing the streak left behind. This can tell them which of the 3Ts is contained in the ore and also something about purity. Diggers can also group together through official or unofficial "cooperatives" to try to set the price or may find a way to illicitly sell to someone other than the owner of the hole. If there is a cell-phone connection, they can use their phones to try to access the "true price" from a friend in the city or even get online to find out something about price; for this reason, négociants will often try to avoid places with cell-tower access, preferring more remote places, which also tend to be more dangerous (vulnerable to armed groups). Depending on the place and context, having this information may not help the diggers too much, but it can still give them some persuasive authority. Diggers can also take value in the form of food, money, water, beer, or other commodities from those above them, expending value "right now" because they're not certain of what they'll get later or when that payment will come. In this way, the digger(s) can turn the table on the négociants, putting the latter under their thumb—the négociants have already forwarded everything they have to the diggers and may be indebted to someone else, so they can now only wait for the diggers and hope for the best or end up being "destroyed" by the digger. Movement can thus overcome, or beat, the machinations and information of those from on high; in this vein, diggers often say that they don't always care what the price is or if they're being deceived, because they can always count on the hole, combined with the movement created by their collective work.

Finally, one of the most important tools that diggers have is their ability to move away; while not all diggers are in a position to do this (especially if it's a small mine near their home), many are, and they are generally less tied down by debt and papers than those above them (of which see below). If they run away after they have indebted a hole owner or négociant by eating or drinking at the hole, it is hard to capture them, as they can use money, minerals, or their ability to work in a hole to pay off the police. A lot of the "bad press" that diggers receive comes from this dynamic: those above them in the chain have some kind of long-term extraction plan or contract to those above them, and those below them are eating value in the moment before the object of all the work has been realized, and they can escape if it never is—hence their association with extreme expenditure and recklessness for others. In reality, they are not actually wasting this forwarded value but using it to prepare the hole by digging while also generating mutually beneficial friendships at the hole, which can later be drawn upon in times of need.

While there is certainly tension, deceit, and acrimony—with those above accusing those below of reckless expenditure (spending too much on alcohol, cigarettes, sex, etc.), while those below accuse those above of selfishness and deceit—this is also a system that operates, when it is operating, on the forwarding of value in both directions but almost never at the same time. Generally, money has to come "from above" *first* for work to proceed below—middlepersons or their financial backers forward money to those below them so that the diggers can dig; if that doesn't happen, work can halt and those in the middle (the négociants) can become indebted to their financiers or those above them in the chain. Sometimes those "above," who buy from those below, want to slow down the process of buying and selling on their end—they may hold onto the product and delay on payment, claiming that their buyer wants to see and test it first. Or they may simply not be able to purchase a certain amount at this point, and so delay forwarding money to those below them, who need to in turn forward the money to diggers so they can eat and drink and generally take care of themselves while preparing the mine, digging the hole, and accumulating material. Those above may also be holding on for a better price, while the diggers are simply digging without a plan and without knowledge of price.

Of all the actors involved in mining, cooperatives like the one Claude heads are among the most complicated, conceptually and in practice (outside of mining concessions, diggers were legally required to be members of registered cooperatives, though it wasn't always the case that they were or that a registered cooperative existed for them to be in). Cooperatives also epitomize the disjunction between visibility and invisibility in the mineral trade and illustrate how these themes relate to movement. In the simplest of terms, these registered, legally and bureaucratically visible organizations were intended to do one thing, but under the surface, invisible to many, they were usually doing something else.[5] Outside of company-owned mining concessions, in ZEAs (legally recognized zones of artisanal extraction), cooperatives are the formal mediators between diggers and négociants.[6] According to official explanations, cooperatives are supposed to empower artisanal miners, who come together through them to manage labor, resolve disputes, and negotiate price with the négociants, who often have their own unregistered cooperatives (some might call them cartels), designed to set the price as low as possible. In practice, so-called diggers' cooperatives usually bring together high-level diggers, like the PDGs, and either négociants or buyers who are not legal négociants—including, in recent times, actors who are not supposed to be involved in mining, such as military and other state actors. In some locations, the role of cooperatives is further complicated by the fact that there are parallel, often unregistered, diggers' associations doing what the cooperatives are supposed to do, sometimes in competition with the cooperatives.[7]

Where there were cooperatives, négociants were expected to buy directly from them rather than from the diggers themselves (this was written into the Congolese Mining Code). Because they could include outside figures, such as politicians and other powerful state figures or their representatives, the cooperatives provided an opportunity for people to enter into business deals with diggers (or, more likely, hole owners) despite the fact that they were not formally recognized as négociants according to Congolese mining authorities and the Congolese Mining Code. Finally, négociants (who were often members of the diggers' cooperatives but not diggers) used the cooperatives to enter into arrangements with diggers in which they would forward supplies, food, and money in exchange for a "good price" or an exclusive buying arrangement that excluded other négociants not in the cooperative. After the gradual introduction of the International Tin Research Institute's bar-coded tags after 2013 (described in chapter 9), some of these cooperatives also worked with state authorities to engineer ways around the system of "traceability" so they and their buyers could sell where they wanted—including outside the province or the country—in violation of what was referred to as the international "law of traceability" (which was not actually Congolese law).

When there are no cooperatives, the artisanal miners organize themselves into committees that interact with the state as infrequently as possible. For example, among the artisanal miners in the company Sakima's concessions in Maniema, there were no registered cooperatives (there were never legally recognized cooperatives within a company concession). Rather, the *mwenye pori* (owner of the forest) interacted mainly with the diggers' elected president (*rais*), who stood for all the diggers and led them; the president was supported by *conseillers*, or advisors, and a *chef de discipline*, who headed a disciplinary committee that imposed judgments and fines with respect to specific cases; secretaries and cashiers distributed and kept track of the value of everything being forwarded at the hole. Every hole, rented out by what was called a chef (or a chief, a colonial designation), had somewhere between three and maybe eight people depending on its size, and if these groups fought with each other over the boundaries of the holes or the overall plan for extraction, the disciplinary committee would intervene. The committee, which transcended and included all the separate holes, also met with government officials in case of any problem that fell outside the arena of mining (e.g., if the police came looking for one of the diggers) or if state representatives showed up. In addition to establishing terms regarding price, the whole setup was designed to keep the actual state as far away from the hole and the other diggers as possible so that movement could proceed uninterrupted.

The People of Documents

Diggers, négociants, and cooperatives have an interest in sustaining movement: they want substances, money, and other commodities to circulate back and forth, and they want forms of value to be converted into other forms of value and for certain ontologies to be transformed into other ontologies (for things that belong to spirits to become liberated commodities, for example). In contrast, state actors derive their income from closing the ways of people involved in the trade for a certain period of time without closing them out completely. While those involved in digging and selling see most of these figures as pests, they all have their own story, which they are sure to tell, about the services they are providing and why they need to be compensated (in part because their formal salaries are either very low or nonexistent). Despite their rhetoric about services, though, most people understand that these are people who could bring trouble (*shida* or *matatizo*) to the workers (diggers and traders) if slighted or left out, so it is best to take care of them even if they're not providing anything. Some of these people are looking for papers and have papers to sell (which are sometimes different from the "real" state papers), while others do not. State officials may also take command of certain holes for certain periods of time as their payment, bringing in their own workers to work for them during that time in lieu of payment from the hole owner; their diggers are often people who are "indebted" to them in some way (in the past, when soldiers were at mines, they were notorious for doing that and in times of war brought POWs to work as slaves in the hole). As mentioned, this practice of temporary forced labor is called *salongo*, and while the term is associated with the Mobutu era, people are clear about the fact that this practice originated during the "time of the Belgians," who institutionalized forced labor for work projects that didn't necessarily benefit those doing the work (during salongo, the hole owners gave up ownership of the hole for a period of time, and often the diggers in the hole were not affected at all; frequently, they would receive less than the normative arrangement, but they would still earn).

There is no single system for how the state emerges around mines; again, "each mine is its own world." Some major determining factors include the size of the mine (big or small); the location of the mine (e.g., near a road or not, easy to access or not); whether the mine is on land or water (it is more difficult to ascertain jurisdiction on a river); whether it is on company-owned land, in a ZEA (zone of artisanal extraction), or in a completely unregistered area (and this can change without everyone knowing); and whether there is conflict in the area. In times of war, state or state-like presence at mines was relatively straightforward and might involve only armed actors, usually cutting down on the number of people that diggers or their representatives had to interact with and pay tax to without necessarily eliminating them entirely. In times of peace, and especially in the years after soldiers were forbidden to be present at mines, the state was more multiple (represented by multiple offices), which is why many diggers preferred the time of war because it cut down on the number of "tax" collectors.

For example, if a hole is on land, eventually it will probably require shaft construction, which brings in different state officials (e.g., SAESSCAM). If it's in a river, it may require the construction of canals, which might bring in the Office of the Environment, because these diggers and their canals are generally understood to be hurting the environment, so that authority has to be compensated and their "salary" paid by the diggers. A small mine in a remote location might be worked mostly by people who already live in that place (who may combine this work with farming or hunting depending on the season), while larger mines attract people from far away, especially when prices are high, creating a vortex that draws in more state authorities. In general, the more people who come to a mine, the larger and more productive it will become, drawing in a wider diversity of state and nonstate regulatory authorities.

Some combination of the following state and nonstate regulatory figures enacting "closure" (*kufunga*) were likely to visit any given mine periodically, in no particular order (they would never be there full time, and only some of them would go directly to the holes where workers were working, interrupting their movement). I am classifying these figures here in the hopes of giving some sense of the breadth of possibilities regarding how the agents of closure, or temporary stoppage of movement, were experienced. Any of these figures could work together, although some were more likely to travel with each other—when they did that, they were understood to be working in collaboration but also watching each other to ensure that revenue was split fairly between them (not necessarily 50/50). Under certain circumstances, they could also come into conflict or at least competition. While the titles were official, the classification of them according to types is my own, based on my and other people's observations of the work they did on the ground (rather than from the point of view of agents in urban offices). At small mines in the forest (probably the majority of mines in Maniema, for example), few or perhaps none of these figures might be present, and there was likely to be only the representatives of descent groups or very local administrative authorities. Still, diggers, porters, and négociants were likely to meet the following state figures on their way out of the forest, at roads or small towns.

First, there were figures concerned with the work of extraction itself, whether legal issues surrounding extraction (whether so-and-so had a legal right to dig or prospect or mine industrially or semi-industrially, for example) or issues having to do with safety (whether the mine was bigger or deeper than it should be according to the Mining Code, for example). Most notably, the representatives of the regional Office of Mines (or simply, the people of mines, *Watu wa Mine*) interacted primarily with négociants and comptoirs to ensure that they had proper papers. They also "sold papers" related to exploitation and exploration. Dealing more directly with diggers was SAESSCAM, an office brought into being by the World Bank–drafted Mining Code of 2002. These agents worked alongside the Office of Mines and were typically housed in the same building or structure in any given town. SAESSCAM was ostensibly established to teach diggers how to dig safely and how to construct shafts while also providing them with tools and materials; in my experience, they did not actually do any of this, but they did impose fines for violation of law or code (including, later on, the rules of the NGO ITRI/Pact regarding tags, which were not actually in the Mining Code). Of all the state officials, others considered them to be the most "like" diggers; they were often referred to as "low" (*chini*) or "dirty" (*chafu*) state actors, or even "prostitutes" (*malaya*) and were seen as relatively uneducated and not cosmopolitan (they were also more likely to be women than the other state actors). Unlike diggers, though, SAESSCAM agents were seen as useless and unproductive, promising things they never delivered or being ignorant of what it was they were supposed to be delivering.

In some places and times (more commonly in company-owned concessions), there were also *police de mine*, the mining police, involved with policing extraction. These state agents policed the company-owned concessions that were occupied by artisanal miners, arresting/"tying" (*kufunga*) people who entered illegally; they also policed within the ZEA in search of fraud or to enforce the orders of SAESSCAM or the Office of Mines over a digger or group of diggers. Finally, there was the Office of the Environment. These agents were supposed to ensure that people were not hurting the environment; since most artisanal miners are "hurting the environment" in some way, they could find violations and "sell papers" related to this (as mentioned, they were especially common at or in rivers, where there were often no other state actors or even representatives of descent groups or concessions and where there was always a threat to the environment).

Second, there were people concerned with what was called "security" (*securité*) in the sense of regional or national security in a context where members of armed groups might be present (or, in many cases, finding real or potential threats to security and extracting revenue from that). The most prominent of these was ANR (Agence Nationale de Renseignements), the national intelligence agency, charged with handling "internal and external" threats to security. They were more prevalent in zones where there was armed conflict or potential armed conflict, and they often worked closely with the army (often via cell phone). Wherever they were found, they might make an argument (whether rightly or wrongly) that a digger was somehow associated with, or working with, an armed group or that the person was an armed actor or was somehow contributing to insecurity. Then there were the DGM (Direction General de Migration), immigration officials ostensibly looking for non-nationals. They often "walked with" ANR (internal security), and together they shared whatever proceeds they could (they operated under the direction of regional administrators). They were expected to only be in places where there were likely to be people from foreign countries—generally large mines during times of great movement; ostensibly, they also helped provide protection to non-nationals who had the legal right to be there.

Last among the category of agents involved with security were soldiers, the one kind of state functionary that diggers sometimes tolerated or identified with (in places where armed groups that were not the Congolese army were near, they did at least provide protection and could be "made supple" through the transfer of work product to them, sometimes even becoming the diggers' soldiers for hire). As already mentioned, during the latter part of my fieldwork, soldiers were not supposed to be involved in mining at all (their presence had long been illegal according to the Mining Code, but this was only enforced after 2010), and it was very rare to find them at 3T mines, because these were too visible and large (they were, however, at gold mines, which were less visible). Higher-level soldiers continued to have a role as buyers (usually via an intermediary in coopera-tives) or transporters in the background. They also operated as hole owners through intermediaries.

In the next category, there were state or nonstate actors representing dif-ferent nodes of political authority, often rooted in different concepts or ontological foundations (say, national legal sovereignty or the power of ancestors). These agents were there to ensure that these authorities were "remembered" and recognized. Among these, outside of ZEAs, were the company officials associated with the company concession. These offi-cials managed the company-owned property and collected revenue from people who came to buy and sell minerals or to rent holes on the com-pany concession. Because the companies (in the Kivus and Maniema, the only 3T company was Sakima) were owned by the state, they were also considered representatives of the state, but they were not administrative bureaucrats. Then there were legitimate tax collectors, meaning they were recognized to be collecting tax for the province or territory and not just for their own office. These people mostly interacted with comptoirs and négociants in towns; but these days, because of the tagging of conflict min-erals, they might show up wherever "tagging" of "conflict-free" minerals is taking place. There were also emissaries representing state-sanctioned local authorities. This varied depending on whether the location was a *chefferie*, meaning a chiefdom, or a sector, an area that was not legally considered to have traditionally "belonged" to any specific ethnicized group. These ad-ministrative offices included the chef de territoire (chief of the territory), *chef de secteur* (chief of the sector), *chef de chefferie* (chief of the chiefdom), and *chef de grouppement* (chief of the smallest administrative unit, or group-ing), all of whom collected revenue from hole owners and others.

Though they weren't state figures, there were also the autochthones of the piece of earth on which the mine was located—the "owners" of the land, whose identity was often contested. This could either be the repre-sentatives of a descent group occupying the land or another rival descent group that claimed rights to it based on an earlier decision, usually going back to colonial or early postcolonial times (e.g., "the time of Mobutu"). These descent groups often sought backing from multiple branches of the state to make their claims effective (in some cases a different "branch" of the state may support a rival group). These are one of the few, perhaps the only, authorities that diggers actually care about, because they are under-stood to be the actual owners of the land—meaning the people who have lived and died there—who can bring problems to diggers through their an-cestors, of which see below (their "customary" rights to land were also protected by Congolese law). These figures were very different from the others because, while they could close mining if they were not satisfied with the rents or tributes they received, they also "opened up" land for people to dig in the first place, and it was widely understood that, without them, movement was impossible. Also, it was understood that these "children of the hill" (*Watoto wa kilima*) or "children of the forest" (*Watoto wa msitu*) were in conflict with state authorities because the more state authorities there were, and the more aggressive they were, the less value in the form of minerals their dead "grandparents" (the *mababu*) would receive from diggers or the company mining concession. That meant that, if the ancestors were unhappy, it was at least partly, maybe entirely, due to the greed of state actors, who were "eating" their sacrifices. Therefore there was a direct and deep conflict between state actors and ancestors—and diggers, the autochthones, were caught in the middle (more on this conflict, and its depth and consequences, below).

As alluded to above, in addition to these regulatory actors, at any given mine the diggers will have organized themselves in some way, perhaps through an official cooperative or an unofficial committee. They are likely to have their own "brigade," or security, which interacts with state authorities, as well as their own tribunal. They will also have their own rules, imposed in conjunction with the owners of the land, and they may employ these alongside, and in addition to, state or international regulations. The digger's "government" (*serikiali ya ma creuseur*) will interact with the government of the state over certain cases (e.g., if one of the diggers is involved in a criminal case outside the mine). Generally, diggers saw their state-like institutions as more "real" than "actual" state institutions, because they actually serviced digger's needs and operated more or less in the way they were intended to.

State regulation, then, consisted of a panoply of authorities representing different spatial and temporal regimes; for example, administrative chiefs were associated with colonialism and the time of Mobutu, ANR and the army with the not-yet-finished time of war, and SAESSCAM with the space-time of global neoliberal governance and regulation. Their pretext for "governing" and their divergent spatiotemporal referents usually had little to do with their actual "function" in any given place in the present, which seemed to be to create movement for themselves by temporarily closing out diggers, who in turn propitiated them so that they might move. While it may seem like a disorderly hodgepodge, the composition of regulatory actors was rule governed but context dependent. It was certainly not "stateless," and it was also never the case that just any state official could show up at any mine at any time and claim the right to be there; assuming they were present for an extended period of time, even wartime soldiers needed to cooperate with other state authorities and allow them the opportunity to collect "tax" from workers, whether in the form of money or minerals. There was, then, a grammar of possible formations of state regulation, and the actual structural organization of the mine (actors representing and enacting movement and closure dancing alongside and with each other) emerged differently in different times and places depending on certain factors and criteria that the actors involved in the work of movement and closure were all aware of (although they often had to argue about it among themselves). While these state and nonstate actors enact regulation with a view to the grammar of regulation, in practice they may come into conflict over who has ultimate authority over a given issue or domain in a given space, and this has to be discussed (see, e.g., the conflict between a company manager and SAESSCAM described in chapter 4).

Moreover, diggers were not simply passive subjects with respect to state regulation. This regulatory system was extremely dynamic, and much of the dynamism came from the agency and power of diggers, who could either conceal themselves or find ways to coopt state officials into their work, compromising and collaborating with them. Some diggers claimed that, even during the war, when the "enemy" RCD soldiers controlled mining and could harass them at will, they still found ways to get around them. As they explained it to me, in those days, after soldiers expropriated minerals from them by force, diggers would use their mobile phones to call the bosses of the low-level soldiers (or sometimes they would go to the office in person), who would then launch an investigation of their military underlings. Typically, the confiscated minerals would be divided among all three parties: the diggers, the confiscating soldiers, and their bosses. Thus the diggers would at least get back a cut of what had been taken from them originally. Wherever there were soldiers involved in mining (whether Congolese soldiers or "enemy" soldiers), diggers and higher-level soldiers each preferred to deal with each other directly because it cut out the middlepersons—mainly, the lower-level soldiers—for both parties, and it placed the decision of how much to distribute to the lower-level soldiers in the hands of their superiors. Also, commanding soldiers found diggers to be more honest than their underlings. The point of these stories about soldiers was always that, if diggers could stand up to occupying foreign powers in this way, it was because they had some leverage over them that others didn't have—they had the ability to work a hole together and thus to transmute potential enemies into allies through their work in liminal, or in-between, spaces. It was easier still to domesticate today's Congolese state agents.

At every mine I spent time at or near, diggers used minerals to "make friends" with state officials in the police department, the Office of Mines, or the army, who would then let them know when other state officials were coming to harass them for "tax." Diggers were sometimes even able to turn police or soldiers against corporations and higher-level state figures (as happened at Bisie, described in part 3). As a result, they figured minerals as being alternate and parallel to the official power of the sanctified colonial and postcolonial social order, and they often specifically talked about how minerals enabled them to "make law" (*kuweka sheria*, or place/set the law) and how their powers resembled but were also qualitatively different from those of educated elites. In this regard, they often referred to their minerals as pens and their shovels as diplomas (parallel powers with different sources). As one digger put it to me (though many expressed this general sentiment), "We didn't study a lot, so minerals are our pen. Minerals are our power, and we make law through them. We force people to respect us and recognize us using minerals. Minerals are everything for us."

This system of movement and closure—produced, on the one hand, by actors in the supply chain and, on the other, by the interaction of state and nonstate actors—shaped an ethical system that was understood differently by actors depending on their "location" in the chain.

The Ethics of Invisibility: The Laws of Diggers and the Paper Traps of Négociants

The conflict between diggers and négociants took shape in relation to the aforementioned regulatory actors and wasn't reducible to the price for minerals. For one, diggers were usually in a better position to avoid state officials and articulated an ethics around their mobility that came into conflict with the patron-client practices of négociants, which would have them stay put. If they sometimes sought to make their worth visible to others (like Claude that day in the bar), diggers also valued their autonomy and invisibility, often likening themselves to snakes, moles, and other animals that hide and burrow in the ground to escape predators. They worked hard to evade state authorities and tended to see every state intervention as an effort to block them. For example, in all the sites I visited they insisted that the de facto illegality of their shafts (most of the time, they were not actually illegal, but state actors acted as if they were) had nothing to do with safety or ecology; rather, it was because shafts made it difficult for state officials to locate or tax them. From their point of view, the informal illegalization of shafts was an effort to make them visible so they could be "taxed" despite the fact that mining "in the open" was more ecologically damaging than shafts.

In a world of paper-wielding state officials, invisibility was an asset insofar as it enabled movement. People throughout the supply chain needed movement to be able to pay their debts and keep money circulating, so blockage was dangerous for them but especially for those closer to the bottom—so much so that invisibility and movement came together in an ethical system that some diggers were explicit about. The aforementioned Claude, for one, liked to refer to the "laws of diggers" (*sheria ya macreuseur*), a set of "best practices" (my phrase) related to visibility, movement, and the promises and dangers of taking on debt. Taken together, his "rules" condemned contracts, bureaucracy, and "transparent" auditing as often-cruel forms of oppression. He laid them out for me and expanded on them often; they are worth repeating at length:

Claude's "Laws of Diggers"
1. Diggers don't fear debt. [Rather], they must be with debt (*Hawaogopi deni. Lazima wawe na deni*). On the other hand:
2. Diggers cannot take on the debt of others (*hawapashwi kuuzisha madini yao kwa deni*), meaning they must be paid in cash for their yield by négociants even if it means sleeping on-site with the minerals they've dug until they're paid. [Also, by sleeping with minerals, diggers held onto the only collateral they had, even if others made fun of their "sleeping with minerals" as an unclean or dirty (*chafu*) practice.]
3. Diggers must sign no contracts nor write their signature (*hawawezi ku tia sahihi*) and preferably offer no real names. [Diggers also said something very close to "what happens in the forest stays in the forest": *byaku pori binashiaku ku pori*.] And this is because:
4. A digger's debt is his alone (*Deni ya mchimbaji ni yake peke yake*). It is a private relationship between individuals based on a principle of trust, and no one else in their family should be liable for the debt.

Diggers often expanded on points similar to these, explaining that they may say, for example, that they will repay their loan after two days but refuse to write anything down because they know that it's possible they won't get the money or minerals in that time. They have to evade other people's linear schedules and the regulatory efforts of state authorities because they know their work is inconsistent, and they also have to eat. At the same time, they limit their liability in the face of rapid price fluctuations and other contingencies by insisting they be paid right away and using their power over extraction to do so. Finally, diggers' refusal to sign contracts was ultimately about not wanting their loved ones to be made responsible for their debts (a situation that often befell négociants, whose spouses sometimes ended up in jail when they couldn't make good on a contract they had signed with a financier; the négociants were sometimes compelled to flee to the forest or even a neighboring country, leaving their families to pay the bill—or they would try to pay the bill for their families while living in exile).

By taking on debt while avoiding the registration and calculation of debt, diggers also tried to avoid the possibility of temporary indentured servitude, which could happen to them if they fell into debt to a state authority or if they were deemed to be in violation of law or custom by the local authorities at the hole. On a more general level, though, diggers resisted the notion that they were truly in debt to anyone, since they were producing wealth of genuine value and were probably being underpaid for their yields (hence their frequent questions about the "true price" of minerals). As some diggers put it to me, "debt is spirit/heart" (*deni ni roho*), meaning that the exact monetary value of a debt cannot be known and that only the diggers know, in their heart, how great or small their debt to their patron is, based on the work they have done for them. Some diggers brought up Jesus in this regard, because Jesus apparently once said that you shouldn't go out seeking to get a debt repaid, because nobody in the world really knows who owes what to whom (see also Graeber 2011).

Paper Traps: The Dangers of Being a Négociant

The "laws of diggers" were extremely frustrating to négociants, who tended to see themselves as patrons (a term they used) since they forwarded food and money, thus enabling diggers' work to proceed. The digger's anonymity, their shape-shifting invisibility, and their power over extraction often made it possible for them to evade the whole state apparatus, but négociants and their families suffered by losing the money they had been forwarded by comptoirs and other financiers. Négociants could even go to jail because of an extraction glitch or other unforeseen event over which they had no control. In some cases, their urban houses were expropriated by banks or judges, which forced them into the humiliation of being "sent back to the village," a fate that some négociants found worse than death (I heard many stories about négociants who killed themselves because of this). As one négociant in South Kivu, put it,

> Even if a digger already has your debt, and he hasn't yet paid you back, if he comes to you for another debt, you have to give him one, because otherwise work won't proceed. You're forced to give him debt because, if you don't, he won't pay you back from the first debt. He'll say he won't be able to get food and tools and so he won't be able to pay for the first debt. Maybe he will run away with everything, go to Numbi [another mine] or elsewhere. Even if you bring him to the police, you first have to give the police money. And then the digger will also be giving the police money, and remember that you don't have any written contract with him. Then the digger comes back home free after his episode with the police, and he's just dancing in front of you. "Hey, look at me! I'm here! I haven't been imprisoned!" Ha! And then you'll *never* get your money back. "Try the police again if you want," they'll say.

Négociants, as this man suggested, were caught up in a web of bureaucracy and surveillance that could leave them destitute, while the diggers they depended on were relatively free; this was another reason why négociants associated diggers with the temporality of the lived moment. Unlike the négociants, they were relatively outside the permanent, encompassing spatiotemporality of the state bureaucracy, materialized in papers.

Some diggers who tried to move up the chain and become négociants found out straightaway how the "government of paper" coupled visibility with expropriation and the potential for indentured servitude. Take, for example, the case of Francis, a PDG in Punia, Maniema; he rented a hole and paid others to dig, forwarding them money, which he usually received from buyers in advance. Francis rented his hole from a mining concession owned by the company Sakima and sold his ore to comptoirs in the provincial capital of Kindu.[8] Technically, Francis was a digger, but I'm pretty sure he didn't have his diploma, or shovel. While his role was more that of a manager of labor and middleman, he had a digger's card but not a négociant's card, and so he was allowed to sell to négociants but not to comptoirs (though he did sell to comptoirs).

In 2013, Francis went to Kindu to meet with a Chinese comptoir there. The comptoir didn't care that Francis had only a digger's card, and he agreed to buy Francis's coltan and also forwarded him money to continue the work of digging, with a plan to purchase from him in the future. Francis signed a contract to this effect, something he later regretted. He then returned to Punia and used the money he had been forwarded by the comptoir to finance the work of the diggers. Then he paid porters with bicycles to transport the minerals to the river (a tributary of the Congo River), where they would be loaded onto a boat bound for Kindu. Francis also had to pay the company Sakima a fee when transporting the materials, because they owned the land where he leased his hole. When Francis and his porters reached the river, he banded together with some other licensed négociants to rent a boat, and they paid people to "clean" (*safisha*) the ore (meaning remove the worthless dirt and rocks) at the river. At some point during this process, the licensed négociants learned that Francis didn't have a négociant's card, and they refused to let him on the boat; by excluding him and his high-value ore, they hoped that they might get a better price for their ore from the same Chinese comptoir. Undeterred, Francis rented his own boat. But when the licensed négociants got to Kindu, they were approached by the Office of Mines, who were looking for infractions and imposing "taxes." The négociants informed on Francis, letting the mining officials know where he was. In this way, they turned the "people of documents" off themselves and onto Francis, who was easier prey since he didn't have the proper papers—"If you want money, just go to the river and get it, it's waiting for you there," they allegedly told the state officials.

The representatives from the Office of Mines went to the river with the mining police and they "tied/closed" (*kufunga*) Francis there, a euphemism for arrest that emphasizes how the state acts like a rope to tie people's ways. The police and the Office of Mines wanted $1,000 in exchange for his release, which Francis didn't have, so Francis agreed to give them one ton of coltan when he got out. The Office of Mines and the police prepared papers for Francis, and he signed a contract promising them one ton of coltan. In the meantime, these state officials took some of the high-value coltan from Francis's yield as partial payment (actually, interest). But when Francis got to the comptoir's office in Kindu so he could sell the coltan that would allow him to pay his debt to the Office of Mines and the police, the comptoir guys used their machines to test his ore. After testing, they told Francis that his yield's tantalum percentage was less than that of the other competing négociants from the river, who had arrived there before Francis. They refused to buy Francis's product. Francis explained that he had been cheated by the state officials with the help of these négociants, but the comptoir didn't care about all that and even brought a case against Francis to the governor. Francis went back to Punia to continue digging to try to earn back what he could (the hole will pay, diggers always say), but he couldn't go back to Kindu because he still owed the Office of Mines and the police $1,000. This left him at the mercy of his enemies, the négociants, who took advantage of Francis's lack of options to "dirty" (*kuchafua*) his minerals, taking the high-value yield and selling it separately on the side while adding dirt to his ore along the way. By devaluing Francis's minerals, they made money for themselves while weakening the fame and reputation of Francis's hole.

In short, as Francis moved up the chain, he exposed himself to a world of paper documents, which were at once dangerous and inscrutable and which made him visible and also restricted his movement. Being without them allowed state officials to extract exorbitant amounts of wealth from him, and of course these papers operated as a force wielded by those who possessed them against those like him, who did not. There is even more to it than that, though, more than a conflict between different state and non-state actors with competing "interests." This brings us to an even deeper dimension of the supply chain, a level that goes unnoticed in politico-economic analyses (but see Coyle 2019; Adunbe 2015). There is a cosmic conflict between documents and the forces of the forest and the ground—really a conflict between the opposed forces that drive and organize the world itself. For it turns out that ancestors and other spiritual entities who ultimately govern mining hate everything about the state bureaucracy (papers and "Frenchness," or *Kifaransa* generally), and they also aren't too keen on the putatively "clean" (*safi*) ways of life in the city, from which the state officials, and the négociants and comptoirs, are imagined to hail (at the same time, indigenous actors can try to appropriate the power of documents for themselves, as seen in section 3, on Bisie).

Ancestors and Documents

At the extraordinarily large mine of Bisie, discussed in part 3, the company MPC (Mining and Processing Congo) at one point had a lawyer who was very suspicious of the *Bagandula*, the "children of the forest," or the ostensible "owners" of the mine. He believed they were falsifying their sacrifices to the ancestors or causing accidents that they could blame on their ancestors, either to rile people up against the company MPC or to persuade people to give them money out of fear. The lawyer would follow these lineage representatives around and try to catch them in the act, describing their work as "charlatanry" without disputing the reality of ancestors or spirits in general. On the other hand, diggers at Bisie enjoyed talking about the time two of MPC's white employees fell into deep holes, cracked their heads, and died after apparently being startled by something they saw, which no one else saw (allegedly one of the Bagandula's ancestors). For diggers, the main takeaway of these stories was that, while the company had documents granting it legal rights to Bisie, the ancestors were still in charge and could drive them away. If the artisanal miners allied themselves with the ancestors through the descent group, they also stood a chance of outliving the company. The conflict between ancestors and documents (*ma document* or simply *karatasi*, or papers, in Kiswahili), or the sovereignty of spiritual forces under the ground and the sovereignty of the state, showed up in virtually every mining context, and it was one of the many ontological and social differences that people in the mining trade had to mediate in order for movement to take place.

Diggers often say that "price is set by the comptoir, but wealth is created by the *mababu*," or the dead. If we take this statement literally, it means that technologies like mobile phones and laptops—at least those containing Congolese minerals—ultimately depend on dead Congolese people for their existence, even if most Congolese are unaware of this connection. There may be "no wealth without the mababu," but for that wealth to be realized in terms of "price," or exchange value, in practice it has to pass through specific agents with specific paper documents from the government, like comptoirs. The contrast highlighted in this statement (price/comptoir versus wealth/mababu) is not the old Marxian one of exchange value and use value but of exchange value and value in the most total and absolute sense—social, cosmological, and ecological.

For example, when state figures come to the mines to harass diggers for "tax" or to "sell" them papers such as diggers' cards, diggers often get upset and may even try to drive them away, saying that their presence "will chase the minerals away" and sabotage the whole project of extraction. Some even say that if someone comes around with nice clothes or speaking French (the language of government and offices), the minerals will disappear. In a related vein, many also say that if they try to dig wearing clean clothes, they won't get minerals, that the "minerals will run from [them]" (*madini yatakimbia*). As one put it, "Minerals are dirty and do not like clean clothes, and ancestors do not like foreign languages." And so, some even refuse to wash their work clothes for long periods, adding to the perception others have that diggers are dirty or that they have abandoned the "clean" ways and dress of the city, or what I would characterize as (post)colonial modernity; some diggers say the dirt, usually associated with disorder and shame, here acts like a magnet, attracting minerals to them. When diggers chase these figures away or complain about them ("leave with your French ways"; *ondoka na Kifaransa chako*), they are reacting to being interrupted in their work and forced to pay "tax," but they are also arguing that extraction will be more difficult because of the very presence of these people and their language/ways (*Kifaransa*, or Frenchness and French language). In a way, the diggers were protecting officials and the movement of minerals because they knew that ancestors and documents were opposed forces, and they didn't want movement to stop—rather, they wanted to hold ancestors at bay and keep them from causing harm, but state officials wouldn't always have it, unwittingly shooting themselves in the foot by showing up (after all, how could the state figures be paid if there was no movement?).

Ancestors and documents are symbolic, in the broad sense of being meaningful and iconic of other things and relationships, and real and agentive, in that it would be inadequate to say that either of these actors (ancestors or documents) are merely instruments wielded by certain people, who one might imagine to be more agentive or real than either ancestors or documents. Ancestors and documents are part of the supply chain, but they are also irreducible to it. The difference between ancestors and documents is experienced in spatiotemporal terms, but it is not only a spatiotemporal difference: documents mark and mediate the passage of things from Congolese soil to the world market, and so their whole purpose is at odds with the intentions of ancestors, which is to retain value in the soil for their living human descendants who live alongside them in a parallel world. There is a fundamental hostility between ancestors and documents/ Frenchness, which comes out in the work of digging and the business of transacting minerals, and in the interactions among the different actors involved in this work. Soil, or *udongo*, is the battleground on which this conflict is waged. The difference and conflict between these realms—the world of ancestors and the world of money and paper bureaucracy—is real, but business requires that they be stitched together through rituals of sacrifice and rituals of bureaucracy so that people can make a living and even go on to be ancestors themselves someday.

In one common formulation, digging was also in tension with Christianity, because diggers were engaging with forces that Christians were uncomfortable with (including but not only ancestors) or that were inimical to Christian modernity, which was related to Kifaransa, or Frenchness. One digger, explaining why artisanal miners put more stock in ancestors and those who communicated with ancestors than in Christian churches, argued that there was an ontological difference between that which was beneath the soil and that which was above it, which had to be bridged. Pastors worked in the light of day and were part of the visible world, but diggers "lived in the dark" where it was "always night." Diggers needed ancestors more than they needed priests or pastors because, as he put it,

A pastor won't help you get minerals or protect you from accidents in the mine! Why? Because the pastor is a person of the world. He travels on the road, above the ground. He doesn't know anything about the soil, about what's under the ground, so he can't help with this. And remember that every digger has his own beliefs, his own religion. There may be some people who will follow the pastor around and insult and criticize him, and so even if the pastor could have helped, now his prayers won't be effective. But everyone believes in the power of the mababu [ancestors] because they have no choice. Because diggers and the mababu are under the ground, just like minerals are under the ground, and they all live together. If the mababu say there's a problem, that we shouldn't dig on that day, we'll all leave. But if a preacher says this, we won't care, because he walks on top of the ground like everyone else.

As this insight suggests, there was something democratic and "global" about ancestors, in that they leveled all the other cultural, religious, regional, and class differences ("every digger has his own beliefs") among people in the mining trade. Ancestors were yet another way in which mining broke down barriers between these people who were forced together by various contingencies; ancestors compelled them to come together across bridges, in collaboration.

Mababu dislike, and also don't understand, the values and behavior that emerge in and through documents and Kifaransa, since documents and Kifaransa imply that agents other than ancestors (the state, a company, a private individual) have rights and authority over earth and the forest. Ancestors also don't like the exploitative and avaricious behavior that emerges through documents—state officials come to the forest and harass their children, expecting something from them when they have nothing to give. Moreover, these state actors eat the money that would have gone to them in the form of food, a manifestation of love. Thus there is a tension between the kinds of power that ancestors and documents represent, in that ancestors are invisible intimates, while visible paper documents are tied up in a remote sovereign regime that challenges the "sovereignty" of ancestors and their descendants over the soil and its products. This is driven home by the fact that, rather than relying on visualization and seeing (maps, documents, satellites, cameras that see underground), the work of communicating with ancestors depends on aural communication, on hearing and naming. You must have a certain voice, and you have to call the ancestors by name, mentioning actual characteristics about them that were positive. This conflict between ancestors and technology emerged in other contexts as well—for example, it was central to the struggle between Mai Mai insurgents, who channeled the power of the forest, and the military technology possessed by the invading Rwandan and Ugandan armies during the war.

Ancestors respond to ethical behavior (as they understand it), and often diggers claimed that this encouraged them to act in an ethical way (or that their natural ethics was rewarded by the ancestors). Throughout the mining areas I visited, diggers insisted on the love they had for their fellow diggers, how they helped each other out in case of need (exchanging food and medicines among themselves or bringing their wives to cook for their friends), and how they didn't account for the price of things when doing so—attributes that other people saw as evidence of fecklessness or the inability to conceptualize the future. They invariably contrasted this behavior to that of those above them in the chain, mainly the négociants. They also insisted that the ancestors, who lived underneath the soil, recognized this generative reciprocity and nonchalance regarding price because it reflected the values of the times in which they lived. For example, in Maniema, diggers talked about how, in the not-so-distant past, people would bring part of their harvest to the *Mwami*, or chief, who would eat some of it with the ancestors. The rest he would leave in a public store for those who needed it, including visitors coming from other towns. This physical manifestation of love through food had brought rain, but the behaviors of négociants and state actors in the mining trade encouraged ancestors to withdraw this rain, through no fault of the diggers, who claimed to be similar in their habits to hunters or farmers (they often used terms from hunting or farming to refer to their work). It was the négociants withholding credit from diggers and the state officials with their documents who together threatened extraction: ancestors responded to this lack of love by making minerals disappear from the ground. Therefore diggers were justified in chasing them and their French ways away, even if they had it in their hearts to share with them.

Again, the existence or nonexistence of minerals is not stable and is dependent on things people do. Selfish behavior on the part of négociants (middlepersons) and state officials, which come out in contracts and the "selling" of documents, drives minerals away. Similarly, if négociants try to undercut the diggers or cheat them by misrepresenting the value of minerals or what the minerals are, the minerals may go away. For their part, some négociants agreed that their presence did indeed chase minerals away, but they argued that it was really their intelligence (*akili*) and capacity for planning, not their selfishness, that did this. One négociant named Gabriel had tried to start off as a digger during the early days of Bisie mine in Walikale—the time of "easy grabbing" (*zola zola*), when each digger was walking away with as much as two tons a day. But Gabriel failed in this business, blaming it on his rational and calculating nature. "Minerals don't want intelligence," he averred, "It's like the minerals are a satellite, and they can see a person with intelligence/the capacity to plan (*akili*) and they won't let you have them. If you have intelligence, and come in with a [linear] plan, you'll lose everything and get nothing." While he was referencing the unpredictable nature of extraction, Gabriel—who was also Muslim—was also suggesting that the entities governing minerals were demonic and so didn't want to release value to people with the capacity to generate genuine, incremental development over time.

Generally, ancestors were described as living in a parallel world and time, a past that was in the present, that was part of the forest, but which most people couldn't see. If you called to them in the correct way (usually this involved communicating with or through water—say, a running stream), they could reveal their world to you: a village might appear, for example, resembling villages of the past. Food, and the act of eating, could connect these two dimensions, and proper communication with the ancestors involved using food and drink in the right way—offering the food that the dead used to eat and drink, for example. By the same token, if you entered into one of the ancestors' usually invisible villages and ate their food, you would join their world for good, disappearing forever from the visible one, so eating their food was something you were supposed to avoid, even if offered.

Always some kind of translation had to take place between the ways of the present and the ways of the past, mediated by somebody who was in a position to do so. Usually, an explanation was in order. In the rainforests of Maniema and Walikale, ancestors had a hard time understanding why the forest alone couldn't satisfy the needs of their descendants, such that they needed to sell the forest, or a part of it, to people who weren't from there. They needed to have money explained to them, as well as all of the different financial responsibilities people have today. There was a technique for dealing with this generation gap: Whoever was "opening" the land for extraction would call out to someone who had died recently, praising that person and the good things he had done for the family. Once that connection was created, that "first-generation" dead person could be persuaded to communicate with the long dead, those who came from a time before mining and money. The recently dead would explain to the long dead about school fees and the need to purchase food, about present-day clothes and medicines, about the basics of extractive capitalism, and anything else that needed to be explained so that everyone would be on board.

Despite this widespread understanding, there was no real consensus regarding whether minerals had any objective existence outside of ancestors. Some insisted that they did not, that ancestors put minerals in the ground for their descendants and could take them away. Others equivocated on this point: they thought that minerals probably had an objective existence, because long ago white people had come and recognized the minerals using maps and other devices that enabled them to see things in the world. Some insisted that minerals were constantly present in the reality experienced by ancestors and other spirits but became visible in "our" reality, the reality of living humans, only when ancestors allowed it. As a result, technologies of visualization and extraction like maps and machines "had a limit" (*iko na limite* or *iko limité*) because, unless the ancestors released these material substances out into the world, extraction would not be successful—there could be accidents, or production might simply stall.

Exactly how minerals were imbricated in relations with ancestors varied from place to place in the Kivus and Maniema, but the only places I worked where there was little discussion of ancestors were those in which the original autochthones had been driven out. For example, in Rubaya, Masisi (North Kivu), the Belgians had brought Rwandans to run plantations, and over time, these groups displaced the Hunde autochthones, who claimed the territory as their ancestral soil. Even there, though, while miners rarely if ever talked about autochthonous ancestors, the Hunde people who lived there insisted that the reason there was so much conflict in Rubaya, and the work of mining never bore fruit for the population, was because they had been displaced from their territory and their ancestors were upset. In other places, some diggers who were constantly on the move, always far from "home," disputed the power of autochthones' ancestors over extraction and claimed that their own ancestors were mobile like them, following them from mine to mine, regardless of how far afield they went. Their own migrating ancestors, not those of the autochthones, ensured their success in mining, recognizing their descendants through distinguishing inherited characteristics.

As Gabriel's comment suggested, there was also disagreement regarding whether these invisible entities were "good" or "bad." Some diggers (like Gabriel) were not certain whether these entities in the ground were actually ancestors or in fact demons. Many urban traders and some diggers insisted that all forest people were essentially witches and that the spiritual agents they imagined to be ancestors were emissaries of Satan. This also came up in the way that many diggers equated ancestors with dirt, or *chafu*—ancestors chased visibly clean, or *safi*, people away because ancestors were neither educated nor cosmopolitan (qualities that could be figured positively or negatively depending on the person or context). For all practical purposes though, it mattered little whether ancestors were good or bad. Diggers, as one pointed out, were "not theologians," (*sisi hatuko ba theologien*), and the "children of the forest," the autochthones, had to be satisfied, regardless of whether their spirit helpers were ancestors or demons. One descent group head at a small mine in the forest in Walikale (North Kivu) explained to me that, while he was in a genealogical position to communicate with his ancestors, his job as a Pentecostal preacher made him unable to have an effective conversation with them. But he knew somebody needed to do this work because without "opening up" the land, nothing would be extracted. Moreover, the mababu would pass through various *njia* (ways), or circumstances in the world, to make sure that whoever was mining there died. Allowing that to happen would be bad for everyone and would also be unethical, a sin (*dhambi*). Fortunately, his brother (who was not saved) had been performing the work of sacrifice, and the ancestors were accustomed to him, whereas a stranger's voice would sound like nothing to them but "noise of the forest." Now, if the preacher tried to take over the job from his brother, the sacrifice would end up killing his brother because the ancestors, not recognizing the new voice of the preacher, would assume that the preacher's brother was dead.

Among Lega and Kumu people in Maniema, there were other gods who were not ancestors but who gave power to ancestors. They also had to be communicated with, and only those who were initiated into the hierarchical and secret organization of Bwami could do it. Within the Lega pantheon of gods, one, called Kimbilikiti, had sovereignty over things of the forest and incarnated in human form during times of initiation, when youth began the life process that would culminate in their becoming immortal ancestors (among Kumu, a similar deity was called Mboyo). Initiated Wami were priests and custodians of Bwami, and they "bridged" these visible and invisible worlds—people said that "Mwami is a bridge" (*Mwami ni kilalo*). These seniors, who performed an offering (*kutambikia*, or the word *sacrifice* in French) to gods and ancestors so that minerals could emerge from the ground, were expected to not be Christian (for Christianity, ancestors are demons, but for Bwami, "ancestors are everything"). At the same time, Bwami was described as parallel to Christianity—outside of it but structured in an analogous way because Bwami was rule governed, hierarchical, opaque, and had a similar system of commandments and principles. The deity Kimbilikiti was an actor that had to be recognized, not only for mining, but for anyone wanting to extract value from, or even just pass through, the forest. During times of initiation, the human incarnation of Kimbilikiti operated a roadblock in the forest with the help of the youth undergoing initiation, during which passersby were expected to contribute "tax" to the initiates and the resource-demanding process of initiation. At the toll, Kimbilikiti sat to the side, announcing himself to passersby in a booming and disarming voice from behind a large wooden structure that had been made for him by the youth in his train.

Minerals, Money, and Documents

Once minerals left the world of the soil and were disentangled from their relationship to ancestors and the forest, they became subject to more intensive scrutiny from different actors (buyers and state officials) and entered into a relationship with papers—particularly, paper bureaucracy and paper money, mainly US dollars. These were also volatile and unpredictable things, materialized powers that acted on people capriciously from above (Kinshasa and, in the case of money, the United States). They were also subject to visual scrutiny, and often all three things were scrutinized at the same time for their content and veracity (inspecting tears in the dollars, figuring out whether a person had the papers needed to do the work, testing the purity of the minerals). This inspection, and the mutual triangulation of papers, money, and minerals—each being inspected in relation to each other as they came into contact with each other—took place as minerals made their way out to the "world market" (*soko ya dunia*) and as they went from being intimately connected to living people to being abstractly connected to people through markets, alienated from the intimate relations that came together in the forest and the ground.

The movement of these things from a sphere of intimate social relations to a domain where they belonged to others was marked by interactions among these media—minerals, dollars, and documents—that each belonged to others in some important way. For example, people said that the reason they inspected dollars so assiduously was that, if there was the slightest tear, *someone else* (never them) might not accept them. If pushed, they sometimes said that US dollars were not Congolese, that they were just being used by Congolese for a period of time, and at some point in the future, those who came with them originally (the network made up of international NGOs, the United Nations, and companies that came to buy minerals) would want them back. If that happened, these powerful and interconnected foreign others were likely to accept only dollars that were in pristine condition and certainly not counterfeits. In contrast, no one cared what Congolese francs looked like or how torn up and dirty they were, because they were and always would be socially and spatially confined to Congo.

Bureaucratic documents, like money and minerals, were also valuable and, because they belonged to others (the state authorities who could distinguish good ones from bad ones, "low" ones from "high" ones), needed to be scrutinized, mostly because they could be falsified, as could money and minerals. A low-level state official might sell you a document/paper that is good only locally or that you don't even need, or he might persuade you that the papers he's selling you makes it so you don't need another paper from a different state official, who may show up later to argue otherwise, when really if you knew better there's another guy, even "higher" than all of them, who can make all these "low papers" go away with a single paper. While documents and money seemed to make the value of the minerals legible and transparent, they too were mysterious and opaque, and so rather than demystifying minerals, they contributed to the mystique around minerals as they passed from the forest to the world market through different sets of actors. Moreover, the work that went into producing and inspecting papers (particularly state but also nonstate documents) was as formally ritualized and symbolically constituted as the work surrounding ancestors—for example, state officials produced these documents with great pomp, creating delays that greatly magnified the importance of papers.

As substances moved from the soil/dirt (*udongo/uchafu*) to the market and were converted into commodities through their contact with money and documents, they became more like money and documents—alien, and subject to scrutiny by experts who knew things that other people didn't, and who had access to the right technology (the testers, which became more sophisticated as one moved up the chain, but could still deliver false information). These three things—money, documents, and minerals—were connected, in a very tangible way, to a world that was outside Congo, like a triangular keyhole that opened the door to that world. Once Congolese substances entered into the domain of visual inspection, they were well on this journey outward, soon to become a thing only foreigners knew how to use, and which could potentially come back to Congolese in a dangerous form (for example, as the bullets Rwandans used to kill them).

Papers were also part of a larger Congolese way of life that was certainly not limited to the mineral trade.[9] They emerged in any context where one was moving, or intending to move, between sociopolitical zones—whether on roads, through paths, or via boat or airplane—or from one place, or kind of place, to another. Karatasi were generally not merely paper documents, but stamped and signed documents, and the stamps were usually much more important than the document itself, because they signaled that a socially relevant actor, usually but not necessarily a node of "the state," was participating in the project of the person holding the papers. These officials took time to invest these documents and stamps with power by making you wait for them and then presenting them with much ado after having done everything in their power to produce fear and anxiety in you, the person "needing papers." On one level, papers were the proof of co-operation between the bearer of papers and "the state," but they were also technologies for including different parties, state and nonstate, into collaboration with one another, which was made visible on the paper. Always there was a sense that the person offering a stamp was agreeing to provide something, usually safe passage, within the limited domain the person's office had control over. These documents became materializations of the cooperation of all of these actors in the movement of the paper-possessing person, and they would ideally include all the people who could potentially bring trouble to, or in some way sabotage, the traveler's projects. The same logic of incorporation of potentially dangerous actors was at work in the system of road and, in some places, forest path tolls (tolls were essentially a technology of inclusion for diverse authorities and one of the main technologies that made "the state" in toto a reality).

We could further complicate this picture of movement by adding another level of contrast to that of ancestors and documents by briefly reiterating the tension between documents and guns. During times and places of relative or potential war, state authorities possessing guns (soldiers) took over in the final instance from those wielding documents, in that soldiers could bypass or make irrelevant the regime of paper by making it so that a stamp from an office or a digger's or négociant's card was unnecessary. When they did this—liberating diggers from oppressive papers regimes and even aligning themselves with diggers and ancestors in tentative opposition to the people of documents—it drove home, for diggers, the fact that the people of documents were nonessential. Rather, diggers transcended, and sometimes were at the center of, all this tumultuous change, the constant force that would endure long after it was over.

Concluding Thoughts

In all the different dimensions of the supply chain that I've tried to draw out in this chapter, there is an underlying theme of moral-ethical contestation and reversal at work. The supply chain is a field of profound argument about the value of certain practices and concepts, many of which are rooted in colonial history. Diggers, in particular, resist or oppose—without always intending to—entrenched urban/modernist Congolese values and assumptions rooted in a colonial worldview embodied in colonial bureaucracy. If others assert that they waste wealth and don't think of tomorrow (an assumed prerequisite for "development"), they are there to remind people that in fact they produce and are ultimately responsible for all wealth and that in some way they epitomize and embody the future. If others see them and their invisibility as immoral or illegal, they insist they are the most moral of actors because of their generative work and their love (*upendo*) for one another, which manifests as reciprocity. Moreover, their tactile work constitutes the conditions of possibility for the state and law (for better or for worse), because the state follows them to mines and would seem not to exist without them (just as law exists for state officials to extract value from them when they break it).

If others see them and their subterranean invisibility as unclean, they know that, while they might have earth on their clothes and bodies, they are also the very source of cleanliness because they are the ultimate reason why any commodities (including soap) are available in many rural areas. And, anyway, dirtiness, in the sense of being with earth, has its own virtues because in the hands of diggers it has become a technology of its own, enabling successful extraction. If others see them as sinful or unchristian because of their profligate habits, they know that ancestors protect them and are more important than anything—and the centrality of minerals to contemporary global political economy makes the importance of ancestors all the more obvious and makes colonial Frenchness (and the papers in which Frenchness is enmeshed) seem old-fashioned and counterproductive. They also know that they're the greatest Christians (*sisi ni Wakristo kuliko watu wengi duniani*) because, while many other groups of people poison and kill each other, they work together with love, even going so far as to split a cigarette or a bottle of beer among five people. If others associate diggers with conflict, they in turn associate everyone else (the conflicting state actors competing with one another and bringing "disorder," the different churches competing for congregants) with conflict. In contrast, their work and the ancestors on which they depend ultimately unite everyone, bringing everyone together in collaboration, even if it is the demand for minerals that compels this collaboration. Their work also domesticates people who are intrinsically dangerous, such as soldiers bearing guns.

Artisanal mining is also a system that relies on the regular forwarding of value—in the form of money and commodities ranging from food to water to beer to cigarettes and condoms—from people "higher up" in the supply chain, like négociants, to people "lower down," often across great distances not tied together by road infrastructure. This regular spatiotemporal arrangement, existing in spite of the sudden temporal ruptures produced by price fluctuations and closures, relies ultimately on trust—not only in people, but in the ability of actors to bring together forces that don't always want to be put together. That is, the moral arguments that emerge out of and in the supply chain relate to an even deeper conflict, which goes beyond social actors like diggers and négociants. This is the conflict between the forces in the ground and the forces of (post)colonial state bureaucracy that mediate the passage of substances out into the world. Actors in the supply chain transform this strict ontological conflict, making collaboration out of it—forcing ancestors and papers to work together when they don't "want" to so that movement can take place.

In addition to being embroiled in a vortex of conflicting forces that they must stitch together, actors in the supply chain are ultimately and unintentionally making it so digital technologies appear to come from Silicon Valley or from the minds of tech entrepreneurs (related terms for this are commodity fetishism and technology fetishism[10]). Indeed, the work of making digital technology and of causing that work to disappear so the technology seems to come from the minds of software developers takes place in many parts of the world in many ways. In Congo, this process of disappearance begins before the materials that end up in digital devices even leave the country en route to the smelters. A strictly Marxist interpretation of this would draw attention to the way markets render this work invisible—for example, how tech companies buy ore from smelters located outside of Congo or how the multiple levels of buyers obscure where minerals come from. But the work through which Congolese agents and forces are made to disappear is nuanced and multiple, at times intentional and at times unintentional; it is not the result of a single mechanism but emerges out of a combination of processes that are thrown together. For example, when the "children of the forest" "open" (kufungua) holes for diggers by performing sacrifices to ancestors, they are, on one level, engaged in an act of remembering, but they are also performing the work of forgetting. That is, these sacrifices are taking place so that mining won't hurt anyone and the minerals can move out to the markets without being cursed or causing grief to people. The work of making digital devices thus includes the work of removing what could be called their mystical but also human power, including the power of the forest and of the leaves. Forgetting that dimension of the life of minerals involves living, dead, and nonhuman agents working together tensely until the substances are cleansed of their pasts. Add to this all the work of the paper-wielding bureaucrats, who mark the movement of minerals from one domain to another, transfer ownership from one party to another, and exact the cut of the state and its various parts from these workers, and we begin to see all that is entailed in the work of systematized forgetting.

Finally, even before the "conflict minerals" narrative, the movement of these materials was set up like a multitiered total "cleansing" (*kusafisha*) process that supported the work of forgetting. As minerals moved to the city, they moved out of the dirt and through places of relative "cleanliness," and the people who engaged with them at these places were supposed to be outwardly "cleaner" in every respect, where cleanliness implied urban (post)colonial modernity. Instead of drinking "dirty" (*kichafu*) beverages in divergently shaped plastic bottles that no one knew the true contents of (often including mysterious herbal stimulants)—with names like Rambo (with his bare-chested likeness on the bottle), The Winner (*Le Gagnant*), Power and Health, or even just Hello!—négociants drank "clean" bottled Congolese beer like Primus and Muitzig or foreign beer like Amstel and Heineken (all the bottled beers were owned by Heineken anyway). Négociants and comptoirs met in "clean" restaurants where they ate "clean" food like *côte de porc avec sauce aux champignon* and spaghetti bolognaise while wearing their "clean" (*safi*) clothes, which some diggers claimed chased minerals away. In the end, the reclusive comptoirs, bounded off in their walled mini-fortresses in the city, performed the final separation of Congolese resources from Congolese soil and Congo in general, using their expensive but deceptive machines to assess the "true" value of Congolese wealth for others. The comptoirs were secretive and difficult to access or have meaningful extended conversations with concerning the trade, partly because they were connected to a network of smelters outside Congo; together they performed the final cleansing of minerals from Congo and all the "dirtiness" Congo is associated with for these others. Eventually, these minerals would go on to power seemingly sui generis devices that epitomize the future and are imbued with an aura of pristine cleanliness. At least from the perspectives of diggers, this enactment of progressively expanding "levels" of cleanliness as one traveled up the supply chain was fundamentally false—a performance played out with a view to the colonial prejudices of urbanites and foreigners. In truth, the work of the diggers was successful because of their ethical but invisible relationships with each other and with ancestors, and if anything, their "true" product was threatened by paper-bearing state actors and "dirtied" by négociants and comptoirs as it moved up the (semiotic) supply chain (in short, they were clean, but those "above" them were dirty).

While this system was set up in a certain way, requiring collaboration between specific entities and forces in specific ways, I never got the sense that there was anything inevitable about the supply chain in its current form or that people felt as if there was. If anything, their experience of the past of mining seemed to underline the fact that there was nothing truly fixed or inevitable about either past or present arrangements surrounding mining. What was consistent was that there existed, in the earth, a powerful potential that could be harnessed and deployed in various ways and that had to be materialized before it could be accessed, conceptually and in practice. This powerful potential emerged most concretely in the forest ruins of mining's industrial past, the topic of the next chapter, and the last major theme I explore before moving on to part 3, an ethnographic analysis of a particular hole that became a catalyst for conflict minerals and conflict-mineral regulation.

FOUR

Mining Futures in the Ruins

Bringing up minerals [out of the ground] brings out everything else—money,
spirits, women, governments, histories.

—A schoolteacher in Maniema

Industrial Mining Ruins as Geographic, Temporal, and Multidimensional Vortexes

In the large rainforest belt of Maniema province to the west of the Kivus, maybe hundreds of thousands of people live in or near the ruins of a few old Belgian mining company towns, where many dig for black minerals in the concessions of a Congolese company named Sakima (a government-owned company that has been granted rights to the land above and below the ground). While it is the inheritor of a legacy of industrial mining going back to the 1930s, Sakima currently leases out space to artisanal miners and collects rents from them and from people who come to buy minerals from the diggers. These industrial ghost towns in the rainforest are difficult to access, cut off from the rest of the country by road, reached mainly by plane or a combination of cargo plane and motorbike. Many of the residents of these places worked for, or are the descendants of people who worked for, one of the colonial or postcolonial mining companies (mainly the Mobutu-era Sominki). Others have come more recently, responding to sudden events and ruptures like the closure of mining in other areas, to dig in the ruins of these large concessions or in the outlying forests where the land is owned by the company. While older people who live in or near the old company towns (the so-called children of Sominki) remember the days when the company (actually, different companies, associated with different times) mined in earnest, youth know only artisanal mining, or a combination of artisanal mining and other pursuits, including farming. Many people, especially seniors, equate the diggers and the ruins of the company with a state of abjection or decline and impurity.

In these towns, industrial mining of the black minerals reached a peak in the 1950s and 1960s. During that time, the Belgian colonial government promoted the region as a miracle of modernization, with roadside restaurants, gas stations, and hydroelectric plants in the middle of the rainforest (little of that exists anymore, although there's still hydroelectric power in Kalima). These companies imposed a rigid and violent colonial racial order—for example, in Kalima, whites lived in large houses with white picket fences at the top of a hill, with a giant social hall and basketball court and swimming pools, while African workers lived at the bottom of the hill in very small houses without plumbing, thrust close together. After independence, the various Belgian mining companies of the east (including both those mining black minerals and those mining gold) came together under the umbrella of a single conglomerate called Sominki, which operated all the mines in Maniema and the Kivus. Sominki never recovered from major drops in global cassiterite prices in 1985, and in 1996, it negotiated a buyout with the Canadian gold-mining company Banro, a contract that President Laurent Kabila contested when he took power after the First Congo War. At the onset of the Great War, in 1998, the Rwandan-backed armies made their way straight to the company concessions in Maniema to expropriate minerals (some in Maniema believe they did this at the behest of foreign companies that were pulling the strings in the background).

These days, while the old company towns still exist, they are no longer centralizing centers, and the borders between them and everything else are not always clear; meanwhile, new settlements emerge at the sites where minerals are actually being mined, sometimes materializing and disappearing again rapidly in the forest. In contrast to the relatively democratic and flexible social and spatial organization of artisanal mining, industrial mining was segregated, vertical, enclaved, and authoritarian. It also appeared fixed, large, and visible—or rooted—relative to the rhizomatic and horizontal social arrangements that characterize the present. In this chapter, I focus on these tin and tantalum mining concessions in Maniema (with ethnographic examples drawn from Kalima, Kailo, and Punia) with a view to showcasing how the ruins of the company attract, catalyze, and resist the democratic concepts and practices of the artisanal miners and their emergent philosophy of movement.[1]

The sense of loss and decline—the feeling that linear temporal trajectories have been reversed—is deeply felt here, as it is in other collapsed industrial mining sites in the region.[2] As in the Zambian copperbelt, for example, the experience of decline has fueled what James Ferguson refers to as a "crisis of meaning" (Ferguson 1999) that is made sensible through material forms like ruins. But these ruins, and the pasts materialized in them, are not just sites of loss or absence but rhizomatic clusters of materialized potentials.

In recent years, a growing anthropological literature has emerged on ruins and rubble, two related but not identical concepts (see Stoler et al. 2013; Gordillo 2014). A great number of themes emerge in this work, but taken as a whole, research on ruins throws into question modernist Western notions of linear progress and linear temporality generally while putting to rest any lingering doubts about capitalism's destructive nature. Importantly, much of this literature dwells on the ways in which the past haunts the present and how materialized pasts engender memories and affects that are irreducible to subjective mind or consciousness. This work also considers how and why people build futures out of ruins, and it suggests that what appears to be the end of the world as we know it (say of capitalism) might in fact be fecund ground for the gestation of new worlds and new ways of relating (Tsing 2014). Thought-provoking work by Yael Navaro-Yashin (2009, 2012) on Greek-Cypriots living, by necessity, in the houses of their erstwhile enemies suggests two phenomenologically different dimensions of ruins that I find particularly compelling when considering old Congolese mining towns in the forest. Borrowing from Gilles Deleuze, Navaro-Yashin refers to these competing forces as the rootlike and rhizomatic poles of ruins: on the one hand, ruins are rhizomatic because they are diffuse and uncontrolled; on the other hand, they are rooted in the past and focus attention on the past while also drawing in unrooted rhizomatic agents. The tension between these different poles ends up being generative in unpredictable and counterintuitive ways that this chapter also explores.

Unlike many other ruins, though, the ones I discuss do not belong to the past in any simple way, in part because they are ground zero for contemporary events and processes, including the invention of new techniques and machines for mining and processing minerals. Also, many are waiting for past companies to return and pay them the back wages and pensions they're owed; they also hope that future investors will be attracted by these ruins and by the machines, infrastructure, and collective memory of mining that already exist in these places as a resource (a less vaunted or visible example of the contemporary "knowledge economy"). Thus it is not just that the past is a potential future for people who live among these ruins but that it is a potential future for others, which incites those who live there to hold on to these spaces all the more. At the same time, different models of the future and of "development" emerge in the practices and discourses of ordinary people in these locations, as they impute divergent meanings to their surroundings in relation to rapidly changing events and circumstances.

"Please, Don't Cut the Grass on Our Account!"

The long, circular driveway leading up to the once-imposing 1930s brick Belgian house we're subletting from the mineral comptoir in Kalima is a shadow of its former self—broken up and overgrown with democratic grass, the house itself surrounded by encroaching forest. I wonder if there are snakes in the grass. Some time ago, in Punia, another Maniema mining town, Raymond awoke in the middle of the night to the sound of groaning in his room at the old Catholic church in the company Sakima's concession there. When he shined his flashlight on the floor next to his bed, Raymond saw the deadly and astonishingly long black mamba coiled up in a ball, its head cocked. Disturbed by the light but not antagonized to the point of attacking, the annoyed snake slowly slithered out of the room through a hole in the door, back into the grass outside. Raymond changed his room the next day to one without a hole in the door, but now it's hard not to think about what lies hidden in the tall grass. On second thought, maybe it would be a good idea to have this guy cut the grass, but we also don't want to communicate that we are taking over the business of buying minerals from the comptoir from whom we're subletting or, worse, that we *are* the comptoir. The self-appointed gardener will end up cutting the grass anyway, despite our protestations, and we'll end up paying him. Like others, he's excited that there are guests, partly because it means that he may be paid for his work. For some, though, our presence also augurs something even more exciting—the chance that investors might return, bringing "friends" from abroad to resuscitate Sakima in earnest so that people might earn regular salaries, as some people did in the distant past.

Sakima, the title owner, does not have financing to mine or, it seems, any real intention of mining industrially in the future ("Sakima is zero!"

people here like to say). Instead, the company "allows" people who live around here, or those who have come from afar, to dig in the concession (the company receives fees from the comptoirs, the négociants, and the hole owners). Comptoirs come and go (these days they are mainly from China and India), usually renting the big houses on the top of the hill, which housed company management in the past—a visible and symbolically loaded enactment of the replacement of the old company system by the comptoir system, in which the comptoir system is inserted into the shell of the company system. Most older people, in particular, hate the comptoir system, because comptoirs don't provide regular salaries or predictable prices, encouraging a "punctuated" temporality (Guyer 2008) that seems the opposite of what the buildings here suggest and appear to have promised—permanence, or at least the company's desire to stay for a long time. A large percentage of these artisanal miners are the descendants of the company's employees, and they also grow crops in the company concession, alternating between mining and agriculture depending on the season. In any given year, they extract more—sometimes astronomically more—than Sominki did in any given year prior to the war.

Sitting on the front steps, I stare at the tilting iron telephone poles from who knows how many decades ago and observe the traces of borders that are now overflowing their limits, as the flowerbeds that once lined the driveway intermingle with the driveway's past. In the distance lie the ruins of a basketball court and a two-story social hall with bar and theatre, now used for weddings and social events by the descendants of the people who used to work here. These days, they are called to church by bells made from the inner wheels of abandoned work trucks. There are maybe forty thousand people living here in the old company town—many more in the "actual" town of Kalima, which lies outside it. Those who live in the company town still occupy the homes built by the company, awaiting unpaid salaries and pensions that will surely never materialize. Their houses are overgrown with moss, held together by leaking roofs and sinking foundations, cut through with meaningless pipes. They continue to decay over time in part because of the cost of repair but also because their inhabitants live in limbo, concerned that someday someone is going to come and kick them off these properties. Most view the decline of the houses and their own collective decline as synonymous with the decline of "the company" (really companies) over time, since the company owns the buildings. As a frequent joke here has it, Symetain (which came in the 1940s) sent us to school and taught us to mine, Sominki (which came in the 1970s) fed us like the children of kings, and Sakima (the government company that came in 2005) is the devil that's come to bury us (*Sakima ilikuya tuzika*).

All of the different stories that the residents of this place have told me about what it was like in the late colonial and early postcolonial past (mostly between the 1950s and 1970s) come together in my imagination as I cast my eyes about, as if I am channeling all these narrated events into a film or a singularly ridiculous Northern Renaissance painting happening in front of me (imagine the ethnographic realism of Brueghel combined with the violent allegories of Bosch): The young Lega men bringing down an elephant while the white managers look on. Families of laughing white kids and women in big flowered hats driving in their brand-new Volkswagens to the town pool, near where the generator still stands. Giant Caterpillar tractors and noisy state-of-the-art mineral-refining machines from Massachusetts, and men shouting orders at one another. White men committing acts of horrible violence on Congolese workers with whips. The groundworkers sharing knowing looks at one another as spouses seize an opportunity to cheat. The buildings are partly to blame for this turn of thought and not just the houses—there's also the silent post office, the crumbling white picket fences, the bad Bruegelesque murals, the giant 1940s mining maps, the desiccated swimming pools, the mocking payroll building, the ghostly children's swing sets, the curious cranes poking above the forest canopy, and all the mysterious cement markers sticking up from the ground here and there (apparently left behind to show whites in the then future the way to mineral deposits). Every conversation I've had with people about the past has taken place in front of these signs and structures and has had them as a backdrop and original motivation to the conversation.

These ruins retain and direct memories, and they act as a kind of catalyst, or currency conduit, for people's imaginations and desires. It's no wonder that presidents, governors, mineral comptoirs, invading armies, artisanal miners, and at least one anthropologist have all been drawn here to live for a time in and around these structures. The company concession may look like a self-contained living history museum of the colonial period, but it is also a laboratory for the most contemporary and fluid permutations of global capitalism. The artisanal mining supply chain flourishes in the shell of the old system, with key operations (like the final weighing of minerals before they're tested) happening in the open air with many different actors present (state officials, négociants, president director generals [PDGs], and the comptoir) amidst the old, now often defunct machines (though some of these machines do get used for certain processes). And when armed groups have stepped in, "just in time" (Hoffman 2011), to meet the demand for "digital minerals," their first stop has been the relatively convenient company concessions. But concealed underneath all this highly contemporary and sometimes-violent episodic flux lies at least one constant—this place, those who live here believe, "is still a Belgian reserve," and the Belgians are still watching over their legacy, even if they can't come back to work because Congolese "eyes have been opened" to the value of minerals (also, the former Sominki employees are still tallying the debts that Sominki and now Sakima owe to them, converting this amount into contemporary currency using a formula they've come up with). Don't you think it's strange, several people have asked me, that this place has been hit by war as many times as it has and at no point was a single building harmed? It means, many infer, that this is still the Belgians' place. Indeed, many believe that, when armed groups have come, they were following the orders of one of the old companies, which is still manipulating events from afar. There is, in other words, a conviction that, underneath the relatively democratic flux of artisanal mining and the come-and-go temporality of the supply chain, there is a more enduring transnational hierarchy that is ready to assert itself at any moment.

To take one poignant example, the house in which we are staying is actually the old payroll office. The comptoir, which currently rents the house from Sakima and pays them for the right to buy minerals from the diggers, is owned by a Congolese guy who lives in Kinshasa who also, we are told, has a French partner. For about eight years, the comptoir owner had been here, forwarding money to artisanal miners, buying their product, and paying grounds people (most of whom used to work for Sominki) to maintain the buildings that they use, but for the past eighteen months no one has been paid. And no one has seen the comptoir owners, though people still await their return. It turns out that this is a major reason why people are so insistent about wanting to work for us—they think we're connected to the company that owes them back wages. A couple of times, we will be awoken by a large group of people demanding compensation in the assumption that we are the comptoir. Through them, we'll come to understand that many people here believe the comptoir has disappeared because it was "fake" all along—the Frenchman's father had allegedly been a geologist for the "real" Belgian mining company Symetain (the precursor of Sominki), and people believed he left a stockpile of something under the house.

Figure 11. Joseph and Raymond walk alongside the white picket fences of the old mining company town in Kalima, Maniema

This is an iteration of a rumor that is by now familiar to us; in fact, at every former industrial mining concession I've visited, someone has thought that I was the descendant of people who worked here in the past, looking for buried treasure. According to this particular rumor, the French comptoir partner was actually working undercover, and his true, secret mission was to recover his father's real, lost stock (implying that the secrets of the old company are of greater value than the ephemeral work of to-day's artisanal miners). He had rented the house we're in with the specific goal of nocturnally extracting that stash from where his father had hidden it, and he had, per the rumor, hired people to dig under the house at night, then to transport his stock out while everyone else was sleeping. That explains why he's been gone for these last eighteen months—he has already taken what he had come for in the first place. People here reason that whatever it was must have been more valuable than cassiterite, coltan, or even gold—their bet is on uranium, which the Chinese are said to be after now. (Some people here say that it was uranium from somewhere around here that enabled the Americans to bomb Hiroshima and Nagasaki with nuclear weapons back when the company was thriving, but I have no idea whether that's true.)

"We Are Living after Development": Histories and Temporalities amid the Ruins

Seniors in this region tell a rather stock story that posits a singular linear trajectory beginning from the time of King Leopold and extending to the time of Mobutu. The short version goes as follows. The Zanzibari slave traders came here first, with their "bad plans" (*mipango mibaya*, which seemed to serve as a kind of shorthand for mass atrocities) looking for ivory, slaves, and gold, but they didn't "see ahead" (*kuona mbele*) because these plans were not long term. Henry Stanley came, and he was different only in that he wrote everything down, cataloging all that he saw, taking notes—this was the beginning of the "eyes of the world" as well as a spatiotemporally encompassing plan of control and extraction. According to local narratives, Stanley brought minerals from here back to Belgium, where they found a market. People saw and then bought Stanley's written work, and a few years later, an "exploratory society" came here. Some say that, to this day, there's a book in a museum in Belgium that Congolese aren't allowed to see, which contains all the old maps showing where this mineral wealth is (the only way a Kongomani can see it is by seeking out a certain renegade Belgian professor). Soon after, a geologist arrived with a "mining research company," and this research company became the first mining company, Simaf, the company before Symetain, which started in the 1940s.

The company, though it ensconced itself in an ideology of reason and science, took what it needed by force. In those days, whites would just go out to the villages beyond the company walls, forcefully and capriciously grab strong-looking young men, shove them in the car, and bring them to work (which must have been horrible, though some talk as if it was a wonderful thing). The company also brought Europe to the forest: Symetain built the church (with racially segregated pews), a hospital, and schools (also all racially segregated), and implemented practices that linked provisioning to standardized, repetitive factory time—to this day, the name for Saturday throughout the east (not just in the company towns) is *Siku ya posho*, the day of the portion (food rations, of which see below; a week is also called a *posho*, or portion). After independence, Symetain was nationalized to become Sominki; this period is generally remembered as having started off reasonably well, but eventually things collapsed as Mobutu drew money from the company until it went bankrupt (as mentioned above, there was also a major global market price decline). When the Rally for Congolese Democracy (RCD) raided the concessions, they also absconded with the futures of Sominki's former employees (including the possibility that they might receive back wages or pensions) because now, with the reserves depleted, the companies have little incentive to return.

Food figures prominently in these memories, linked to collaboration and provisioning. People talk at length about the comparatively cheap food that the company used to fly in on airplanes (tomatoes, mayonnaise, potatoes, beer, etc.), all of which was deducted from salary. This sticks out for people, in part because these days food is always much more expensive in areas where people are mining artisanally, and prices and availability are highly erratic. On the one hand, food provisioning is seen as indicative of the sybaritic luxury in which some people are imagined to have lived, as well as the comparatively close, if not exactly good, relationships between Africans and whites (they ate the same food at the same time, even if not together; some people dwelled on the conviviality of white women teaching African women how to prepare certain kinds of foods, for example). At a most basic level, those who ate the company's food are imagined to have shared in and been included in the community of the company, materialized in now-ruined building and infrastructure. But many also interpreted this imported food as a form of bribery, or combined temptation and deceit, saying that the company "used food to close people's eyes" to the value and type of mineral wealth in the ground (all of which was *presque*, meaning "almost" but used in the sense of "near") and the value of their work. Moreover, nearly everyone claimed that provisioning created a long-term condition of food dependency that continues to afflict people today: they are now unable to plant crops in a sustainable way, the story goes, because they can't get accustomed to the temporality of agriculture (the waiting), and some can't abide its lowly nature—the task is simply too common for the once-salaried children of Sominki. This is but one specific version of a more general tension between agriculture and mining, communicated in the common observation that "mining destroys agriculture." People in mining areas usually frame this in terms of learned habits—people have become addicted to the fast, right-this-minute temporality of mining and the rewards that can be derived from it and so can't go back to the waiting entailed in agriculture, a condition that leads to exorbitant food costs and dependence on food from elsewhere.

In the ruins of the company towns of Maniema, this was a locally resonant morality tale, but it was not exactly true. After all, aside from the small group of people who worked for the company (about five hundred people had a "permanent contract" with the company in the town of Kalima, for example), most Congolese in and around the company towns *did* grow food, even during the time of industrial mining. And most people here do shift back and forth between mining and agriculture depending on the season. When people choose to mine instead of grow food, it has more to do with inflation and the need for consistent money than laziness or ignorance about farming.

If these stories reference the comforts and abuses of the company's provisioning, a history of horrible violence and exclusion can also be read from the bodies of those who walk among the ruins, interrupting the nostalgia that the large, permanent buildings tend to produce. Take, for example, the case of my aging friend, Diamond, who literally cannot touch others because he once tried to eat some of the company's food. Back in the 1970s, the Belgian management of Sominki cut off both his arms at the shoulders, and he has lived each day for the last nearly half century by the grace of others, who assist him in the most quotidian of tasks, including that of relieving himself. These daily humiliations are compounded by the fact that he is all alone with no family to care for him. Diamond's father had worked for the Belgian mining company Symetain and had continued working with them into the independence period, the time of Sominki. By that time, it was after independence and Mobutu was president, but those Belgians who had remained still ran things at the mining company concession in much the same way as they had during the colonial period. When cassiterite prices fell in the early 1970s, the company downsized, and Diamond's father lost his job. Diamond reacted to the fact that his family was no longer sharing in the food of the company by engaging in some acts of adolescent resistance: stealing food from the company canteen.

When the company police caught the young Diamond, they tied a rope tightly around his body, which dug into his arms. Then they chopped some logs and chained them to his legs, immobilizing him. They left the agonizing, pleading Diamond like that in prison for five days, and when they finally removed the rope, the flesh on his arms and legs was swollen and rotten. The main doctor had been away on vacation, but he returned just as his underlings, operating under the direction of the company's management, were about to chop off Diamond's legs, having already removed his arms. The doctor was infuriated because, he insisted, they should have simply removed the rotten flesh from Diamond's arms and legs and not amputated the limbs. Maybe he was also infuriated because the removal of Diamond's limbs exposed the violence of the company and the violence that undergirded and suffused his seemingly benevolent healing profession (Hunt 2016). At least the doctor was able to keep those butchers from going after the legs too. Diamond is confident that the negligence was purposeful and that the company was trying to punish him in whatever way they could—they wanted to take him apart, limb by limb. Diamond was able to find a Congolese lawyer to file a case against Sominki, and in the end, the company agreed to give him two kilos of fish every time they catch fish, if they ever catch fish. Sakima, which now owns the company, has a fish pond, and when they open it up for the season to gather fish, Diamond gets a kilo (a little more than two pounds).

Who Brought "Development"?

While local historical narratives about the lost "time of development" seem to place whites and their habit of writing things down and "seeing ahead" at the center of history, there are other dimensions of the story that complicate this, instead highlighting collaboration between opposed groups and forces. Seniors, in particular, are quick to point out that it was senior men, leaders of the Bwami society, who "opened up" mining for the company and who continue to open holes for people to mine through their communications with ancestors. Moreover, while it may seem like the Europeans brought electricity and other symbols and conduits of "development," the company has been able to have electricity only because the senior men of the Bwami society "opened up" the water source where the dam is decades ago by communicating with ancestors and spirits, and they continue to do so today (indeed, the reason electricity doesn't always work now is because confusion was created during the RCD occupation, when the RCD government mistakenly recognized a different family as having the ability to sacrifice to ancestors). This tension over who really brought electricity and who really has rights to it also played out in Punia. There, when Symetain was nationalized and the Belgians were forced to leave, the whites tried to take the main power generator with them, but the Bwami society leaders organized all the initiated youth into a battalion. When the whites landed in their helicopter and tried to take away the generator, the young men surrounded it, bearing spears and arrows. Punia still has electricity to this day because of that act of defiance, which also sprung from a recognition of the fact that senior men, ancestors, and forest deities—not the company—were ultimately responsible for the area's electricity.

While seniors see themselves and their ancestors—and, in particular, the "opening" work of the Bwami society—as the precondition for all "development," in the long term the Faustian bargain that the seniors had entered into destroyed their authority and their ability to mediate effectively with the forces of the forest. Seniors complained that the onset of mining ultimately weakened the power of the Bwami society by making it so young men could earn money and get married without passing through seniors or undergoing initiation into the society. Moreover, the annihilation of animal species brought about by large-scale hunting irrevocably changed the Bwami society—first, with the selling of game meat to the population that expanded around the mining town and, later, with the closure of Sominki and the onset of war, when the invading armies hunted animals to near extinction. According to those who were around at the time, when Sominki finally shut its doors in the early 1990s, it created a huge problem for locals because all these laid-off people went into the forest to kill "all the animals" (or at least as many as they could) for food. Bwami society leaders no longer had access to the materials used for making masks (ivory, certain animal hairs, shells), which had enabled gods and spirits to incarnate in the form of Bwami society mediums, who in turn had provided hunting opportunities and knowledge to young men. With that source of connection to an alternate dimension destroyed, the only work left to Bwami seniors was opening up mines, so they too were now dependent on mining, as were the young initiates who found themselves without animals to hunt.

Also, most of the people who were brought to work for Symetain and Sominki had come from far away, including Kasai and Orientale provinces. When they left, they had planned to send for their kids later, but then other companies in other parts of Congo also closed, and after that, the war came, and so these children were left behind. Local seniors see these two things—the death of animals and the emergence of a landless, unemployed population that were not autochthones—as being the nail in the coffin of their already diminished authority, which had been based on communication with ancestors and the use of ritual and masking technology to mediate relations between people and the forest. To this day, the collapse of the authoritarian, hierarchical company and of seniors and "customary authority" (the authority of colonial and postcolonial chiefs as well as the authority of Bwami society leaders) are understood to be tied together. For example, when President Kabila came to visit Maniema in 2016, there was no space at the table and no food for the chiefs (*Wami*), which seemed to sadden the directors of Sakima as much as the chiefs, as their authority was interconnected.

The United States of America and Kigali 2

When the war came, the tenuous continuity between past and present that the company professed to mediate was replaced by rupture, fusion, and simultaneity, as a new temporality emerged in the mining concessions. Wormholes opened up in the forest when the armies came to buy minerals, and even the United States of America appeared, for a time, in the middle of the Congolese rainforest. From a distant, hard-to-reach forest in the middle of nowhere, the United States of America grew into a nation of immigrants almost overnight, and all in the midst of one of the most destructive wars the world has ever seen. As one man remembered, "When you were approaching the United States of America, before you even saw it, you could hear it. It sounded like a great torrent of rushing water, but that's not what it was. It was the sound of people at work, all the people digging. The movement! You'd find workers from Senegal, from Nigeria, even from Kenya."

In the beginning, there was no government to speak of in the United States of America, but over time the Mai Mai army "brought law" (*ilileta sheria*): "The more people worked, the stronger the Mai Mai army grew. They grew rich from the 50 percent that they earned from the diggers." But even then, like its namesake, the United States of America was known as a relatively liberal place with many freedoms, which helped to make it a land of opportunity: "A person could go from nothing to something, and no one cared who you were or where you were from, what your ethnicity was, so long as you were strong and could work." Because so many people wanted to go there, and there was a constant threat to security, when a newcomer visited the United States of America, she had to apply for a visa: "They gave you an identification card, and the Mai Mai told you, 'Welcome. You are in the United States of America now, another country. You have left Congo behind.'" There were customs officials there, inspecting goods, assessing value, and placing tax.

Diggers from the United States of America joined diggers from other mines to sell their wares at Kigali 2 (named after the capital of Rwanda), a town controlled by the Rwandan-backed RCD, the "enemies" (*adui*) of the Mai Mai (though both armies depended on trade and collaboration between the mines—in fact, the armies staged "fake" battles so they could engage in business with each other). As one man recalled,

The market in Kigali 2 was lined with the tents of prominent Rwandan generals. They had with them all of the things they had looted from Congo and the many things they had brought with them from Rwanda, like mobile phones. They gave these things to traders who were on their way to the United States of America.

As many as five planes could land at the village in a day, each carrying goods and diggers from Kigali (all of them Hutu prisoners) and returning to Kigali with minerals. The Rwandan military figures had their own network of buyers, a market, and they had come with motorcycles, televisions, radios, money and, most importantly, cell phones, which ended up being put to use in the trade. Nowadays, the United States of America has returned to being a forest, and in the village of M. (formerly Kigali 2), sometimes one can't even buy a bar of soap, the loaded icon whose absence many eastern Congolese use to signify absolute poverty. Its residents are left to wince at the name Kigali 2, the only moniker outsiders know their village by, since the term implies that they were traitors who assisted the Rwandans during the war—the smear is the only thing that remains of that time.

During the war, the mining concessions were relatively uncontrolled and ultimately fragile openings, or portals, to other dimensions, but in the wake of war people sought to make incremental futures in the ruins, tying together the past, present, and future through their daily practices. In the process, they threw into question, and sometimes imploded, inherited institutions and practices—not least the vestiges of the (post)colonial state. In what follows, I will present some ethnographic moments from the field that each bring out a slightly different dimension of how mining concessions remained rooted in the past while also drawing on and combining new rootless, or rhizomatic, forces. In the first episode, I write about the work of company directors as a kind of collaborative practice that stitched together different times, forces, and sovereignties. In the second, I write about the story of a famous local figure, Lagome, or the rubber sandal (flip-flop), a man widely associated with the act of going to the toilet, who was rumored to have become the state. I contrast the mythology of Lagome with what I came to understand as a story closer to the actual truth in order to arrive at a deeper understanding of the symbolic tensions and struggles at work in this region. And in the third, I consider a new group of workers—the divers—who made their living in the liminal, watery spaces of the company concessions, refashioning waste into artisanal machines and ultimately bringing the state to them. All three ethnographic moments depict challenges, or even alternatives, to state authority as state-like figures emerge from the ruins.

Working with the Ruins: The Spatiotemporal Practices of Company Managers (Kalima and Kailo, 2011)

Old man Sebastian, a company director for Sakima at one of the large colonial-era mining concessions in Kailo, was proud of his ability to navigate between *kiasili*, the language and ways of tradition, and *kizungu*, the language and ways of Europeans. He saw this as central to his work of managing the mining concession, and he had come to this mediating position, in part, through anthropology. In the 1970s and early 1980s, a French anthropologist, probably trained in French structuralism, came to this part of Maniema to conduct ethnographic research on Kumu religious ritual, employing Sebastian as an assistant. Sebastian hasn't seen the anthropologist since the early '80s, but he has continued to do historical ethnography on his own, developing a full manuscript about a nineteenth-century Kumu prophet named Nkoonga, who is said to have predicted the coming of "Arab" slave traders from Zanzibar and Belgian colonials and who preached about the need for everyone to "come together under God." Sebastian sees this prophet as a harbinger for a possible future that now exists, if incompletely, in which formerly separate and conflicting "worlds" (e.g., Belgian, Arab, and African) are bridged through the work of mining.

Sebastian, who identifies as an educated Christian, knows about the power of ancestors firsthand because his grandfather is now a leopard in the forest near his home who visits him whenever Sebastian goes home, protecting him and his house. But Sebastian doesn't engage with the ancestors directly: rather, his job is to determine who the "correct" descent group representatives of a given area are and to make sure that they are satisfied. He needs to do this because otherwise there will be accidents in the mines and declining output. Because diggers come to this place from far afield, sometimes they encounter people who claim to be these figures when they set about opening a hole for exploitation. Like many others, Sebastian holds that, if a person promises to be able to bring you wealth from the ground without being a "child" of that area, then that person is a witch, manipulating appearances and convincing others that something morally meaningful or valuable is happening when it is not. The difference between witchcraft and communication with ancestors is that, as Sebastian puts it, "you can't know where the power of the witch comes from"; it may be satanic power or magic purchased with money, whereas the ancestors are loved ones, and their power is a manifestation of their love. In contrast, the witch's power entails altering the appearance of something without actually affecting its value or intrinsic nature.

One time, a Congolese doctor visited Sakima's concession from Kinshasa with a gold detector. He requested permission from Sebastian to use it, and after paying a certain fee and receiving the requisite papers, Sebastian agreed. But when the doctor went out to the forest to use his machine, he was confronted by a group of senior men who requested the sacrifice of one goat. The doctor bought the goat and beer, and he informed Sebastian, who went with him to perform the sacrifice. When they arrived at the agreed-upon location, Sebastian couldn't help himself from breaking out into loud laughter: the guys who were performing the sacrifice were not from this area at all but from Punia, a good eight-hour motorbike ride through the forest. Through fits of laughter he remembered, "I know . . . I know all the people here, and . . . how . . . can someone be communicating with ancestors from *here* when they're from *Punia*?!" For Sebastian, the ancestors were the living but invisible conduits of the past of a specific location in that location in the present, and so he saw these practitioners and their spirits as fake and therefore probably witches. He was, I believe, also inclined to think this because he was a trained informal anthropologist wedded to a fixed idea of tradition *and* the manager of a company whose predecessor had "purchased" rights from a specific group of people, and the company's claims were legitimated through that exchange. But if ancestors were as mobile as people, as some diggers thought, then the idea of multiple ancestors and different mediators among them was conceivable and perhaps not as ridiculous as Sebastian thought.

Another time, Sakima entered into a contractual arrangement with a German NGO that was working on a road project in Maniema, and they had a lot of accidents in a short period of time; one man had his head cut clean off by a falling iron bar. After listening to his Congolese workers, the German asked Sebastian to go find the seniors from there to see if anything needed to be done. The seniors told Sebastian that he had to give two chickens and beer to the old men and the *mababu* (ancestors), but when Sebastian explained this to the German, the latter agreed to only part of the offering, believing two chickens and beer was too much for a few old guys (suggesting he didn't really believe the ancestors were eating too). Sebastian went to the seniors with the offering the German had given him, and a sacrifice was performed. During the sacrifice, another man, much older than the others, happened to walk by. He stopped momentarily to lean upon his stick, silently considering. The other, younger old men ate the chicken and drank the beer, but the oldest man didn't have any of it and continued his walk into the forest. Sebastian followed him and, when he caught up, asked the old man why he hadn't participated or eaten the chicken. The old man simply said that the sacrifice had not been accepted, but those "youngsters," the other seniors, just didn't know it yet. Sebastian went back and reported this to the German, who seemed indifferent. But in the ensuing weeks, "death was everywhere"—fifteen people died from trucks turning over. Eventually, Sebastian went back to the very old man who had told him "the truth." Now the old man told him that with all these people dying, the sacrifice couldn't just be about chickens anymore—Sebastian and the German had to give a couple of goats. They did, there were no more accidents, that old man never asked for anything for his trouble, and Sebastian got his salary scale adjusted from a level 3 to a level 4.

Clearly, the work of the company manager was not just about collecting rents but about mediating temporalities and ontologies. In addition to dealing with entities from the past living in the present, company officials also had to make accommodations among ever-changing state authorities in the here and now. Working in the mining concession, whether as a company manager or other authority, meant actively navigating these changing contingencies and accommodating different stakeholders while also crafting persuasive and compelling stories regarding peoples' rights and expectations toward one another. For example, in Kalima in 2011, I accompanied the Sakima director there, Lucien, on a small excursion to visit one of the artisanal mines on Sakima's concession in an effort to ascertain why he wasn't getting any revenue from the artisanal miners on the site. Lucien had written up a contract agreement with a representative of the lineage who owned a hole and was collecting revenue from diggers, some of whom were from there and some of whom were not. From Lucien's perspective, everything under the ground belonged to Sakima, which should have been collecting rents, regardless of who the PDG, or hole owner, was. But, for some, the fact that the PDG was a representative of an autochthonous lineage meant that his obligations to Sakima were somewhat mitigated, because Sakima was also expected to compensate indigenes for land use.

After lunch, we drove our motorbikes on trails toward the construction site (*chantier*) until we reached a point where we were forced to walk. Along the route, we happened across the representative of SAESSCAM, the small-scale mining authority, who made a show of how unlikely this run-in was since, he insisted, he's really never here. At that time, in 2011, SAESSCAM was a fairly new state authority in this area, charged with ensuring the safety and legality of artisanal miners and their work, and people were still figuring out what it was they were supposed to be doing and whether they should be located in company concessions at all. The SAESSCAM representative's performance (acting as if he just happened to be there) was probably purposeful because, if he had been there for some time (say on a daily basis, for weeks or months), as Lucien believed he had, it would mean that SAESSCAM was collecting revenue and monitoring mineral extraction in Sakima's concession without informing or sharing it with Sakima, the state-owned company. This probably would have been considered a legal infraction, and it would certainly not have been collaborative or kind.

Our visit out to the mine ended up being a meeting between three men: Lucien, the SAESSCAM authority (a man I'll refer to here simply as SAESSCAM), and the hole owner, or PDG, who was the head of a family that could trace its ancestry back to before the time of the company. The conversation began as a friendly but pointed argument between Lucien and SAESSCAM, two men who ostensibly represented "the state" in different iterations (the state in the form of the government company and the state in the form of the small-scale mining authority). Lucien wanted to know why he wasn't getting reports from SAESSCAM, as this failure, if it were purposeful, would suggest that SAESSCAM did not recognize Sakima to be the owner of the land and the legitimate state and business authority in this context.

"So," asked Lucien, "you've refused my requests [for a report]?"

"No," the SAESSCAM official responded, "I just haven't been to the mine for a while."

Lucien, addressing the PDG, asked where the toilet was. "Where's a house that people can enter into when it's raining? Where's the water?"

"There's nothing," SAESSCAM almost moaned.

Lucien continued his litany: "Where's the drain so the rainwater doesn't go into the hole [where the men are digging]? I said all these things have to be here before you start working. This is necessary to stop accidents. Where is all this stuff I asked for?"

"You know," the SAESSCAM official needled, "sometimes you say there will be an accident, but you have also said you don't care if there's sickness or accident."

"I said that?"

"Yes, sir," affirmed the PDG, chiming in. "You said that."

Lucien, now angry, dug in: "Who is it that can answer questions about the affairs of this place and explain them to others?"

"It's you, sir," uttered the PDG, meekly.

"That's right," proclaimed Lucien, triumphant. "If the government starts asking questions, I don't want issues or conflicts."

The rain began to pour, so we retreated to a makeshift lean-to, where Lucien and SAESSCAM discussed these issues in more detail in front of the PDG. Throughout the conversation, Lucien tried to undermine SAESSCAM by continuing to ask questions about safety and what Lucien called "development," which is the ostensible job of SAESSCAM (making sure that mining is safe and that holes are dug and built in ways that are likely to endure over time, into the future). A disagreement emerged over which authority, the government-owned company Sakima or SAESSCAM, had the right to extract rents and direct papers at artisanal miners. Some weeks earlier, Lucien had written up a contract for the PDG, in which the latter agreed to pay a certain amount of money every month. Per the contract, some percentage of the ore mined by the diggers would pass through the PDG to Sakima. From Lucien's perspective, his letter allowed the diggers to dig on this land, which belonged to Sakima, and for the PDG to organize their labor; he was still awaiting confirmation from the PDG, which would take the form of a signature on paper—a contract.

The PDG now told his version of events. When SAESSCAM had first found his way to the hole and started asking people questions, the PDG had brandished his letter from Sakima. SAESSCAM saw the letter and told the PDG that "this is not a letter" (*Hii haiko barua*), meaning that the contract was not valid and that the only people who could grant permission to dig were SAESSCAM representatives, not Sakima. SAESSCAM's justification was that his office alone works directly with artisanal miners and administers papers related to their work. After rebuking Sakima's letter, SAESSCAM continued to visit the site from time to time to collect revenue of his own from the diggers (he also had his own hole). SAESSCAM chimed in here to explain that he had indeed written a report about the work to the Office of Mines, his bosses, implicitly ignoring Sakima's and Lucien's request for a report. Lucien, clearly upset with SAESSCAM, turned now to the PDG and insisted on knowing why he hadn't shown up at his company-owned house on the hill with a signed contract. The PDG politely explained to Lucien that he hadn't gone to see him because the SAESSCAM representative had forbidden it.

Lucien was angry. Sakima hadn't been getting revenue from SAESSCAM, the diggers, or the PDG, and Lucien also hadn't received a salary in months—after all, he and the company depended on money from diggers, négociants, and comptoirs to survive. When Lucien asked the SAESSCAM representative why he had behaved in this way, SAESSCAM wondered aloud why Lucien was talking to "lowly people" (*watu wa chini*), meaning diggers and autochthones, when he was right there, and they were both "adults." At this point, SAESSCAM began rhetorically devaluing the mine, as if to show that its yield was so small that there was no point in SAESSCAM's or the PDG's reporting anything to Sakima anyway. SAESSCAM suggested that the mine would close soon. Lucien responded by saying that he would close it himself tomorrow, since he had the authority to do that, and SAESSCAM reiterated that it would just close itself. In the end, the two state figures came to an agreement about the distribution of rent revenue that was sure to end up "taxing" the diggers more than they had been taxed previously.

Both Sebastian and Lucien tried to discursively forge a sovereign spatio-temporal order, placing themselves and the company at the center of linear time. In his work with "customary authorities," Sebastian worked to create continuity between the historical past and the present, in which his expertise and the sovereignty of the company created a seamless connection between the time of the ancestors and the present day. He resisted, and found plainly ridiculous, any alternative claims to a historical connection with spirits involving persons not included in the company's historic contract relations. Nor did he believe that the ancestors of a place and the "past" that was relevant in the present might change over time. By stitching together the past embodied in ancestors and the present, he was able to fix the company at the center of local history and create incremental temporality for himself—by advancing a pay grade, for example.

Similarly, in his dealings with the local autochthone/PDGs, Lucien employed bureaucratic papers (contracts) to encompass people who had lived in the concession prior to the company's arrival within the enduring and overarching space-time of the sovereign company. At the same time, Lucien consistently returned to a concept of linear development and historic rootedness, which he associated with the state in the abstract (not necessarily SAESSCAM). According to this rhetoric, the state is concerned with miners' well-being and with the sustainability of the hole over time, and one is only a legitimate state authority to the extent that one is similarly concerned with duration (SAESSCAM's fleetingness thus meant he wasn't legitimate).

These official actors tried to present themselves as seniors and rooted sovereigns through the deployment of papers that carried the trace of a larger spatiotemporal order, extending across historical time. In contrast, in the mine of Chamaka outside of Punia, a young man came to prominence who seemed to purposefully explode the idea of incremental temporality, instead recreating sovereignty around practices of rapid expenditure.

The King of the WC (Punia, 2013)

Lagome, meaning "the rubber," a local colloquialism for cheap flip-flops, was a man of a little less than forty years of age who leased an artisanal cassiterite mine in Chamaka (in Maniema province), a full day's walk from the former industrial mining town of Punia, through the forest and across a tributary of the Congo River.[3] Like Kalima, Punia and Chamaka were once company mining towns operated by the Belgian mining company Symetain, which was later taken over by the joint venture of mining companies and government called Sominki. Now the area is owned by Sakima, which leases out space to people like Lagome. As in Kalima, in the town of Punia and the mine at Chamaka, the halcyon days of industrial mining and the formal employment that came with it serve as a baseline for most everything that is said, thought, and done today. At least in senior people's memories of that time, the company and seniors collaborated to create and manage a moral economy in which "traditional" values (Belgian and Congolese) held sway and in which no moment was unaccounted for.[4] In Punia and the forest town of Chamaka, people would get excited when they saw me because they hoped that a real company might be coming back, one that would bring jobs and set about production. What could they do or tell me, they wanted to know, to make this happen faster?

Moreover, the state officials in Punia were notoriously aggressive in their pursuit of miners' income, and most people felt that the presence of a company might keep this in check. It seemed like every single state official was trying to make money off miners by "selling" them papers to do business. Since in Punia mining was the only real game in town, state officials didn't let go of diggers just because the price went down because of the decline of outside demand—rather, they increased the pressure. Some diggers felt that they would have to take their cue from the diggers in not-too-far-away Kasese: in 2012 and 2013, miners there took to hiring a Mai Mai group called Raia Mtomboki ("the infuriated citizens") to protect them from state officials and to help them find an alternative route for their minerals so that they wouldn't have to pass through the expensive legal channels, including the distant and difficult to reach provincial capital of Kindu.[5]

It is against this backdrop of nostalgia for the company that many people in Punia and nearby Chamaka understand artisanal mining as a chaotic, unpredictable, and even immoral (if also necessary) activity. And it's against this memory of industrial mining that Lagome is understood in Punia. Lagome, whom they say used to bathe in beer just because he could, made all his money from mining and selling cassiterite. Like the earth Lagome mined, there was never anything glamorous about him—he was a simple, uneducated person—but he was for a time the most fabulously wealthy man in Punia, and that's why he gave himself that self-effacing praise name, Lagome. You might be ashamed to enter the city of Kisangani without good shoes, and your friends might look down on you for having to wear them, but at the end of the day everyone has to go to the toilet, and so everyone has to wear lagome. And so too anyone in Punia who wanted anything would sooner or later have to come to Lagome, the simple, looked-down-upon thing that everybody needs and that is proof of the basic sameness of everyone. People also called Lagome *kikombe*, or cup, because everybody drank from him, and little kids used to sing a song about how Lagome was the father of the WC, the king of the toilet.

It was rumored that Lagome made his wealth through a sacrifice. They said it was his mother who helped him get rich by sacrificing someone's life (exactly who no one knew) so that Lagome could find seemingly endless cassiterite. His mother never got anything enduring from that sacrifice, and she must have been angry because all the cassiterite Lagome found he gave away to women or sold to buy things he now no longer has. Lagome made so much money from his lineup in Chamaka that he didn't know what to do with it all—people said that it was his mentality, the fact that he "couldn't see far," couldn't plan, or make good investments. Like all artisanal miners, he didn't know anything beyond what he immediately experienced and couldn't imagine a future beyond the here and now. He was just like Lungwa 4, another local antihero who had made a fortune out of gold.

The story of Lungwa 4, an artisanal gold miner during the 1980s (the time of Mobutu), is worth telling in brief because many people's understandings of Lagome were filtered through their understanding of Lungwa. In the 1980s, Lungwa 4 rented an area in the forest outside Punia from its autochthones. People say that with the money he made from gold he bought a huge stereo and hired a guy to carry it behind him when he walked through the forest. Then he bought a car battery for the stereo and paid for another guy to follow the guy with the stereo, carrying the battery. And all of these guys were surrounded by another fleet of guys with umbrellas, as well as guards to protect Lungwa 4, his stereo, and the battery when they walked through the forest to his mine. If Lungwa 4 ever needed for anything while he was living in the forest, he would simply take a cup and put it in his always-present basin of gold, exchanging it for whatever he wanted at a "price" far less than it was worth. Afterward, there would be uproar (*kelele*) in the forest for days, as other people participated in the overflowing expenditure. One day, Lungwa 4 decided that no longer would he bathe in water ("From this day forward, I will only bathe in beer!" he's rumored to have said), so he asked one of his now numerous wives to fill a basin of beer for him to bathe in. After that, he decided that he wanted to sleep in the governor's house ("I must sleep in the house of the governor!" one person recalled him saying). So he went there with his basin of gold, where he was received by the governor, staying in his house for a whole week. The rumor has it that the governor "gave" Lungwa 4 one of his wife's sisters to stay with him for that week as his companion and that she was able to use the money she made after that one week with Lungwa 4 to buy a house in Bukavu.

One of the implications of the rumor was that artisanal mining could dirty (*kuchafua*) the state and elites, bringing them down low (*chini*)—the governor becomes a pimp and his sister-in-law a sex worker. At the same time, some people were able to convert gold's potential for permanence and spatiotemporal extension into enduring futures—note, for example, the contrast between Lungwa 4's rapid expenditure and the solidity, visibility, and permanence of the house in Bukavu. These days, Lungwa 4 is said to be indigent, a condition that many blame on his mother (just as they do with Lagome today): they say that he had used magic that she had prepared and that this magic had "conditions," the main one being that its user would never be able to create anything of enduring value with the money acquired. Meanwhile, other people, including several women, rented holes from Lungwa 4, staffed them with diggers, and used the money to educate their children (including one young man who ended up attending graduate school in Brussels to study the potentials of mining for "development").

As in the days of Lungwa 4, when people came to Lagome's place up in Chamaka to buy cassiterite, they would come with things like motorbikes and stereos and leave them outside the door to his place. Like Lungwa 4 before him, Lagome never inspected the things—he would just say, "leave the stuff there." Or he would let people walk away with five hundred kilos of cassiterite at whatever price popped into his head (e.g., fifteen dollars per kilo), usually far below the actual price. He was, it seemed, recklessly indifferent to the prices of these things, but for a while it didn't matter. So many people wanted access to his mine, or the cassiterite that came from it, that even to talk to him you needed to bring two goats. So if twenty people came to talk to Lagome today, there would be forty goats in front of his place at the end of the day. Eventually, Lagome had accumulated a whole lot of motorcycles, and he needed to get rid of this stuff, so one day he had all of these bikes transported by truck to the town of Punia, and when that happened, the police confiscated everything. This, after all, was a visible form of wealth upon which they could levy "tax." When Lagome got the cell-phone call saying that his things had been confiscated, he invited the police up to his *carrière* (quarry), but first he rented a generator from Sakima so they could talk in the light. The police and Lagome sat down for a discussion, and Lagome agreed to pay a certain amount of money to the police, after they had established the appropriate "price."

The story went that Lagome wanted something meaningful from the police in exchange for this money: mainly, to effectively shut down the state for a period of one month. He allegedly demanded that all drivers in Punia be able to come and go as they please without buying licenses or paying tickets—that there be no police on the road at all. This was agreed to, after which he earned copious praise, as all the young motorcycle taxi drivers went around singing his name—"Hey, Lagome, Lagome!"—and there were parades when he came to town. But making the state disappear for a month was only the beginning for Lagome. So many state officials were still coming to him, looking to collect tax, that he was said to have decided to become the state himself—it was said that Lagome went to Kisangani and "made friends with" some bigwig in ANR, the not-so-secret Congolese intelligence agency, thereby becoming the director of ANR in Chamaka. Once that happened, Lagome *was* the state—the other state authorities no longer had any authority over him and so mostly left him alone.

But at some point, the stories went, things went downhill for Lagome—most said it was the fact that his wealth was the product of a sacrifice, the fruits of which are always ephemeral. In any event, suddenly Lagome had, or seemed to have, nothing. He lost all of his things and all of his women, and he became an example of what happens when you can't "see ahead" (*kuona mbele*). In general, the discourse about Lagome was ambivalent: Yes, he was great and worthy of praise, and he did manage to sequester the state, but he was also stupid and shortsighted. He had money, but he never built anything that lasted, and now it was all gone. One young woman talked about how she met him on the way to school one day, and he said that she was beautiful. She responded that she had no time for him, that she was going to school. And he asked her why she should go there in the first place: he would buy her, the school, and the teachers, whisking away the old established structures with a sweep of the hand.

When I was hearing all these rumors about Lagome while in Punia, I was wary of taking them as literal truths. After all, living in the moment is at once a strategy of dispossessed peoples and a quality that is imputed to them by those higher up in the social hierarchy (Day, Papataxiarchis, and Steward 1998). To learn more about the man behind the stories, I decided to walk, with Raymond and the local Sakima director, through the forest and across the river to see Lagome, informed by all the rumors that I had heard about the young man who knew nothing but the ups and downs of artisanal mining and who therefore was incapable of seeing far and of conceptualizing the day after tomorrow. On the way, we either passed through or came fairly close to a number of other mines whose names spoke to the new, otherworldly, extreme, and risky character of artisanal mining and the mineral trade: Modern Times (Les Temps Modern), United States of America, Kuwait, Last Chance, Morocco, Leave the World Behind, Find Your Own, Grab Your Own, End of Life. All these names for mines signify a rare rupture in space-time, a quickly emerging opening that is likely to close up just as quickly.

When we finally made it up to the town of Chamaka, we were, as usual, confronted with state authority, particularly DGM, the official representing immigration. I won't delve into all of the various rituals of state-making that Charles, the DGM official, went through with his stamp and pen and uniform, or the lengthy discussions that took place afterward about what his job was, or what DGM was for, and what the state was—except to say they took place in a French that grew increasingly eloquent, as the state became ever more reified and imaginary, and that the conversation included miners and traders, SAESSCAM, the Sakima representative, and my assistant, Raymond. At bottom, the discussion concerned what the appropriate fee, or price (*bei*), of an immigration official should be in the forest, when there is no border anywhere nearby. For his part, Charles was convinced that I was the grandson of one of the former managers of Sominki and that a store of cassiterite had been hidden long ago and no one knew where, except for me. I had come to find it, and Charles wasn't going to let me out of his sight (it seems he wanted his cut). The next day, we walked to Lagome's mine, and as soon as we arrived, Charles rushed to Lagome and asked him for "rice" and "cooking fat" (meaning money) for having brought me, a potential business partner, to him. Charles called it his *frais de mission*, the price of his traveling costs.

Over the next few days, I got to know Lagome a little better, and one of the first things I came to know was that his father was the chief of engineering when the company was there and had taught him everything he needed to know about that particular site (he had maps), as well as how to go about mining. Lagome had been studying this since he was a kid. His father had long since passed away and Lagome, coming into adulthood in a time when there was no company, abandoned any hopes of becoming part of it and instead struck out on his own, negotiating with state figures and easily finding cheap labor—people who had been tossed around by war and ended up with nothing.

Lagome turned out to be nothing like his image of a loud, self-aggrandizing fool. He was soft-spoken (even shy), self-effacing, and detail oriented. And, rather than having his sights on every woman in the world, as per the rumor, he seemed particularly upset about an old flame who had allegedly grifted him in a mineral deal and left him hurt. Lagome was very proud of his loyal and well-oiled operation, which, he was quick to point out, was more productive and faster than a "machine" (his word). And he was equally proud of how he had worked to keep the state in check and of the money that he gave to ordinary people to build roads and bridges and pay for kids to go to school—acts of reciprocity and community-building that everyone confirmed he did in fact do.

Over time, he explained to me the organization of his mine, which varies depending on a variety of factors: Lagome has to be prepared to expand and contract when necessary. When I met him, they were in a state of expansion in a new mine and were expecting to be producing in earnest in a couple of days—hence the delays. Lagome explained how he, the head of the carrière, negotiated with the senior representatives of the land, as well as the company Sakima. His director general was concerned with the whole administration of the carrière. And then there was the director of finances, a treasurer, as well as a *mwandishi*, or secretary. There was a chef de camp, who inspected the entire camp and where people were working. The *commissaire ya madini* were nonstate police who protected people in the carrière and were concerned with security and regimentation, making sure that work started at 8 a.m. and ended at 6 p.m. The chef de chantier coordinated work among all the different holes. The chef de secteur worked with the chef de chantier to observe the organization of work. And so on, and so on. The point was that it was highly organized and that this organization interacted with the state but was not subsumed by it or created by it. As elsewhere, the terms were borrowed from the old company days and drew attention to those days, remembering and reproducing them.

These days, it is not trendy to point out the difference between people's beliefs and reality—we anthropologists are often expected to drop the
concept of belief entirely and treat everything as an equivalent "force" in
the world. Leaving aside the value of this, I couldn't help but be drawn to
the radical difference between the stories people told about Lagome and
what I saw in Lagome. So, when I returned home from Chamaka, I put
people in Punia to task about Lagome: There was no way Lagome sacrificed
people to find that vein of cassiterite. His dad was an engineer. And he
wasn't some caricature of the artisanal miner, concerned only about the
present moment—he was in fact quite the planner and architect, not only
of mines but of social forms, which he cultivated as if they were, in his own
words, machines. And really, wasn't it possible to see a positive plan in
Lagome's practice? Whenever he could, he tried to collapse or absorb the
predatory state and supplant it with something else. And his practice also
envisioned a social future in which the state as it currently existed was no
longer necessary—didn't he give out gifts to people who built roads and
bridges of their own volition, without any help from the state?

Invariably, my interlocutors agreed with me—Lagome's dad was a chief
engineer. The mines are organized. And, yes, Lagome turned out to be a
pretty nice guy. But the stories about Lagome continued, and I was surprised to discover, some of my very friends who confirmed everything I said
about Lagome to be true also persisted in telling these stories about how,
for example, Lagome's mother had sacrificed a loved one so Lagome could
have that cassiterite. Why, I wondered, was there this discrepancy between
what I saw as the reality of Lagome (a man who outmaneuvered the state
while trying to develop sustainable social networks outside of it) and the
narratives that people spun about Lagome's destroying incremental time
so that he could enrich himself at the expense of others without even realizing that the fruits of such momentary sacrifice are always unsustainable?

The answer, I think, is this: Lagome didn't sacrifice the lives of others
to get ahead, but there was a sacrifice happening in Punia and Chamaka.
Stories about Lagome were acts of sacrifice, and what was being sacrificed
was Lagome and the possible future that he represented. In these stories,
the possibility of a society without oppressive state authorities, in which
people voluntarily contribute to the improvement of their world, was acknowledged (after all, Lagome is praised) and then foreclosed (he is also
ridiculed) in favor of a society that values education, jobs, and functioning state institutions. Ultimately, these stories about artisanal miners sacrificing the future valorized formal employment, steady wages, and secure
futures, and in them, the artisanal miner was the scapegoat, the sacrificial

lamb, for all that is wrong in the society. The true potential of the artisanal miners, and of artisanal extraction broadly conceived, was acknowledged and then disavowed in favor of a future based on formal employment and a rehabilitated state. But through it all, Lagome continued his work, day by day. And he continues to enact a certain kind of state-like network by providing for others and modeling a potential incremental future—even if others see him as being synonymous with the "punctuated time" of here and now neoliberal capitalism (Guyer 2007).

The Invasion of the Divers (Kalima, 2016)

Tupac, or perhaps 2Pac, far from home, is laying down a rap about his condition. The young Luba man from distant Kasai rarely stops moving, using his whole body to tell the story of his forced evacuation from the diamond mine near his home in Mbuji-Mayi, where the diamonds were "all white, 100 percent. Not 80 percent!," and his long, circuitous journey here to Sakima's mining concession in Kalima. His hands and arms rapidly perform rowing, diving, digging, running, flying, being stopped, and even fighting as he becomes, in turn, a speedboat, a motorcycle, machine guns, a car. Tupac explains how the governor of Kasai "gave a contract" to "the Chinese," or some Chinese contractors who wanted to mine diamonds semi-industrially, and then he brought in Congolese soldiers to chase out the artisanal diamond miners. Tupac clearly misses the high-stakes world of diamond mining:

> The work of [digging] diamonds is higher than the work of the governor! The governor's just a ticket, he has a ticket he got from Kinshasa, and he uses his ticket to get diamonds. He has an office filled with diamonds, like a museum in Europe! Heh?! The governor has three phone numbers on his mobile: his office, the diamond business, and his family! If you call him and say, "I have diamonds here," it doesn't matter what work he's doing; he'll come and follow you there, no matter who or where you are. He'll close his doors to all visitors. That work [of diamond mining] leads/organizes everything (*kuongoza kila kitu*).

But the governor had "made partners" with the Chinese so, with no other choice in the matter, Tupac and his "colleagues" (*ma colleague*) left their home: first they "crossed the water," using the river to get to Kisangani (formerly Stanleyville), the capital of Province Orientale, far to the north. From there, they took motorbikes on to Bunia and Beni in North Kivu, to the southeast of Kisangani. When they were in Beni, these former diamond miners heard there was gold in Maniema, to the southwest of where they were, so they went into the forest to Punia, but by the time they got there they had absolutely no money, so they walked all the way through the forest here, to the old company mining town of Kalima. And why not? Movement across space and time is what being a digger is all about: "I'm a digger!" proclaims the displaced-from-his-place Tupac, "Congo is our country! Every town is ours! Every place is ours! Every place belongs to the digger. We're bringing globalization (*mondialisation*) to this place!"

One of Tupac's friends from Mbuji-Mayi went to Shabunda, South Kivu, to the southeast, where he made enough money to buy an artisanal drag (drag is the name given to any machine used to suck ore from the ground, usually from underwater) and came back here where work is safe. The gold mining is good in Shabunda and the price is higher than here, but unlike here, there's no security: there's a group of Raia Mtomboki, a kind of Mai Mai, which controls certain gold mines and attacks others. And the Chinese are mining there with their "robots" as well, of which see below. They say that last week the Chinese murdered a Congolese miner in Shabunda and threw his body in the river—perhaps, the diggers muse, because he knew the value of what they were pulling from the ground, and they feared he might tell the world.

In 2016, there were about five thousand *plongeurs*, or divers, and six thousand *dragueurs*, or drag operators, working in Kalima, so roughly ten or eleven thousand of these workers overall. Using scuba gear to submerge under the water (they ventured as many as twenty-five meters down), they tied sandbags around their waists to keep themselves in position while there. Ropes connected the plongeurs, who worked in two-hour shifts, to the boats. Meanwhile, those above the water employed machines powered by car engines to suck up ore from the riverbed through a pipe. While underwater, the plongeurs shoved the sand (*mchanga*) from the bottom of the river into long plastic pipes about six inches in diameter. Above the water, the coordinator, called *mutiste* (the mute, or man who says nothing because he's working too hard to talk), took the stuff out and passed it on to the *cyaneur*, or cyanide man, for cleansing (the term *cyaneur* comes from industrial mining, but they were actually using mercury rather than cyanide). The work of the plongeurs was particularly dangerous, in part because of the risk of being underwater for so long and in part because of the different entities that live in the water (both water spirits and ancestors, who come into conflict with one another through the divers). About twenty-five of them died each year, mainly from the bends, though the plongeurs tended to blame these deaths on the conflict between ancestors and these underwater spirits (if an ancestor from the area sees you making an offering to an underwater spirit or wearing a certain kind of amulet, they will decide to take your life because they don't approve of this kind of "witchcraft").

The emergence of the divers was related to a couple of seemingly unrelated factors coming together in time: mainly, the return of industrial companies coming, in the wake of war, with contracts from the "government in Kinshasa," and the effective barring of military personnel from mining, which forced high-level soldiers to get creative in how they profit from mining. Around 2014, when the diamond miners were being expunged from Mbuji-Mayi, Tango 4—an infamous, well-connected, and now quite rich general from Maniema—allegedly financed the purchase of some drags for some of these arrivals. Tango 4 was from Kindu, the capital of Maniema, but he fought for the Rwandan-backed RCD during the war, waging a war on villagers across the river who he claimed were helping Mai Mai; later, he was accused of selling weapons to the Rwandan-backed CNDP and M23. As the divers used their cell phones to call their friends, their numbers increased, and the number of drags started to increase as well, with different financiers emerging as the drag owners (most of these people seem to have been state-connected: politicians, high-level administrators, and high-level soldiers). While you can order the drags on the internet if you have the wherewithal, over the course of a couple of years many people learned to make their own drags, replacing imported ones, which cost close to $30,000. An "artisanally" made drag cost about $15,000 to purchase; it was made from found auto parts and local metal, including the remains of the Caterpillars and other abandoned vehicles and machines found around the mining concession and in the forest. These new inventions, called *tia mai* ("add water") because they have a tendency to overheat, were forged anew from the scavenged ruins of industrial mining. As one local man put it, "So all this abandoned waste has found a use! Congolese sure have a lot of intelligence!" (*Wakongomani wako na akili sana*). Originally used for mining gold, the diggers in Kalima had also adapted the machines to the 3Ts: tantalum, tin, and tungsten.

While many older people glorified the days of industrial mining and looked down on the diggers, "ordinary" diggers were, nonetheless, mostly "children" from here; they were more familiar than the divers and somewhat less associated with rapid expenditure, fluctuating prices, and interdimensional travel. Most importantly, they were less likely to move away permanently, especially given that they also grew crops on Sakima's concession during the appropriate growing seasons. In contrast, the divers came from far away, spoke a "foreign" language (Lingala) that most people in this Swahili-speaking region associated with the army, and had brought with them different food, music, and dress. Compared to other artisanal miners in this area, these people generally had money, and in 2016 they were the main group contributing to movement in this town and its immediate surroundings. Or, as one of the plongeurs put it to me, "We don't build houses, which means we rent—that means circulation of money (*circulation ya makuta*)—in order that others may drink, eat, and feed others (*kwa ajili ya watu kunywa, kula, na kulisha*)." Indeed, when they first came to town, they were buying so much and seemed so flush with money that shop owners nearly doubled their prices (the going rate for sex workers went up even more—allegedly from as low as 2,000 francs to between 15,000 to 20,000 francs). Eventually, different state officials intervened to persuade the plongeurs to bargain more over the cost of things. Many said this was just the way of all diggers: "Diggers," several people complained, "have no time to haggle," implying that the fast pace of their movement drove up prices, causing blockage or closure for those who couldn't keep up. In this regard, rumors had it that the plongeurs' rapid expenditure had ended up breaking up a lot of marriages, and in several cases, bridewealth had to be returned ("They've destroyed households!" some lamented [*wameharibu nyumba*]). According to this view, the divers' rapid movement, in terms of expenditure, caused disorder and the rupturing of other people's incremental, long-term plans.

There were other aspects of the plongeurs' work that differentiated them from ordinary diggers. While ordinary digging was relatively egalitarian, with groups of friends or relatives renting out holes together, this kind of semi-industrial mining was more transparently hierarchical and regimented. Most of what got extracted went to the owner of the drag, whose social and economic distance from the other workers was marked by the term they used for him: *Mzungu*, or white man, a term also implying "patron," which people around here hadn't heard in the artisanal mining business in quite a long time (long ago, gold buyers who would extend credit to diggers were also called Mzungu). The Mzungu had "the means of production," as one plongeur actually put it to me (*Mzungu ana moyen de production*), and supplied them with food, medicine, and anything else they needed. He might operate the drag or remain in the background; under him were two kinds of workers: the dragueurs, who operated the drag and took care of everything except for going underwater (cooking, cleaning the minerals, and taking care of the machines) and the plongeurs, who dove. In many cases, no money exchanged hands: rather, once the "dirt" was cleaned, the plongeurs received a certain percentage and the dragueurs another; the majority of the final product (as much as 50 percent of the original ore or 90 percent of the "cleaned" ore) went to the owner of the drag. Often, the dirt left behind, which contained cassiterite and other ores, remained with the divers, and they sold this at rather cut-rate prices to the négociants.

While democratic, the plongeurs' organization gestured at state authority, reproducing some of its forms, and some of their titles also suggested that their work was dangerous, like that of soldiers (and many of them were former soldiers, as their permanently crooked index fingers indicated). At the top of the plongeurs' association was an elected "supreme commander," underneath whom was the vice president, the secretary general, and the OPJ (*officier de police judicaire*), which was also the title for a local judge (this was a name directly appropriated from the state's judicial system). The OPJ presided over the BD, or brigade, meaning nonstate police, who resolve conflicts at the camp. In addition, the plongeurs and dragueurs called their journeys onto the river "missions," giving a certain percentage to SAESSCAM and the *mwenye pori*, the owner of the land (the autochthones), for "every mission." And their cooks were called S-4, which is also the title for cooks in the Congolese army. In addition to mirroring state officials, they were also much more directly connected to higher-level state authorities. A few times a year, the jointly elected committee of plongeurs and dragueurs went directly to Kindu, the provincial capital, with gifts in envelopes, which were distributed to all the requisite high-level state officials, including military figures. Other, lower-level state authorities, like SAESSCAM, "got theirs" by coming to them in person at the camp.

The divers' powerful connections with state figures allowed them to break the rules and transgress, placing them in a liminal position that had other ontological implications. For one, they lived on the margins of the company, often employing their purchased connections with high-level state authorities to bypass or evade the tenuous sovereignty of the company's management (they were inside the concession but under the water). New regulations did not affect them as acutely either: for example, in 2016, diggers on land were subject to the rules of traceability (they had to have digitized tags for their minerals and deal with state agents if they didn't), but divers and the dragueurs fell outside of this system (which meant that they were also technically dealing in "blood minerals"). Their liminality had other implications: Several people suggested to me that the divers had, in their interdimensional travels, journeyed beyond the pale of the human. Humans, several put it me, are not meant to spend so much time deep under the water, an alternate dimension populated with dangerous spiritual entities ("God doesn't like people to be like crocodiles and fish and snakes under the water—he created people to remain above").

The divers needed these strong state connections in order to stand alongside their powerful foreign competitors, who also had secured the right to mine in the river from the governor. For, once they got to Kalima, the divers again found themselves competing with the Chinese, this time in the form of what they called the Chinese robots (*les robo za ki Chinois*). They referred to these larger, semi-industrial drags as "robots," but it was a bit of a misnomer because each one employed thirty people and had four rooms; the Chinese workers slept in the rooms. Artisanal miners hated the robots, in part because they could strip a whole area of gold in "a minute," and they work throughout the day and night. The robots and the divers worked alongside each other, often in the same space. While the plongeurs and the robots were competitors, their relationship was also somewhat symbiotic in that the presence of the Chinese gave the plongeurs a certain amount of leverage. In brief, the government had shut down diving in these parts in 2014 after a government geologist issued a report claiming there was uranium in the water. But when the government "opened" a certain area of the river for the Chinese robots, the plongeurs and dragueurs protested to the governor and other state authorities. As one put it to me, "How can you shut down our brothers who buy food, rent houses, and bring movement here, and you leave things open for the Chinese, who have their own food that they bring from elsewhere?" So, in what struck me as a particularly science-fictiony, if also dystopian, resolution, the plongeurs earned the right to dig in the potentially radioactive water alongside the Chinese robots.

Back in the bar, Tupac's rant touches, whether intentionally or not, on the ambivalence of movement—that the diggers' displacement and dispossession from home is connected to their circulation and expenditure and therefore the movement of others; in this way, the lives of artisanal miners are important sacrifices, even if they're not taken seriously by those around them. He continues, "Ever since we started working here with our drags, two years ago, there's been movement, and now this woman [the bar owner] is putting down pavement. Why? It's us who caused her to put down pavement, because we like it here and we're her only customers! This is not a good day, we're about to go back to work, but if it was a good day, we'd buy you all beer. Because that's what life is. I'm not going to be buried with gold. God will help me to get more because I used what I had for others!"

"Tupac, you're a prostitute/slut!" teases Sarah, a young businesswoman who sells gasoline to the plongeurs.

"Slut?! Don't bring disorder here, my love!" Tupac ripostes, "They call me Tupac because I can't stand disorder. I have a lot of strength/power (*nguvu*), and people depend on me. And all eyes are on me because I'm the solution. If someone's bringing a problem for others, I'll say, 'Rather than bring a problem for this person, come and fight me.' If the soldiers come to the construction site [mine], I'll say, 'What business does a soldier have at a mine?' What is a soldier going to converse about there? I'll start with fists and in the end I'll swallow his blood! Yes! Me?! I'm a vampire! I'm like Jet Li. If I see the person has a lot of strength, and he's beating me, I start biting. Yeah, Jet Li—I travel via the screen (*nasafiri kwa ecran*)."

"You're a Rwandan! A clown (*mkalamusi*)!" Sarah retorts.

"A Rwandan?" Tupac smiles and laughs, feigning offense. "Why I'm a good Lega boy from this forest right here!"

"Well, you sure talk like a Rwandan."

"I'm a Lega! I'm best friends with your parliamentarian!" (Laughter.)

"You're a Rwandan. If you try to touch me, I'll tell the police I'm 17."

"But, my love, a diver must be a diver everywhere . . ." (Laughter all around at the innuendo).

"See," Tupac continues, "but that's your habit. Rwanda. You think if a person talks fast, he's from Rwanda. I speak fast because I'm a speedboat [*caneau rapide*; to show what he means, Tupac makes the "broom broom" sound of a speedboat while moving his body about quickly]! If you have a speedboat, it's broom broom, and no one can stop you. We're already teaching you Lingala, showing you globalization. Yes, we bring globalization to this place! We came here, and most of the houses were small, and they had no color [paint], and people didn't know how to dress. If you speak Lingala, it means you're the same as a soldier, but it's the national language! Look, here's a professor from America. Do you think he'd be here if we divers weren't? A real professor. His job is to give out grades to all the people of Maniema. If you pass, you pass. If not, tough luck. He doesn't need your minerals, so go play with your Chinese!"

"Oh, diver, you're lucky," says Sarah. "You've come here just in time to meet the president [he's visiting the provincial capital in a few days], while all your other diver friends will be underwater and won't get to see him!"

"The president?! Why, I've seen the tribunal in Kindu! Have you ever seen that?"

Sarah has come to this bar on a weekend to meet with the plongeurs and dragueurs because she knows this is the day they come into town to take advantage of the town's infrastructure to access the internet, googling the price of gold and using mobile money to send money back home (they claim they send home about 70 percent of their earnings). Today, here at the bar, there is also going to be a meeting of the joint plongeurs' and dragueurs' association. Sarah's life has converged with the divers, at least for this moment, but as one of the many "children of Sominki," her personal history roots her to this place; like Lagome, she was born into the mid-high-level echelons of Sominki and grew up in its ruins. The daughter of a former Sominki employee, this astute businesswoman now provides loans and necessary materials like gasoline to the plongeurs and dragueurs. She is one of the primary people with whom the divers from Mbuji-Mayi have been dealing since they started arriving here a couple of years ago, and without her work, they wouldn't be able to mine at all.

The rooted past of industrial mining provided Sarah with some of the foundation that she needed to enter into this new work with the rhizomatic artisanal miners. With the salary that he garnered while working for Sominki from the '60s through the '80s, Sarah's father was able to put her through high school and university in Bukavu; afterward, she worked as a nurse for hospitals in Kindu, Goma, and Kisangani. During the war, Sarah continued her nursing work for the United Nations, working with refugees. She was able to leverage her connections there to access cheap gas, which she sold elsewhere, and eventually she ended up being the main supplier of gasoline for the governor of Maniema; in other words, she quietly brokered the transfer of resources from the "international humanitarian community" to state actors who were in turn "fending for themselves." Three years ago, one of Sarah's woman friends told her that she could make even more money if she left her work of bringing gasoline to state officials and instead brought it to the plongeurs, who needed it for their machines (another example of the diggers being "stronger" than the state). So she came home to Maniema and started venturing out into the surrounding forest. When I came to know her, she had been doing this work for three years: selling gas, mercury for separating gold, and beer, medicines, and condoms to everyone. In that three-year period, Sarah managed to earn enough money to build two houses, one in the North Kivu capital of Goma and one in the Maniema capital of Kindu.

Over time, Sarah brought in other women who were unable to get jobs for NGOs or the UN, the main formal employers in the postwar economy. They came from all over the country, including Kinshasa, but she often complained that "most of them forgot the work they came to do," selling materials to diggers. Instead, they either became sex workers or entered into relationships with men who took their money or things, usually on credit. Without money, the women couldn't continue their businesses, and the end result was ultimately the same: their bodies became "boutiques" (as in the common phrase, "I'm going to the construction site/mine with my boutique"). Sarah was bitter about the way diggers used women, and she blamed them for women's dependence on sex work instead of other businesses; as she put it to me, "The main plan of diggers is to eat the money of these women, and then they have no recourse but to become prostitutes (*maraya*)." She has also had bad experiences extending loans, or *mkopo*, to diggers and getting grifted in the process; typically, she had lent out food and cigarettes to diggers in anticipation of a certain amount of minerals in the future, but almost every time she did so, the diggers left, switching their SIM cards on her so she couldn't find them ("They used the cell phone to escape!" as she put it). One time she lent $1,000 worth of gas to a dragueur; when he disappeared, Sarah went all the way to Shabunda in South Kivu—a great distance—just to look for him, hiring men and searching the forest for a whole week. In the end, she "grew tired" and gave up.

Overall, though, Sarah prefers working with dragueurs and plongeurs to "ordinary" diggers because they tend to be more solvent, and they have a centralized workers' association that governs them, which she could interact with if she needed to. But she also finds their small town in the forest risky and intimidating, hence her visit to the town on Sunday. "The problem," she tells me in the bar, "is that there is every kind of person at their town [in the forest by the river]—former soldiers, teachers, and even students on vacation." "All of them," she claims, are "bandits" (*wabandiya*) at heart, lacking manners and undependable, unlikely to forge ties and quick to move: "Most of them marry here and quickly have children, but then they disappear!"

Eventually, Felix, the supreme commander of plongeurs and dragueurs, calls the joint meeting of the plongeurs' and dragueurs' association to order. The main issue at hand is that one of the plongeurs has been accused by the local ANR of possessing a gun that he claims he found while underwater. The ANR is using this find to claim that the plongeur is working with an armed group, probably Mai Mai or the Hutu-backed Forces for the Democratic Liberation of Rwanda (FDLR) or both. The plongeur and his dragueur claim that this is ridiculous; they have a particular history with this ANR representative. He had been coming to the forest for some time, approaching the drag operator, who has been running the drag for its owner in the shadows, in order to borrow small amounts of money. After a while, these small amounts of money had grown into a large amount, which the ANR guy didn't want to pay. According to the plongeur and the dragueur, this is why the ANR officer decided to bring a case—to run from his debt (*anakimbia deni yake*). After some consideration, Felix proposes that, rather than paying the ANR, they should push the case back against him in order to clear it entirely and domesticate this particular ANR officer. This will mean bringing in higher-level authorities—probably a delegation of "higher" ANR from the city and local administration. Some kind of *kope*, or cooperation (money), will have to be paid, but it will settle the matter for good and keep this from happening again so they can continue their work. This proposal is voted on and everyone agrees. MP4—a young man named after a digital multimedia storage device because he knows so much music—raises the point that the state authorities have become more aggressive lately because they know the plongeurs have money. He recalls how, some time ago, he was stopped by police and "tied" (*kufungwa*) for a day, even though he had a digger's card and a card from the plongeurs' association. He ended up giving them $300, and upon his release from the jail, he was so mad that he went across the street, bought a bottle of champagne, drank a sip, and threw the bottle to the ground in front of the police ("crack!")—a demonstration of his disrespect for them and his nonchalance about money in general (for them, the display no doubt reaffirmed the commonsense understanding of diggers as being reckless, childlike spenders with no thought for the future).

In the end, a few of the association members suggest that it may be time to make another visit to the governor with a gift. Felix, the supreme commander, agrees but adds that, given all of these cases and the ongoing cold war with the Chinese robots, "We need our own parliamentarian [elected public official], like the [motorbike] taxi drivers in Kisangani got."

Concluding Thoughts

To this day, the memory of industrial mining extends far beyond the geographic limits of the old mining towns, influencing the terms artisanal miners use in their craft and the organization of their labor, the names of days of the week, culinary tastes, and the general sense of the kinds of futures that Congolese minerals make possible. In the infrastructurally remote province of Maniema, where the ruins of roads, bridges, and industrial towns are splayed out amidst the rainforest, these memories are all the more palpable and sharp. Because they attract highly contemporary, ever-changing people and forces—from armies to state officials to artisanal miners to the new category of semi-industrial miner known as the plongeurs and dragueurs—these places and the minerals that they contain are much more than congelations of competing and conflicting memories. Rather, these vortexes are active agents, even sources of unbridled energy, that people who live around them are compelled to come to terms with in various ways.

There may have been a time when the old company towns, enclaved from the rest of Congolese society, seemed able to contain all of that energy, channeling it and even slowing it down so to produce incremental, linear time, the kind that materialized in salaries and pension plans, a time without rapid price fluctuations and long closures. For many people, life after that time seems like an aberration—"living after development." But, as much as people living in these places like to talk about it, that experience of regular, repetitive time was made possible by repetitive violence and racial apartheid. Beyond that, it was also an illusion, since real value was being siphoned out of the country such that, to this day, if you really want to know what's in the ground you have to talk to some (possibly imaginary) professor in Brussels. Certain people—for example, the directors of Sakima—sometimes gesture toward the notion that the company is still invested in a project of long-term development in which past, present, and future are sutured together through the company and through the work of the company management. And older people recall a time when there seemed to be a moral economy forged by the company in collaboration with senior men. Most people tend to like the idea of the company in the abstract but see current iterations of the company (and the comptoirs that rent out space from the government company) as fake because they come and go so quickly. However, for quite a while now, the decaying structures of the company have been the retreating mise-en-scène for the most current and, often volatile, forces and events. The uncontained energy of concessions now explodes in here-and-gone-again wormhole mines like the United States of America, which seem to capture, for a moment, something of the essence of the faraway places for which they're named.

The old mining towns are not just fortuitous or symbolically potent backdrops for these new happenings, because the old companies have let loose the social material that makes the new social arrangements possible.

There is some kind of crazy dialectical magic in the daughter of the company manager coming back home with gasoline from the UN to finance a new generation of semi-industrial artisanal miners who have been kicked off their old mines by "Chinese robots." Perhaps even more in the son of a company engineer using his father's knowledge and expertise to make his way under new conditions, only to find himself transformed into his whole hometown's ur-symbol of the coming undone of the world—the very opposite of the old economic and moral order out of which he was born (All hail the self-effacing, heartbroken King of the Toilet—at once the King of Dirt and the one who makes soap and cleanliness possible—who can help turn the destitute into the rich and convince the police to withdraw). At which point it is almost trite to point out the fact that uprooted artisanal miners like MP4 give themselves the names of the digital devices their unrecognized labor helps to make without themselves realizing it.

For many people, these figures were symbols of collapse—or, more precisely, of the instantaneous expenditure of energy, resources, and opportunities. Lungwa 4, who bathed in beer every day and seemingly couldn't think of anything more important to do in the world than sleep with the governor's wife's sister in the governor's house (maybe there really wasn't). But, while the energy potential of the concession may not be as contained as it once was, these people were not the mushroom cloud of chaos and doom they represented for others because their work was channeled into futures as predictable as futures can be. They used the cell phones their work helped to make to send money back home to their parents and to pay for their and their kin's bridewealth—thereby making possible marriage, children, and their own future transformation into ancestors themselves. As always, houses were the main visible proof that all of this expenditure was actually materializing in something enduring and collectively useful (even if the houses weren't theirs). Moreover, they were building incrementally from the reservoir of their personal histories to produce a kind of regular, incremental temporality for themselves. Sometimes, as in the case of Lagome, they even helped build actual bridges.

In sociological terms, this drama of collapse, transformation, and rebirth took the form of generational and gendered conflict. For its former employees, the company was a paternalistic father that had fed its children well only to abandon them to a world of uncontrolled young men and loose women. All of its figures were men, who rested their legitimacy on agreements with other senior men, who belonged to a secret society run by senior men (the Bwami society). All these people tended to engage the artisanal miners as if they were uncontrolled children. But the collapse of the company, based on gerontocratic authority, and its transformation into a staging ground for artisanal mining and global supply chains, opened up an opportunity for youth and women to become more powerful and influential than they had been. The artisanal miners often took on the form of a motley youth movement, which often seemed to others like a revolt against the elders and of the social order—breaking up marriages, introducing new languages, and seeming to mock or dissemble established religions. The seniors in Chamaka clearly distrusted Lagome, for example, blaming him for the corruption of Chamaka's daughters. But the diggers' radical egalitarianism and reciprocity among themselves and their circulation of money were also regenerative, revivifying a region that had seen the collapse of the company and the theft of its future by invading foreign armies (their practices were also deemed to be closer to the ways of actual ancestors). These artisanal miners created a division of labor out of those displaced by war, and their decision to remain mobile and *not* buy homes was a kind of unintended self-sacrifice, allowing their money to circulate throughout the society so that others (especially the women shop owners and businesspeople) could build homes that endured and paint the ones they already had. When these diggers did become powerful, they also inclined toward organized state-like and even state authority—often taking the opportunities presented to them to supersede the state and become actually functioning state (or state-like) forces that were also antistate in that they defused the actual state.

The three chapters comprising this section have ranged fairly broadly throughout the Kivus and Maniema, drawing out major themes in the mining of the digital minerals—mainly issues having to do with different dimensions of movement. In part 3, which follows, all of these different themes emerge and come together in a new historical and geographic place that, I argue, was historically transformative, ushering in a new moment in the history of mining and "conflict minerals" in eastern Congo (and arguably the whole world), in which foreign mining companies with contracts from the central government came into direct conflict with the army of artisanal miners brought into being by war. This event, I argue, helped give birth to the US Dodd-Frank Act and the corporate-sponsored "bag-and-tag" tracking schemes that followed and whose implementation is discussed in chapter 9.

PART THREE

The Eyes of the World on Bisie
and the Game of Tags

FIVE

Bisie during the Time of Movement

Bisie had been the security of everyone here.

—Artisanal miner reflecting on Bisie before it was formally closed in 2010

Bisie means something close to "it will never end" in the language of the Bakumu, the people who live in the part of the Walikale forest in which it is located. A giant artisanal mine in the North Kivu rainforest in Walikale District—at one point one of the largest artisanal mines in the world— Bisie became regionally and even globally famous between 2002 and 2010. The place was a vortex that sucked in all kinds of actors, things, and forces, and it was ultimately made possible by cooperation and collaboration among diverse groups, many of whom had a tense and even conflictual relationship with each other. As a result, for many it became a model for peace, prosperity, and the good life emerging out of multidimensional collaboration in the context of dispossession brought about by war. But for the world outside of Congo, Bisie was, or became, ground zero for conflict minerals, and its history strongly shaped the international response to conflict minerals in Congo. Bisie was, among other things, a large hole in the ground—a particularly exemplary hole, one containing many holes— that drew in many people from all over to be transformed by it. Drawing from speculative physics and science fiction, I refer to it at various points as a wormhole because of its capacity to connect far-flung worlds and space-times. The wormhole is a physics concept that has had a rich life in science fiction, and I feel comfortable with the science fiction analogy because there was something "Afrofuturist" and "science-fictiony" about the way eastern Congolese were thinking about and engaging with new forces, many of them technological, during and in the years following the wars.

With respect to Bisie, my use of the concept reflects my understanding that people were coming together in and through Bisie not only to improve their lives but to improve their lives by altering and compressing space-time and that this compression was at least partly the outcome of their actions, desires, and intentions.

Enter the Wormhole

Going to the gargantuan and world-historic artisanal mine of Bisie was one of my first experiences venturing to a mine or really going any significant distance away from the eastern Congolese cities of Goma and Bukavu. It was also my first visit to the Congolese rainforest. And while I only visited Bisie mine one time for a little over a week (it became an illegalized mine soon after I left, making it more difficult to visit), I would return several times to the proximal towns and village of Walikale in North Kivu (Walikale town, Mubi, Njingala, Biruwe, Osokari, and others) for several visits of a couple of months at a time to interview people who continued to depend on Bisie for their livelihoods. My first visit, though, was in July of 2009, during what I considered preparatory fieldwork; I had virtually no Congolese contacts at the time (this was before I had met Raymond or Joseph), and it was years before I would put together a conference on artisanal mining in Luhwindja and begin to learn more about how artisanal mining worked. The year 2009 was also still a time of war in many places, as the National Congress for the Defense of the People (CNDP) insurrection had just come to an end, and a reintegrated battalion was stationed at Bisie.

I was in Goma with Ngeti Mwadime, my Kenyan friend and collaborator from my Kenya fieldwork, and I was looking for a future field site closer to the cities of Goma and Bukavu (we visited Nyabibwe and Numbi for example, places I had also visited with Jeff Mantz). I had heard from people in Goma that if I wanted to know about coltan, I should go to the rainforests around Walikale—no one actually said anything about Bisie. Somehow we decided, rather spontaneously, that this would be a good idea. To get there, we had to buy space on a cargo plane, so I had my first experience doing what would eventually become a pretty routine thing. I already knew people who had strong enough connections at the airport to get Ngeti and me on the plane, but we had to show up two or three times before the plane was actually ready to go. Once we got through to the plane, we waited until the Russian pilot with the querulous look on his face finally gave us the go-ahead to get in. Aside from one elderly woman sitting on a pile of boxes, we were the only passengers. After enduring the takeoff, I looked down in awe at the forest canopy that stretched below us, realizing for the first time how close this different world was to the city of Goma. When we touched down about half an hour later, it was on a road adjacent to the forest, with people on bicycles scurrying out of the way and other people selling bananas; naturally, we had no idea where we were.

Although it's only about one hundred miles from Goma as the proverbial crow flies, getting to Walikale continues to be difficult, with the route and mode of travel changing according to a host of circumstances, including road conditions (the main road connecting Goma and Walikale hasn't been passable by car for years, but there are roundabout routes), plane availability, and what eastern Congolese refer to as *securité* or *insecurité*, referring to the absence or presence of armed groups in various, changing locations. Knowing which way to go at any given time requires good communication networks with as many people in as many potentially relevant locales as possible (and, of course, cell phones). For example, on a subsequent trip in 2015, the government had shut down the cargo planes, apparently in an effort to reduce smuggling to Rwanda, and the only way Raymond and I could travel was by road with an ill-equipped Toyota RAV4 that we had rented: this meant driving 403 kilometers north from Goma to the city of Beni, then another 820 kilometers northwest to the city of Kisangani, then 200 kilometers southeast to Lubutu, and another 230 kilometers east to Walikale. Along this 1,653 km circle through what was almost entirely forest, we passed through five provinces (North Kivu, Ituri, Orientale, Maniema, then back into North Kivu), a national park that was home to at least one militia and seemed to be protected only by "scarecrow" soldiers (sticks with uniforms and helmets), a stretch of river where the bridge had collapsed, and over thirty barriers where various state officials collected tax and tolls. (Our headlights also went out, and at one point, we drove quickly through the forest at night on bad roads for hours, holding flashlights out the windows in front of us to show the way.)

In 2009 I knew little about traveling far beyond the city—I'd assumed that, with my visa, I'd be good to go wherever, and I had little knowledge of what state authority looked like in Congo's sylvan worlds. As soon as we landed at what I soon learned was locally referred to as the "International Airport of Walikale" (again, because of the many planes that flew back and forth, mainly to Rwanda), we were detained because we didn't have papers from various government offices in Goma: DGM wanted papers from the DGM office, ANR from the ANR office; someone else represented the Ministry of Interior, another the Ministry of Mines. They argued a little with each other about what kinds of stamps and papers were the ones we needed. All of this involved storytelling about the nature of the place—that it was insecure, that it was in the interior, that there were valuable resources there—and also what each state figure was for. Eventually, the state officials made me get motorbikes for all of us—they didn't have transport of their own—and we all drove together from the international airport to Walikale town, the seat of local government, almost an hour away (the state officials wanted to be able to talk to their superiors back in Goma).

Once we arrived in Walikale town, they were on their cell phones talking to their multiple bosses back in Goma, who insisted we had not been in to see them, which we hadn't (we never had any idea that we ever needed to). After a few hours of being threatened and going back and forth on the phone, they made us purchase what they and their superiors understood to be the necessary papers for being there. At some point, they also called up a Congolese soldier (FARDC) attached to the UN (Monusco) named Patrick, who quickly showed up; they seemed to think it would be good for us to know him. Major Patrick, a middle-aged Azande man from a neighboring province, had been stationed in Walikale during the Second Congo War, at which point he trained Mai Mai insurgents and also learned some of their techniques.[1]

As soon as we were done, we were all friends, and the various state officials looked up at us smilingly; one of them said, "You'll go to Bisie, right? You have to go to Bisie!" That was the first time I had heard of it, and they seemed a little disappointed.

Ngeti and I got a room at the Catholic church in town, where I came to be somewhat derisively known as "Bwana Mchuzi," or Mr. Sauce, because I once asked for a food that they associated with the Swahili coast and used a word for it that is uncommon in Walikale. Part of the message was that, unlike Goma, this was a domain and region that was part of a different ensemble of geographic relationships. The relations that connected Walikale with other places were also, I would come to learn, not all straightforwardly "geographic." Walikale was certainly connected to the rest of the world, especially through Bisie and other nearby mines, but in a way that was fundamentally different—specifically in the sporadic and explosive temporality of those connections, which differed from the long-term, long-distance connections in which terms like *mchuzi* were imbricated.

Ngeti and I started casually talking to people in town, and we slowly learned what Bisie was and a little bit about what had happened there. In general, the people we met in Walikale were happy that the people of Bisie (who exactly was not yet clear to me) had fought, seemingly successfully, against a company headed by whites who wanted to take their livelihoods away—the name "Bwana Yves" came up a lot, but at the time I didn't know who he was; most people weren't sure if he was Belgian or South African (they were talking about the company MPC, or Mining and Processing Congo). Not that all people were diggers or involved in the mining business, but whatever their livelihoods were, they knew they ultimately depended on the money earned from artisanal mining and, specifically, Bisie. Some said that these people had come with the "mentality of whites," bearing papers from Kinshasa and thinking those papers gave them the right to take the mine and close everyone out, even people who had been heavily invested (and indebted) in their mining businesses. These whites had been forced to leave, and now Bisie belonged to the people again. There was a celebratory mood, tempered somewhat by anxiety that those whites could return. Some were suspicious about who I was—was I one of them, one of the company guys? I, of course, told people I was an anthropologist, and I explained the concept, but I'm not sure to what degree it registered. We also learned that just a couple of weeks earlier a small film crew had come through (I think it might have been the television show *Vice*, because they came out with an episode about Bisie soon afterward, but it could have also been the makers of the movie *Blood in the Mobile*, which also came out soon after we returned). While we obviously weren't a film crew, there was some thought that we could be with a humanitarian organization. I remember getting the sense from people that film crews, humanitarian organizations, and the white people's company were somehow connected, but at the time, I didn't yet understand how, being inclined to see them as separate and even opposed forces.

At the time, the FARDC was fighting FDLR not far away—we saw FARDC troops driving out of Walikale town in trucks to places that we were told were not far away. We also learned that, about three weeks prior, the FDLR had attacked Bisie and robbed nearly everyone at gunpoint, then disappeared, apparently without harming anyone. There was some sense that the attack was "fake," that it was staged by the FARDC battalion that was stationed there so that they could continue to stay in Bisie (already there was some pressure from politicians and NGOs for them to leave Bisie because, according to Congolese law, the army was not supposed to be at Bisie). In any event, we were assured that there was no problem up at Bisie now.

As the days passed, Ngeti and I started thinking about how to get to Bisie—some people in a bar had told us it really wasn't all that far, and since we had a lot of experience hiking in the Taita Hills at fairly high altitude, we figured this wouldn't be that big a deal. Ngeti had REI hiking shoes that I had bought for him a while back, but the thought of going to Walikale and Bisie had just come up spontaneously, and I hadn't actually anticipated hiking into the rainforest on this trip or even going much further than Goma. Walikale had a market, and I was able to find one pair of boots that fit me. Sometime the next day around noon, we followed the directions of some people who pointed us toward the forest, assuring us that we'd have plenty of time to make it to Bisie by nightfall.

Ngeti and I made it perhaps a hundred feet before we arrived at the first of what I would later learn were four tolls on the way to Bisie. A group of soldiers, representing diverse battalions, and other state officials wanted to see our "papers" (really stamps on paper—my "papers" never consisted of anything more than a letter of introduction from a university department chair). Rather than the ones that we had just gotten from the airport, it seemed they wanted stamps from the army and the administrative chief. By this time, I had come to know Major Patrick a bit better, so I gave him a call. With Patrick, we went to see Colonel Manzi, a Tutsi colonel whose CNDP regiment had been integrated into the army; Manzi had just replaced Colonel Sammy, who had recently been deposed as the military commander at Bisie. I had learned that Colonel Sammy's 85th battalion had been in charge when the guys from the white-owned company, MPC, were attacked, and Patrick explained to me that he had helped MPC during that time, when they were challenged by Sammy, but I didn't understand how. After some discussion, Colonel Manzi signed my papers without asking for any payment, and in a few days, Ngeti and I started out for Bisie with a small group of others. Patrick, who wanted to check up on how work was proceeding at his hole, decided to accompany us as well (at the time, people said that anyone who was anyone in Walikale had a hole at Bisie—certainly soldiers and state officials).

As we prepared to leave, I developed a slightly more realistic understanding of what it was like to get to Bisie: a thirty- or nearly fifty-kilometer hike (depending on the route) over difficult terrain on an at-times barely visible path in thick rainforest. I learned about how people sent money to Bisie before they started off so that it was waiting for them when they got there; without such trust-based practices and procedures, Bisie probably would not have been possible (this was before M-Pesa and other digital money-transfer systems had begun operating in this area). Before entering the forest, a middleperson would bring the money they wanted transferred, along with an additional 10 percent, to the owner of the cell phone, who communicated with his or her partner in Bisie, who would then give the money to the traveler once they had arrived at Bisie. I was wary of this system, though, so I just shoved $5,000 in an envelope under my belt. Everyone had said that $100 in Walikale was like $10 in Bisie, so I knew prices were inflated there, but I didn't yet understand just how much.

The next day, Ngeti and I went to various stalls in Walikale and stocked up on water, canned corned beef, canned sardines, and a ginger-based beer called simply Tangawizi (ginger) that diggers liked for its rejuvenating qualities, and which we also liked. When the day came to leave, we set off first thing in the morning with Major Patrick. Leaving early was important because the forest was not safe at night—there were FDLR, Mai Mai forces, and random armed actors. We took motorbikes to Biruwe, the jumping-off spot for entering the forest. But, as was always the case in any kind of trip like this, we ended up being delayed by various state actors from Biruwe, so we didn't get quite the start we wanted.

I hadn't realized that, once we hit the forest, we would practically be running, through thick brush and over streams, with no time at all to stop and observe our surroundings. The forest was beautiful, and darker than I had expected, thick with mushrooms and noisy with the sounds of monkeys and insects. I was told most of the animals had "left." As we walked, we joined in and were joined by other people along the way, and the atmosphere was one of mutual encouragement, interspersed with occasional warnings—"Watch out for Ekonda!" "What is Ekonda?" "It's an insect. If it urinates in your eyes, you go blind. On your head, and you'll be bald." For me, the hardest parts were the slippery, rain-soaked log bridges—there were two long ones, one of which stretched over a dramatically deep and rocky ravine.

Then, about halfway through the journey, my feet went right through the cheap boots I had bought at the Walikale market, and I was barefoot. Someone lent me a pair of gumboots that were about two sizes too small, and I suffered through them for the rest of the trip. Patrick used my camera to snatch a video of me crossing slippery logs, and he provided politicized running commentary about it: "This road of ours is very hard for the white man. . . . The white man is shocked by the conditions in which we are forced to live by our leaders. . . ." A couple of times, lone women appeared, smelling of perfume and dressed as if they were going to a nightclub (I remember it was Ngeti who used that term because he thought it was funny that we were going on this very long hike through the forest to potentially arrive at a nightclub).

On the way, we were also passed by groups of porters walking quickly past us in both directions. Being a porter was the easiest job to come by at Bisie, and it was the way many men who had no money to start with generally broke in, often with the hopes of moving up the ladder to become merchants or hole owners. An average of two hundred people, each carrying 120 pounds (50 kilos) on their heads over more than forty kilometers, set out from Bisie to Njingala each day, with probably the same number moving in the opposite direction. People told me that about one of these porters died every week, usually of stroke; some also suffered from compression of the spine.

While everyone said that I had to see Bisie, when we finally got there after what was pretty much the hardest thing I've ever done in my life, the sound of it struck me more than the sight. As we got closer, we could hear what sounded like rushing water—the hum of people working. Up and around a bend, peer down, and there it was—a town in the forest. I wasn't able to see the ballot boxes that were put in for the 2006 election, but I immediately caught a glimpse of the satellite dishes that people had carried in through the forest, in pieces. We descended the hill and went through one more small swamp until we were in an area with deep pits with many diggers in each one. People called out to us—the atmosphere was friendly and jovial. Happy to have finally arrived, I relaxed my guard a bit and didn't think anything of the large puddle that lay before me. I stepped into the water and fell like a stone into what turned out to be a mining hole, twenty feet deep. I was underwater for a while, and it took a couple of guys to help me get out, and they had pretty concerned looks on their faces when I finally did alight.

Once I was dried off, we had to sit for several hours in a small government tent while people tried to figure out who my Kenyan friend and I were. These people were mainly soldiers, and they wondered if I might be a representative of the ousted company Mining and Processing Congo (MPC)—the fact that Major Patrick was in my group added to that suspicion. If not MPC, a rival company? A humanitarian NGO? What exactly did I want? An anthropologist (*anthropologue*)? What is an anthropologist again? Oh, yeah . . . OK. . . . Eventually, we made it through that station, but it was a couple of days before all the state officials at Bisie were done with us. Our first stop was the Bagandula, who also stamped our papers (they had their own stamps, from the Bagandula *groupement*, or grouping) and presented us with the story of how they had come to discover Bisie and all the trouble it had brought them. Their voices rented the air via loudspeakers as they warned people not to accept just any papers from anyone for renting holes ("Not all papers are true!" the voice intoned).

I soon learned that Bisie was actually a complex of sites discovered at different times. From a single mine, called Manyore ("black stuff," from the French *noire*, referring to the high-value black cassiterite there), the area had grown to include at least four main mining areas: the ever-expanding Marouge ("red stuff," referring to the red color of the cassiterite there) and the mines known as Forty-five Minutes, Fifteen Minutes, and Ten Minutes, so called because of the distance it took to walk from them to the "camp" where most people lived, near the river, which here took the name "Bisie."

The process of getting to Bisie was not just a big deal for me, but for a lot of people—its distance and the difficulty of getting there was for many a kind of rite of passage, allowing entry into a community of people who shared in the experience of what Bisie was and who felt they had experienced something special through it. This collective effervescence and other-worldliness are, I believe, what people really wanted me to see there. The power of Bisie was augmented by the tolls one had to pay to get to it, and the system of tolls gave a sense of how diverse state authorities collaborated and negotiated their respective rights and responsibilities from the bottom up. I was told that this system didn't come from on high, from Kinshasa or even the province, but had been negotiated over time by state and non-state actors on the ground based at least in part on their understandings of the purposes of certain kinds of state and state-like officials, and what they were worth in the context of where they were (although, as I unpack the history of Bisie in the chapters that follow, a somewhat different picture emerges about how tolls came to be staffed in the way they were). I learned that the toll system actually began with the RCD occupation around 2002 (they had come to Bisie briefly) and was carried on by the Congolese state authorities who inherited it after that period ended, although the composition of personnel collecting tax changed somewhat over time. Again, there were four separate tolls on the path to Bisie: the first at the entrance to the forest from the roadside towns (Biruwe and Njingala); the second in the middle of the forest, where paths diverged to two separate mining zones at Bisie; the third at the entrance to the Bisie market area; and the fourth at the entrance to the main hill (Mpama), which was the site of the richest cassiterite mine (Manyore, or black stuff).

These sites were strategic, and they were also symbolic. At the entrance to the forest, people paid tolls to the military and all the other services of the state. This acknowledged the fact that, at that boundary between "ordinary" society and the forest, the state had two major dimensions, the "government of paper," or documents, and the "government of guns." Each of these was legitimate, although the perceived importance and power of the government of guns over the government of paper was clear from the greater percentage of revenue they received at this toll—the military receiving half, while the governments of paper split the other half among themselves. In contrast, the toll in the forest was 100 percent military, reflecting not only the armed power of the military but also the fact that, in the forest—a place of danger and "insecurity"—the military was the only de facto state authority, and the government of paper was neither present nor relevant. At the entrance to Bisie, the border logic of town/forest repeated itself once again, with the government of guns and the government of paper "equally" represented once more. Finally, there was a toll at the border to the main hill (Mpama), where revenue was shared between two authorities only: the military and the autochthones (the Bagandula). The logic here was that the Bagandula were the "children of the hill" and its rightful owners, as well as the mediators between the living and the dead, and were therefore the final say on individual and collective production yields at Bisie. According to this logic, military figures received toll fees here because the mine was also a valuable natural and strategic resource that had to be protected. And so the revenue collection system at the hill, the main mining site, also reflected the different qualities minerals have at different moments—they are substances in the forest, and they are part of the forest, which is presided over by ancestors, but once they emerge from the ground, they are a potential commodity that must be protected as a resource.

At different moments in Bisie's history, other actors were included in these tolls, reflecting the changing ownership of the mine, which increasingly came to be a commodity in itself, as powerful interests wrestled for the papers to the place. For example, there was a period when a comptoir was included along with the military and other state authorities, reflecting the fact that they had, they argued, acquired exclusive buying rights to the mine, which they hoped to one day own, ostensibly in partnership with the Bagandula. Later, after a combination of actors expelled the rival company MPC from the mines, new and different military authorities would be included in the toll system, as Colonel Sammy struggled to hold onto his position against increasing pressure to quit and hand it over to the company MPC.

Movement at Bisie

Having passed through the barriers on the way to Bisie, we could now experience the movement that had made the place famous. As discussed in chapter 1, the concept of movement had a few major senses, and at Bisie all of them clearly came together: the literal movement of people, things, and money across great distances; the totality of energy and activity that was created by this movement when it attained a critical mass; the unpredictable consequences of this totality (undesirable things, conflicts, and people come in, creating "disorder," which can sometimes put an end to movement) and the uncertainty created by it; and, finally, the capacity for self- and collective transformation, through which certain types of things and people are converted into other types of things and people, for better or worse (e.g., a porter becomes a PDG, a négociant becomes a militia leader, cassiterite becomes food becomes cassiterite, flashlights become a house, a group of women traders turn into packs of wild dogs at night and attack their competitors, and an unknown place in the rainforest becomes a global city connected to the world through satellite dishes and to the nation through polling booths).

The Office of Mines representatives at Bisie, as much in awe of the machine of artisanal miners as anyone, gave us some figures, some of which I supplemented over the years through interviews with other Mines officials who had been there at different times. They pointed out that, in the year 2004, still the time locally known as *zola zola*, or easy grabbing, Bisie's workers produced eighty thousand tons of cassiterite (they reckoned it was actually much more because of how much unregistered coltan was smuggled to Rwanda). To put this in perspective, the Office of Mines officials showed documentation indicating that the industrial mining company Sominki had extracted thirty-five thousand tons of cassiterite in the entire North and South Kivu provinces over the course of the company's thirty years. That means that these artisanal miners, working by hand and without a production plan or schedule, extracted more than twice as much material in one-thirtieth of the time in a single mine. Because of the volume of production, events at Bisie following its discovery in 2002 have had a direct and dramatic impact on global markets, with news of fighting in the region or of a shift in military command at the site leading to global price fluctuations in tin of as much as 40 percent (Magnowski 2008).

While some estimated that there were twenty thousand people at its height and that the population of diggers was around twelve thousand, it is impossible to get an accurate number, especially because of the high mobility of the nondigging population of porters and traders. At its peak, Bisie had about eighty large holes, each with a single owner, often financed by outside investors and employing between 100 to 140 people. There were approximately five hundred smaller holes, mostly spread out in areas other than the main large work site known as Manyore, or black stuff; these holes employed an average of closer to forty people each.

Extraction was only one element of the work that took place at Bisie—the other main jobs were the buying and selling of all kinds of food and goods (everything from plastic bottles of water to cows, all of which were exchanged for minerals, as traders moved among the many holes and zones at Bisie hawking their wares); moneylending (also carried out at holes); the transportation of money and goods over the forty-kilometer forest trail (give or take); cross-cutting interventions, by state and nonstate officials, ostensibly aimed at regulation; and different forms of security and conflict-dispute resolution (some organized by the diggers' cooperatives and others by the mining police or the army, with different systems prevailing at different sites and different times). There was a great deal of social mobility, an aspect of movement, between these groups, and it was common for someone to go from being a porter or digger to a hole owner (later I learned from state officials in the Ministry of Mines that between 70 percent and 80 percent of hole owners had started off as either porters or diggers at Bisie or somewhere else).

In addition to the rate and extent of extraction, what was stunning about Bisie was the availability of "exotic" commodities from far-flung places that you couldn't find elsewhere, even in more traditionally urban locales—chocolates from Belgium, beer from Germany, meat from Brazil. This diversity of commodities reflected the diversity of Bisie's population. There were two main types of merchants: those who had purchased land at Bisie from the Bagandula and owned shops at the central market area at the base of the main hill and those who came to Bisie to sell something specific and then return, usually with a view to doing it again. The latter group was large, highly mobile, and nearly impossible to count (it seems that no one ever did). There is little doubt, though, that this mobile population contributed more to their families back home, and to the larger regional economy, than the lower-level diggers, who ended up spending much of what they had right there in Bisie. These people also earned more consistently—while miners' fortunes were up and down depending on production (particularly, what phase the mine was in at any given moment), restaurateurs and bar owners earned consistently because, as one woman restaurant owner put it, "I always eat because everyone has to eat."

While "movement" in the form of extraction and transportation was almost exclusively the work of men, "movement" in the form of circulation and the conversion of things and qualities into other things and qualities was almost but not quite equally the work of women: perhaps as many as 30 percent of Bisie's traders were women. For a long time in the life of Bisie, women were also involved in the "washing" of minerals, the first stage of refinement, but the army restricted them from going to the river that bordered the market town and the main hill, Mpama, when NGOs started coming to take pictures of women and children digging at Bisie after the Battle of Bisie in 2006 (there had also been a time when women were involved in digging and some continued, with some difficulty, to rent holes; see subsequent chapter). The army was responding, in their military way, to the threat of closure posed by the "international community." Despite this formal exclusion from extraction, women's work provided an important source of credit to miners, as women sellers gave food and fluids to the hole owners on loan, who in turn provisioned their workers in lieu of cash.

Bisie was a prime example of the regional aphorism that, if you go to a place and find that most of the people are not originally from there, you can assume they have been displaced by war. By some estimations, about 90 percent of the population consisted of people who had lost land and property or had been separated from their families during the war. These statistics are impossible to determine for certain, in part because, as mentioned above, while some of the population of Bisie was relatively stable, a larger number was itinerant, moving back and forth to sell things to the people digging at Bisie. But it is certainly not a stretch to say that people went to be transformed, even reborn, by the forces and potential that materialized suddenly at this mine. While outsiders associated it with conflict, that certainly wasn't the main theme for people at Bisie. For one, people (including women) generally saw it as relatively secure because it is a very difficult place to get to: it was hard for "enemy" groups to reach, especially since soldiers controlled the pathways to the mine and provided security at the mine itself.

Some holes became famous as sites of rebirth, where people went down into the ground to tangle with the forces to be found there and came out anew. These were holes whose managers were able to successfully navigate the difficult work of organizing and subsidizing labor, acquiring loans or profits sufficient to enable that work to continue when it got harder, paying off multiple state and nonstate authorities, and securing a good price from outside buyers. The fame of these holes circulated far and wide through names like Safina (ark, as in Noah's ark), and Maternity, as well as Modern Times and Modernity. It hardly bears mentioning that Safina suggests doomed people being rescued from a storm, and that Maternity implies that the mine nourishes and raises up its children, while turning former strangers into kin. As one former digger at Bisie put it, "There's nothing like the love that us diggers had for one another, because we were together in a time of trouble and if one of us got a problem, the rest of us helped in any way we could." Names like Modern Times and Modernity were explicitly intended to convey the fact that these dynamic mines had put this remote place at the forefront of the world.

Later in its history, Bisie as a whole came to be known as "the Arab world," partly because of the multiple conflicts over property and authority and the competition among certain companies trying to possess it, but also because of the great significance that was attached to it by the largely dispossessed population that now lived there. At one point, this name attracted UN forces in Goma, who for a time believed that Al Qaeda might be using Bisie to earn and launder money. They sent a delegation there to investigate but never came up with anything useful about Al Qaeda. The name "the Arab world" suggested something else important about Bisie, though—that the place and the seemingly wretched people who made it were so powerful that they reached out to the world, grabbed it by the collar, and pulled it in.

Many people were deeply affected by the collective effervescence produced by movement, and they had faith in Bisie's capacity to transform them. For example, I got to know a twenty-nine-year-old Tanzanian man who claimed to have come to Walikale from Dar es Salaam with $17,000 in cash he had borrowed from his politician father to attend university in Canada. He hadn't seen or spoken to his father in two years, but for all his father knew his son was well on the way to earning his bachelor's degree in physics at the University of Montreal. Instead, the son had given all the money to négociants in Njingala, the main market town you reach before entering the forest, to go to the forest and buy cassiterite. The négociants disappeared, so now he had climbed up to Bisie, where he was trying to recoup his losses by digging. I was overwhelmed by how happy and optimistic this Tanzanian who had just lost $17,000 was, how sure that it would return to him in a matter of months, even weeks.

You wouldn't know it from the reports about it, but people at Bisie weren't just digging and making money—they sat around and talked with each other, watched movies at small movie houses, drank in bars, went to one or more of the many churches. They talked a lot about the future of Bisie—everyone was impressed by the multiplicity of people there and by how far they had come (Kenya, Senegal, Nigeria). They were proud of the name "the Arab world" and even proud that the UN had come looking for Al Qaeda. I suspect some were, if not proud, at least impressed by the dog-witches and the prices of things—the originality, fame, and extremeness of Bisie. (Male diggers insisted that women traders, who competed with each other in a friendly way in the new market town during the day, turned into wild dogs at night and attacked one another. This was blamed not so much on them, but on their recent overconcentration in the market center because, before they were relocated, they had been spread out around the different mines and competition was less intense; more generally, it may have reflected the idea that women in mining towns had been "unleashed" from patriarchal control.[2]) But what they were especially proud about was the fact that all these people from all over got along and, for the most part, didn't fight with each other—that they all found a place for themselves at Bisie regardless of who they were or where they were from, and in many cases they were positively uplifted by it.

In the churches (there were about fifty), the pastors talked about God's plans for Bisie, and what they had in mind was that collaborative, peaceful cooperation would lead to a kind of utopian city on a hill (there was always the idea, though, that things could go one way or the other depending on the behavior of people at Bisie—like all wormholes, it was fragile). I remember getting into a friendly argument about evolution that was, at its core, about the moral implications of implying a basic sameness between humans and animals. For my digger interlocutors, God had a plan for humans, and the power of Bisie to resurrect the lost and give them a new life proved that to be the case. Among the enemies in this story—in some versions, agents of the devil—were the ecological conservationists (NGOs), who violated God's will by trying to close people out of the forest.

Movement, though a source of salvation, could also be a kind of trap, especially for those who were stuck in their holes, depending on minerals for their livelihoods. In part because of uncertainty about the future (Would MPC return? Would the diggers be kicked out if they did?), people were not growing food or building schools (at the time I was there, there were no schools). The refusal to grow food also had to do with the fact that, when minerals were moving, people wanted to take advantage of that and "didn't have time" to stay and grow food. Obviously, this was one of the factors that led to the increased cost of food. Meanwhile, a moving population took advantage of the population of diggers, who often used ore as currency and sold at a rate far below what it was worth, even in local terms. The direct exchange of minerals for food and other goods led to an extreme on-site devaluation in the price of minerals, putting diggers at the mercy of négociants. For example, in 2009 one can of beans exchanged for three kilograms of cassiterite, which in turn sold for $6 in Bisie. Meanwhile, the price for a can of beans in Goma at the time was 80 cents. Paper, which was crucial to the innumerable daily transactions and contracts, was even more egregious: four sheets of paper exchanged for one kilogram of cassiterite, or $2 (the price that value of cassiterite fetched at Bisie). An entire box of paper containing five hundred sheets exchanged for 125 kilograms of cassiterite, which in turn exchanged for $250. Meanwhile, a box of paper in Goma was around $4. It's no wonder that the counterfeiting of US dollars was at one point an underground business at Bisie. Moreover, since people were moving back and forth all the time, no one actually knew who owned what. As mentioned, the broadcast towers never stopped advertising the uncertainty: "Not all papers are true! Not all money is real! Make sure you go to the Bagandula before buying land so you have the right papers!" With the counterfeiting of money, this uncertainty entered into the buying and selling of minerals, undermining trust-based relationships, the first prerequisite for movement.

Some diggers were stuck because of investments they had made in the land or because digging tied them to one place while they consumed highly priced food, water, medicines, and alcohol. These people were also more vulnerable to various external shocks, from price dips to closure, than were the hit-and-run entrepreneurs who were able to "flexibly accumulate" in response to changes and contingencies. And it didn't help that digging was getting more difficult and more expensive all the time. Many did acknowledge the need to find some way of transitioning out of mining to another economic activity. Others were more cavalier, though,

Figure 12. A section of Bisie mine in 2009 (North Kivu)

clearly feeling empowered by all they had done already—they lifted up their hands to the vast ocean of forest that surrounded us and said, "When we're done here, we'll just go over there, and take down all of that." Others talked about planting food, about bringing in machines, and even about sharing the space with a responsible company that might hire some people, work alongside artisanal miners, and help them by establishing a bank and/or grocery store (mechanisms for generating incremental time), so their money wouldn't just disappear. What most seemed to hope for was long-term collaboration with powerful others, the kind of cooperation for which Bisie was already a model—anything but being closed out of the forest completely and plunged into debt, forced to go elsewhere or nowhere.

The Eyes of the World on Bisie

Several people told me that the "eyes of the world" were on Bisie—it's where I first heard the phrase—and a couple said that everything that happened there had to be understood in that context. At the time, I didn't really understand what they meant. What I hadn't realized was that Bisie was already well on the way to developing an international reputation for war and conflict minerals. In the three or four years after the Battle of Bisie in 2006 (described in a subsequent chapter), Bisie's fame spread far beyond the world of Congolese artisanal miners and traders. As already mentioned, the television show *Vice* did an episode about Bisie. A Danish filmmaker made a full-length documentary about the mine called *Blood in the Mobile*, which had such a profound effect that some Congolese NGOs linked it to President Kabila's decision to close mining in the east in 2010. The first two chapters of Kevin Bales's excoriating diatribe on "modern slavery" are devoted exclusively to Bisie (Bales 2016). There have been hosts of NGO reports. However, the more that has been said about Bisie, the less is known, because everything that has happened there has been seen through the prism of "blood minerals," "illegal actors," "dangerous artisanal mining," "failed states," and forced labor at "the barrel of a gun." Today, this demonizing storytelling about artisanal mining is foundational to the self-presentation of the mining company that is there (Alphamin—more on them later), who profess to see themselves as an engine of salvation for the region.

Kevin Bales, a "professor of modern slavery" who coauthored the Global Slavery Index and cofounded the organization Free the Slaves (an NGO that was active for some time in Goma), has one of the most dramatic takes on the place. For him, Bisie in the time of artisanal mining was the heart of darkness, every day spilling out all the horrors of the world and the soul from a hole in the earth. The workers there were "trapped in this Hell" of "rape, slavery, and murder" from which they could not escape, bonded as they were to military overlords (Bales 2016, 49). Over the years, Raymond and I questioned many people who had been there about slavery at Bisie, and we never had anyone agree that this existed; they did point out the intermittent use of the salongo system of short-term forced labor by various military and state officials, often adding that this colonial-era practice affected the hole owners more than the lower-level diggers. Rather than seeing the multiplicity of people at Bisie as generative of peacebuilding, Bales sees only criminals of the highest order; indeed, for him everyone at Bisie was either a slave or complicit in a heinous crime:

> Equally guilty [as the soldiers] are the parasites feeding on this crime—the officials, tribal leaders, corrupt police, moneylenders, brothel keepers, and mineral dealers—who levy taxes, harvest slaves through false arrest, enslave through debt, trade in the flesh of children, and buy and sell the ore in full knowledge of its ugly human cost. (2016, 52–53)

It is revealing that one of the main victims in Bales's account is the foreign mining company, MPC, which he portrays as the rightful but helpless owner of Bisie, egregiously dispossessed of their property by armed thugs: "MPC still owns the legal right to Bisie, but all they can do from their office in Goma is watch a constant relay of old Russian airplanes smuggling *their ore* out of the country" (51, italics mine). It takes a staggering amount of faith in a company to accept without any skepticism or critique such claims to ore and land, especially given the fact that even the state's ability to grant land ultimately goes back to the land policies of the colonial period (and, according to law, when the site in question is on the land of indigenes, even the ability of the state to grant land is attenuated).

What exactly was going on here? Why this extreme difference between the stories people at Bisie, and Walikale as a whole, told about it, and the globally famous story that would come back to determine the future of Bisie? What forces underpinned such staggeringly different understandings of war, peace, and prosperity? To get at this, we need to take another look at the history of Bisie and how the efforts by state and corporate actors to create a mining enclave depended on a reframing of artisanal miners as violent actors—even though, for those at Bisie, the exact opposite was true.

SIX

Insects of the Forest

Having assembled together in the compound of a member of the Bagan-dula lineage, the oldest man there, who has never laid eyes on me before, looks up at me and chokes, "Why did you close the forest?!" He can barely get his words out he's so angry. His sons, seemingly embarrassed, try to tell him that I'm not with the Canadian mining company Alphamin, which has purchased the papers from MPC and its mining rights, but the senior man doesn't back down immediately. "People think we're just insects of the forest," he warns, but if they communicate their anger to their ances-tors (the *mababu*) in the proper way, with the right rituals of communica-tion and the proper offerings, my machines won't do me any good, and I'll be crawling back to them looking for help because "we" won't be able to get anything of value out of the ground. Once he calmed down, the old man went on to explain how closing them out of the forest was a long-term state effort, begun during the colonial era, to seize the very source of life. In the old days, the authorities had tried to do this by literally going into the forest, pulling them out, and relocating them to the roads where they could be seen and taxed. He explained the long history of the colo-nial gaze: coming to the road, he said, "brought us into the light and away from darkness, exposing us to things we had never seen before." But it also meant that they were subjected to an oppressive and violent regime predi-cated on visibility: "When the road came, that is when we got the stick and the tax. That is when we started having to carry white people around. Even now, if you sell bananas, they see you. If you trap and sell animals, they see you. If you fish, make palm oil, anything. Everything has a tax."

Nowadays, being closed out happened through conservation initiatives and extractivism, which for the Bagandula were alike efforts to dispossess them from their world. The government's and the company Alphamin's closure of the mine at Bisie was especially hurtful because it was the Bagan-dula who had set in motion the events that would bring the company to Bisie. While his sons tried to separate me from the larger colonial project of closing the forest, the young men backed their father up on his recognition of dispossession and the magnitude of the problem. As one of them put it, "The forest was our entire existence—wildlife, farming, minerals, ancestors, trees. Now it's like we're in the jail at the Hague. You're sort of free, but re-ally you've been imprisoned."

I can't stress enough that Bisie, one of the largest tin mines in the world, a place whose movement and activity has reverberated throughout the world on registers financial and ethical, came into being because of the intentions and actions of this family of farmers and gatherer-hunters with little or no formal education who were painfully aware of the fact that oth-ers saw them as "insects of the forest." As they put it to me, they had seen an opportunity to create "development" (*maendeleo*) on their own terms— roads, electricity, and hospitals—by using their historical and spiritual con-nections to earth and to the past to their advantage. At last, roads would no longer be something that oppressed them, a tool for government to tax and watch them, but something that they controlled and benefited from as the guardians of custom, the owners of the hill, and the mediators between the living and the dead. The irony is that now they can only sit and wave when Alphamin's trucks pass by their compound on the way to Bisie on a road that the company has built on top of the old path that the Bagandula have long used to reach Bisie. Their efforts to continue and expand their way of life by opening up Bisie to the world on their terms have ended up closing them out, probably for good.

When, during the coltan boom, these people set out to look for coltan in the forest, they felt they had little choice but to pursue the path of min-eral exploitation: as they put it, the war had "ruined the forest," and the animals that they once hunted with ease had either been killed or moved somewhere else. (They liked to think they had moved somewhere else, and some stuck to this idea, but no one was really sure.) Regardless of who dis-covered the cassiterite that would bring roughly twenty thousand people to their place (and I describe the controversy around this below), the Bagan-dula family helped promote it to the world, and they actively set about developing connections with powerful figures, local and international, to protect their interests against state authorities who would eventually try to exploit them and drive them from their patrimony. One might say that, in reaching out to the world in this way, they ended up inviting the wolves that would pit them against one another and ultimately devour them, but that would neither be fair to wolves nor would it give the Bagandula the credit they deserve. How did they come to arrive at the point where they are now, closed out of their own forest in a way that remains inconceivable to them, on a scale that makes colonial-era efforts to drive them to the road pale in comparison? The answer is not simple because, while the end result was dispossession, the circumstances and cross-cutting intentions that set this chain of events into motion were not only multiple—they were surprising.

Patrilinies, Precedents, and Peacemaking

For years, a Kumu clan named Bassa lived and hunted in the area that includes but is not limited to the hill now known as Bisie. While they seem to have lived there for some time, I never was able to determine for how long, but I was informed that at some point, probably in the 1920s or 1930s, the Belgian colonial government recognized the Bassa clan as the customary authorities of that area, and it subsequently came to be known as the Bassa grouppement. Most people don't recognize the Bassa claim as having originated in colonial times, however, and instead simply refer to them as the traditional authorities of a large swath of territory.

Later, perhaps in the 1940s, a Kumu hunter named Olema Mtokotoko, the head of the family now known as Bagandula, came to the forest to settle there from Loso. The story has it that Olema was looking for a place to farm and hunt, probably because the colonial government was relocating people to the roads and "closing them out" of the forest. When he arrived at the Bassa grouppement, Olema came to an agreement with the Bassa clan that he and his family would live in a part of the forest that the Bassa were not using. In exchange, they were expected to bring a portion of their harvest or hunt to the Bassa lineage, in accordance with colonial customary law, which most Congolese simply recognize as African tradition. Over time, more people from their clan came over from Loso, and these people and their descendants began extending into other areas of the forest to farm and hunt. They also intermarried with Bassa, and eventually they came to be interconnected with one another as affinal kin. The hill that is now known as Bisie was located between the area that had originally been given to Olema and the area where the Bassa clan resided. The river that passed through there made it a rich and desirable place for planting, and years later the same water would be used to dig and wash minerals.

In 1978, during Mobutu's presidency, the Bassa filed a grievance against Bagandula with the authorities for the district, claiming that they ultimately had a right to the hill and that the Bagandula were merely guests. There was at one point a violent conflict between the two descent groups, at which point the term *Bisiene* (not Bisie) stuck; at the time, the name referred to the idea that the conflict between the descent groups would never end, although it now refers to people's hopes that the mine will never stop producing minerals. After conducting research among those who lived in the area, an official tribunal decided that the Bassa were indeed the "traditional chiefs" of the area according to colonial and postcolonial law, but Bagandula were "customary guardians," since they lived and died there, and their ancestors also dwelled there. According to this understanding, Bagandula possessed a territorial and "ontological" claim to the area (by which I mean that they were connected to the place through the dead), while Bassa's claim was tied up in the history of colonial administration. (I also came across a handwritten document, dated 2009, stamped by a number of chiefs and signed by multiple "clan" representatives, concerning the conflict between Bagandula and Bassa at Bisie, which employed similar terms: it said that Bagandula had to give the Bassa respect and "his rights of traditional sultan-ship" [*haki yake ya usultani ya asili*] and that the Bassa had to in turn recognize that Bagandula were the residents of the forest "because they sacrifice there" [*kwa sababu ni yeye anatambikiaka pale*].) The tribunal reiterated that Bagandula were expected to give a portion of their hunt over to Bassa, who were understood to be their "seniors," as if Bagandula were the junior siblings of Bassa.[1] Afterward, the conflict was declared to be finished, and the peace was cemented by a marriage alliance between Bassa and Bagandula. Bassa offered a wife to the family of Bagandula, and the couple bore a child, who they named Esa Yongo, meaning the peace bringer, because they hoped his birth would create collaboration between the families.

While this agreement was cemented through local practices and relationships, it was backed by the power of the state under Mobutu, represented by the official authorities of custom at the level of the district and the province. To this day, those who possess the official documents created during this time, or their copies, treat them with great care and reverence, keeping them in secret parts of the house, and presenting them to visitors with a combination of caution and drama that may seem strange to an outsider, since these papers have long been superseded by the new documents of new regimes. Indeed, people show the same reverence for all of the many documents that I will bring up in this chapter, regardless of whether or not they would seem to have any real power or relevance now. Each signature and stamp is discussed, and its history rendered—"See, that was signed by so-and-so, and I saw him do it, face to face!" "And that one—do you recognize that one?!" There is a whole world of something in this, a recurring ritual that recalls the original ritual when the documents were made. In these moments, ephemeral existence becomes concrete, visible, and permanent through documents of different orders and types and through signatures and stamps that were backed up by real people who may be gone now. All this is proof that others, who were often remote and powerful and recognized by even more powerful others, in turn recognized these humble people's significance in space and time and their command over some piece of the world and of all the times and happenings that came together in that piece. At that moment when that stamp was fixed on that document, these people, who hadn't always known or respected each other, were together, in agreement, and they expected that this recognition, and the collaboration and collusion that undergirded it, would last forever through the power of the documents and their preservation by living people and their descendants.

This is what makes it so painful when all that beleaguered ritual and history gets swept away by another set of unrelated documents "from high" (*kutoka juu*), from Kinshasa. Why should these arrangements made from on high be more important than the documents that come "from below" (*kutoka chini*), from people who were actually there, who had actually done the work of finding out what had taken place, who had lived and died there, and who knew what had been said and promised by those long gone?

Anyway, despite the peace that was achieved in the 1970s, the conflicts between the families never disappeared, and the presence of gold in the forest exacerbated them. Both Bassa and Bagandula had known about gold in different parts of the forest, and at Bisie, for a long time, at least from the 1970s and probably earlier. But the only people who could legitimately mine from the colonial period until 1982 were the state-sanctioned mining companies, which had stopped mining in the 1970s. Anyone who did mine artisanally did so at great risk, and so while they had occasionally traded in gold, it was never all that significant in comparison with hunting and farming. Even after the liberalization of gold mining in 1982, they never sold a great deal of gold to others, because there were plenty of places in the forest to get gold, especially in the open deposits that were eventually abandoned by Sominki, which still held these areas as part of their concession.

After the ousting of Mobutu in 1996, the presence of Laurent Kabila's army and, two years after that, the Rwandan-backed RCD and various Congolese Mai Mai groups greatly expanded the market for gold and diamonds and the frenetic activity around it. Meanwhile, the armies further eradicated the local animal population, making the herds of elephants, leopards, and antelope that once flourished there a thing of the past. Bagandula and Bassa participated in this gold trade and in the trade in animals, but they were never in a position to control it—again, because high-value/low-volume gold and diamond deposits were spread all over the hills in small enclaves (note again how the materialities of certain mines and minerals— mainly the black minerals—enable centralization and state-like formation, while others do not). The discovery of high-value cassiterite in their area would expand their ability to control the forest, but the extensiveness of the claim and the magnitude of the labor needed to extract these heavy minerals were among the factors that would also make it difficult for them to completely control this industry in the face of multiple other state and corporate authorities.

Local Knowledge versus the Map

In Walikale, people tell two main stories about the origins of the cassiterite mine at Bisie. The first is that some of the Bagandula men were hunting together when they lifted up a stone to set a trap and noticed it was particularly heavy. There is some disagreement as to who among the Bagandula was the first to identify it, or who was there at all on that day—some say it was the young man Fikiri, who later became one of the main representatives of Bagandula, while others emphasize that it was the whole group. Overall, this story emphasizes the somewhat accidental nature of Bisie and, while it gives some credence to the Bagandula's agency, it downplays their education and knowledge. A slightly different version of the same story is that, not only were they actively searching for minerals, but some of their grandfathers had worked in Lubumbashi, where they conducted "research" and had passed this knowledge to their descendants. In using this term *research*, some Bagandula suggest that science and geology were artisanal crafts that they had developed over time and which had become their heritage.

In any event, according to this story, they took the stone to two different comptoirs—Sodexmin and Saphir Inc., the latter run by a Congolese Munyamulenge who other Congolese simply saw as Tutsi with the name Alexis Makabuza. The comptoirs never told them what the mineral was, and even later, when many comptoirs started coming to buy, neither diggers (who during that time of war doubled as négociants, or middle-persons) nor Bagandula knew what this material was and simply called it coltan. Later, when some of the diggers went to Kigali, Rwanda, they found a comptoir with a tester who broke down the ore into different parts (coltan, cassiterite, bauxite, etc.) and offered different prices for each one. Even much later, when people came to know this material as cassiterite, comptoirs would claim that they were only purchasing one of the minerals present in the ore and that the rest was garbage, which they claimed to throw away. Middlepersons and diggers suspected this was not true but could do little about it. In any event, thanks to the cell phone, word got out to indebted coltan miners and traders in nearby Itebero, and a rush ensued. Négociants went to gather diggers from other locations and paid their way to come to Bisie and begin digging for them, either leasing holes from the Bagandula or paying in the form of ore.

The second major story about the origins of Bisie, told by people who were more connected to what was happening in official circles, is that the discovery was made by a young Nyanga man in Walikale who had studied geology at a university in the city of Bukavu. While he was undergoing his training in the Office of Mines, he saw an old map from the time of the Belgians in the office there. From the map, he discerned that the area known as Bisie had been identified as a giant reserve by the colonial mining company that was the predecessor of the current government company, Sakima. According to the story, this Nyanga man stole the map when he was left in the office alone one day in 2002 and went with it to Walikale. At that point, he visited the senior men in the Bagandula clan, informing them that he wanted to go to the river and prospect for gold. They consented to this request, but instead of gold he uncovered this very heavy stone, which he thought might be coltan, so he brought it to the Office of Mines after passing through to see the Bagandula family. It turned out to be cassiterite instead. It is worth mentioning that, in general, this trope of people walking around the forest with colonial-era maps they acquired either from offices, or from seniors who had worked in colonial-era mining companies, recurs over and over in stories about the coltan rush, and it is an example of people's drawing on the history of colonial-era exclusion to take advantage of the opening that the war and the price hike made possible.

According to this story, the Nyanga man never got anything for his troubles, and he spent the rest of his life going to the Office of Mines and the Bagandula, back and forth, saying that he deserved some portion of the rents and income being generated at Bisie. The people at the Office of

Mines in Goma tried to get the map back, but he spitefully refused. They say that the Bagandula eventually gave him 150 kilos of cassiterite worth about $160 to avoid any trouble, including the possibility that his emotions could manifest as witchcraft. The story has it that the geologist died from depression and resentment, and the map was never seen again. In addition to minimizing the knowledge and labor of the Bagandula while recasting them as selfish parasites, this story also positions the Office of Mines as inheritors and rightful guardians of secret colonial knowledge. This knowledge takes the form of the map, which epitomizes transparent vision and the seeing eyes of the state and the world community, which always has its eyes on Congo and its wealth. The story also points to the conflict that would ensue between the Bagandula and the government (mainly the mines office), which would eventually drive some of the former to seek out powerful allies to defend their interests.

Another aspect of both stories is that, while they privilege the actions of certain people over others, they also stress the importance of cooperation and of networks while showing how those networks are fraught with conflict and deception. The Bagandula profit by making and managing connections with certain people who can help them, like comptoirs from Goma and négociants, while those who fail to manage these connections and rely only on techniques or knowledge (the Nyanga guy with the map) are left out. As one trader put it, remembering that time,

> In those days, the main problem the Bagandula faced was transportation—they might find a two-ton, even a five-ton, rock and lose out because they didn't have a way of transporting it. When the Bagandula made it to the comptoir, the comptoir initiated the process: in no time, the comptoir was on his cell phone, and the carriers came out of nowhere, like a line of ants. First the porters (*wabebaji*) and then the négociants, together with the diggers. Later they were followed by the armies and the state authorities.

Unfortunately, the Bagandula were unable to manage these relationships indefinitely, as the "opening" that Bisie created and embodied brought in hosts of players with their own competing networks, crystallized in papers. Similarly, diggers and lower-level middlepersons would struggle against government officials who insisted that they couldn't just "go here and there" buying and selling minerals: they had to have cards, and they had to follow a system, laid down in the Mining Code, of buying from cooperatives and selling to licensed middlepersons, or négociants.

Customs and Governments

Regardless of who found what, there is no doubt that a rush ensued, which would ultimately culminate in the rapid transformation of the social and geographical landscape at Bisie, with mines expanding in the various locations that I stumbled into when I went to Bisie. In these early days, the Bagandula, the so-called children of the hill, were at once customary authorities, porters, and hole owners, essentially leasing from themselves. People who came to Bisie during this time of war knew nothing of the earlier agreements between Bassa and Bagandula arrived at during the Mobutu era. They met the Bagandula living there and, by and large, seem to have accepted that they and their ancestors required recognition in order for digging to be successful. For diggers, the Bagandula were the only state-like authority that really mattered, because they alone could ensure the movement of minerals through the mines and their safety at the mines through their relationship with spirits. This would eventually become a source of conflict between the Bagandula and other state authorities, especially the Office of Mines, who would go to great lengths to "sensibilize" diggers about the secular nature of mining, and the fact that success was not dependent on spirits. They did this, in part, to legitimate their existence and the power of papers while minimizing the power of the Bagandula (recall the tension between papers and ancestors discussed in chapter 3).

An early example of Bagandula's acting like a state involved a collaboration between the men of what later came to be called the Bagandula Association and the male miners at Bisie. They apparently came to the mutual understanding that the spirits of the ancestors disliked the presence of women at the mine, which allegedly threatened extraction. They prohibited women from coming to the mine and restricted them to the camp known simply as Bisie. At this point, though, women still could go to the river that separated the camp from the mine called Marouge to wash minerals and to carry them to the border between Bisie and the forest (that changed too, later on, when the international community drew more attention to Bisie, and the army intervened to force them out). This collaboration of men made it more difficult for women to be effective *wenye shimo*, or hole owners, because they couldn't manage labor when they weren't around, and this began to relegate them to other forms of work, including trade and sex work.

From the point of view of Bassa, now that all these people were showing up, they actually deserved more revenue than before, rather than less. After all, this was no longer about who had rights to reside in a place:

new people were coming, and the Bagandula were giving out land to use, essentially subleasing from the permanent lease they had from Bassa. Bassa contested their right to do this, and in the early days they showed up at people's holes to offer the sacrifices that Bagandula in turn insisted wouldn't work, because Bagandula were the actual autochthones. The vast majority of diggers seemed not to buy into the Bassa's legal-bureaucratic argument, rooted as it was in colonial bureaucracy, and favored the social and spiritual claims of the Bagandula, which were the only ones with practical implications for them. They also claimed that the Bassa sacrifices indeed didn't work and that the Bassa were endangering their lives while also threatening the possibility of extraction from new holes. The diggers seem to have nearly unanimously resisted the Bassa and, after the RCD left, allegedly even subsidized the Mai Mai under Sammy to help remove them from Bisie. In the meantime, Bagandula had in turn developed arguments for why the Bassa claim to territory had always been illegitimate. They argued that Bassa were never the original autochthones and that the Bagandula had been there at the same time. According to this argument, the eponymous ancestor of Bagandula was in fact senior to the eponymous ancestor of Bassa, and the Mobutu-era authorities had muddled things up, wrongly placing the clan of Bassa in charge. Some argued that the error preceded Mobutu and went back to the colonial period—according to this narrative, the Belgians had given the Bassa a large swath of land that included the area in which Bagandula were living as appreciation for the role the Bassa had played in the Belgian war against "the Arabs" (*Waarabu*, actually the Swahili under the nineteenth-century warlord Tippu Tip).

When the RCD was routed at the end of the war, the Mai Mai under Colonel Sammy from Walikale "did rotation" (*walifanya rotation*), coming into replace them and also recognizing the customary authority of the Bagandula. In 2005, Sammy's Mai Mai militia was integrated into the Congolese army along with most other militias; this was part of the larger plan to integrate all the warring factions of the east into the Congolese army. As mentioned above, Sammy was from Walikale, and it was his success in routing the RCD from there, allegedly with medicines that his mother prepared, that led to his being stationed at the profitable location of Bisie.

In the old days, new arrivals had asserted their claims to land based on informal agreements they had with "some Kumu guy," usually but not always someone in the Bagandula clan. "It was simple," one Shi "manager" (a middleperson between hole owners and négociants) remembered, "You give money to the Bakumu, and you dig." But by 2005, a middleperson trying to sell one ton of cassiterite worth $2,500 might have to pay $100 to each of five different state authorities on top of money to the Bagandula. By 2005, then, Bisie was no longer a world unto itself, outside the government of paper. There was a larger shift underway, and I don't want to portray what was happening as a natural process of state authorities simply proliferating and becoming more powerful. When the mixage, or blending of armed groups, began, independent militias like Sammy's Mai Mai group were expected to act as, if not exactly extensions of, then at least collaborators with other state bureaucracies. Following the integration of Sammy's militia, other state officials gradually started showing up and asserting themselves at Bisie, but they needed the backing of the army. This was a tricky thing, because other state authorities, such as the Office of Mines, had nothing to give the army in terms of money, whereas hole owners (*wenye shimo*) and the Bagandula were in a position to pay the military to do what they wanted (e.g., to back the Bagandula against the Bassa, despite the fact that the Bassa was bringing a claim against the Bagandula in the court, demanding inclusion in the rent-collecting Bagandula Association). With the integration of militias into the army, state authorities were able to use their connections with other, "higher" state authorities in Goma—the governor, for example—to assert themselves. If the Office of Mines met with resistance from the army when they tried to impose tax on diggers, they could now complain to higher-ups, and Colonel Sammy, or any military figure, would be in danger of losing his position and being placed somewhere else (which is what eventually did happen to Sammy). With time, the control of the army and the Bagandula decreased together, mostly because, even if the on-site army had weapons, other state officials could now complain to Sammy's higher-ups in Goma, if indirectly.

It was in this changing context that the Office of Mines and other officers, such as the chef de territoire, started to assert themselves against the Bagandula, sometimes using the army to help them in this process. They also took to being seen alongside soldiers to demonstrate that there was now an alliance between the "government of guns" and the "government of paper," even if privately soldiers often expressed a disdain for these ever proliferating and expanding "people of lies." It was also in this context of pressure from the government to get papers that the Bagandula started seeking out companies and thus opened the door to their own final dispossession.

Meanwhile, some tensions were emerging within the Bagandula Association, which companies would later exacerbate and exploit. The Bagandula had organized their association through patrilineal descent, with some pushback from the women in the lineage, who encouraged the men to incorporate other criteria into the structure of the organization, including matrilateral ties. The lineage founder Olema had died long ago, and his first son, Mtokotoko, had gone missing just before the discovery of Bisie; Mtokotoko's first son, Mtokotoko II, had also passed away. (Some wondered if the Bassa had somehow caused Mtokotoko to disappear because he was the oldest Bagandula and the only one who knew the history of the Bagandula's claims.) One of Olema's other sons became the collector of revenue, but there was some concern about how he used and distributed the money. With the first son and grandson of Olema out of the picture, the remaining children debated over who should hold positions of authority, in view of the fact that those who did would be in a privileged position to collect and distribute rent and toll revenue. Marcelin, a firstborn daughter of Mtokotoko II, and her aunt Vumilia, or Struggle/Endure, were passed over because of their gender, but Vumilia, the firstborn daughter of Mtokotoko, insisted on her right to be represented as firstborn and nominated her son, Fikiri. She was also concerned that Fikiri—who it was widely held discovered the claim in the first place—would otherwise be passed over because his father was from elsewhere. Moreover, Fikiri was the only person in the family to have much of a formal education, being the only one with a high school degree, and the family hoped that he would be able to engage with the *Kifaransa*, or French language/ways, of the state bureaucracies, whose officers were now showing up at Bisie to harass them.

The Bagandula Battle the "People of Lies" and Search for Papers

Today, when the Office of Mines officials remember that time, they speak of themselves as being in a constant struggle with backward people of the forest to get them to "understand" the "need for papers." Their repeated refrain is that the Bagandula believed that the forest and the hill were theirs even though they didn't have papers, and so it now fell upon the Office of Mines to "teach" them about papers, which also meant teaching them about a larger spatiotemporal reality, which they depict as being synonymous with "seeing ahead" (*kuona mbele*). As one Mines employee put it, "In the beginning, Fikiri didn't even have a digger's card. They just believed the forest was theirs and would always be there. For them, that mountain is like a totem. If it dies, they die. . . . their intelligence was very low, like an animal's." Another continued in this vein, elaborating on how the Bagandula's oneness with nature and the ground threatened to drag state officials down to their level so that everyone would be living together in the darkness of a hole:

God gave us His intelligence, but these people are like pygmies, living in nature, unwilling to use their minds, and living only for this moment. Humans sleep in good houses, they grow and cook food, they prepare for the future. But when we come to the Bassa and Bagandula, it's like humans meeting animals. They want people to bring them things, like pygmies, eating whatever fruit falls to the ground. So when you leave Goma and come to Walikale, really you are traveling down into a hole. You go there with your education, and they beat you with their rocks.

Note how, in this framing, being under the ground, in a hole, is opposed to being in the light and transparency, as well as to being human, as the state officials reiterate a set of colonial associations that diggers often inverted. Part of the idea here, also, is that forest people and artisanal miners alike live in a constant now, unable to conceive the future. And so, for these officials, the Bagandula's major defect was their spatiotemporal orientation, epitomized by their alleged ignorance about papers, in which were materialized predictable futures and national space. According to this (post)colonial, state-centered worldview, this inability to conceptualize the future was sure to eventually backfire on the Bagandula, and so they had to be taught a lesson about their proper place in space and time:

They thought this place would always be theirs. They just couldn't imagine that anyone else, like a foreign company, would ever go there to stay. And now [Alphamin's] helicopters fly in several times a day! But back then I told them that if someone comes around later with questions about who you are or what you're doing here, you better not complain about it. Without papers, you'll find you've been beaten.

The government of paper's efforts fell on Vumilia's brother Ramazani and his nephew Fikiri, who the officials helped install as the PDG of Bisie, in accordance with the Mining Code. The Office of Mines officials certainly intended to emasculate the two men when they called them "bad women" who offered themselves and their services to this random group and that random group—this was also a reference to the rival companies that they would eventually court and who would court them. These officials urged them to instead behave like "good women and stick with one husband," the state; and diggers' papers were the formal contract that would demonstrate their marriage to the male state as represented by the Office of Mines. One Mines official described how he came upon Ramazani and Fikiri when they were collecting "tax" from porters on the walking path to Bisie. Finding that they still had no papers, he forced Fikiri and Ramazani to take off their shoes and kneel on the ground. Then he had a soldier beat them with a stick in front of everyone and send them to prison:

> I gave them the stick (*fimbo*) to teach them about tomorrow, about building, so they could see ahead. Wealth is not meant to be eaten [meaning consumed and so extinguished]. [Laughing] It is not for breaking bottles [a stereotypical behavior of miners after consuming an alcoholic drink, performing a lack of concern about money] but for building preparing/planning (*kutengenza*).

According to this mining official, coercion was necessary to teach these people—who he imagined to be living in a hole, like diggers—about linear time and planning, which should replace the destructive spontaneity shared by Bagandula and diggers alike: "They thought I was doing a bad thing, but I was teaching them the importance of papers." After all, he continued, "Paris wasn't built in a day, but they forget this."

Fikiri finally got a digger's card but, with the thousands of people coming and going from Bisie, the folks at the Office of Mines deemed it insufficient. As one put it, extending his arms as if to compare the weight and size of Bisie to a digger's card, Fikiri and Ramazani's puny papers were like "a bottle in the sea." The Mines official continued, shaking his head at their shortsighted stupidity and indolence, "How can you compare the movement and value of all of Bisie to this card? You can't!" The Bagandula's failure to grasp this self-evident fact was just more proof, if any was needed, of their lowly, animal nature. Now the Office of Mines was telling Fikiri and Ramazani that they needed incredibly expensive exploitation papers. They were treating them as if they were a company, translating their putative "ownership" of the hill to something that made sense to them in terms of their reading of the Congolese Mining Code.

My friend Major Patrick, who became close to the case, and who for a time defended the Bagandula against the Office of Mines, explained,

> The Office of Mines conned and persecuted the Bagandula! The Mines people were saying that Fikiri and Ramazani needed $300,000! For exploitation papers! And they needed two sets of papers, one from Walikale [the district] and one from Kinshasa, which meant that Fikiri and Ramazani would have to travel to Kinshasa in an airplane, as if they were a company like MPC! These guys, they live in thatch homes. There's nothing there! Where are they supposed to get this money?!

In lieu of paying this money, which they were now alleged to "owe" the Office of Mines, Fikiri and Ramazani developed a "habit" (*tabia*) of giving the mines guys a hole to dig for a period of time whenever they showed up trying to sell papers (the salongo). Over time, the amount they received was equal to the value of the papers; ironically, *there were no papers to account for any of this extraction.* Moreover, the Office of Mines guys would then give less than the customary 10 percent to the Bagandula, who in turn were supposed to give a certain part of that 10 percent to the "discoverers" of new holes, as well as to the Bassa. This meant that, because of the Office of Mines, the Bagandula were becoming increasingly indebted to other diggers and risking their ire.

The Office of Mines staff liked to talk about how Fikiri and Ramazani drank away all the money they earned from Bisie, but those in the know said otherwise. Bagandula weren't shiftless and lazy. They were furious: "Sure, Fikiri and Ramazani drank a lot, but it was their anger at the system of papers that led them to drink the small amount of money they received, to never do anything with it." Even more consequential was the fact that the Office of Mines' demand for "papers of exploitation" drove Fikiri and Ramazani to go looking for money to buy these papers. As the Bagandula found their claims to territory undermined by the government of paper, which in turn took value away from the ancestors, they searched for more powerful connections further afield. They sought out the soon-to-be notorious coltan middleman Sheka and his friend Pastor Raymond to facilitate their acquisition of papers of exploitation by forming a new company, the subject of the next chapter in this history of Bisie. It was pressure from the state that encouraged the Bagandula to try to become, or merge with, companies, which ended up indebting them to powerful people and fueling tensions that these companies exploited as they in turn sought the support of these indigenous people.

SEVEN

The Battle of Bisie

By 2005, there were more than thirty comptoirs, or buying houses, represented at Bisie, but the two largest and most well-financed were MPC, managed by a Belgian man named Yves, and Saphir Inc., owned by the Congolese Munyamulenge Alexis Makabuza, from Masisi, North Kivu, the same province as Walikale. Both of these comptoirs had offices in Goma and in Mubi, the main market town between the district capital of Walikale and Njingala, the jumping-off town to enter the forest en route to Bisie. Makabuza's brother is believed to have been one of the main financiers of the Rwandan-backed CNDP insurrection, which evolved out of the older Rwandan-backed RCD (Global Witness 2009). MPC, headed by Mr. Yves, sold its ore mainly to South African smelters and at least appeared to have connections with Western buyers. These comptoirs were, then, each elements of a larger set of spatial and temporal networks which they also have indexed for others: Makabuza's company was touched by its association with Rwanda and the invasion of Congo by Rwanda-backed forces at different times in the recent past. In contrast, people often associated MPC and its workers with Belgium (and so the colonial-era Belgian Congo) and apartheid-era South Africa, particularly after they tried to employ papers to exclude people from their livelihoods at Bisie. At other times, MPC, which sold the company to the Canadian Alphamin in 2012, has been able to benefit from its possible European and American connections by suggesting that it can connect miners at Bisie to larger markets or open up markets to them after these markets were closed owing to international concern about conflict minerals.

In other words, for Congolese "on the ground" in Walikale, these companies were never merely companies but were understood to be material and financial embodiments of particular spatial geographies and histories, and this had some influence over the way events played out at Bisie. I place the emphasis on representation—rather than the actuality of the networks—to make the point that, regardless of the actual relationship that these comptoirs had to other foreign companies or political actors, people constructed their own narratives based on their understandings of who these people were and who they were likely to be connected to, and this had more importance with respect to people's actions on the ground than the "ontology" of the actual social networks.

Makabuza's Saphir Inc., with its Rwandan connections, had a somewhat longer history with Bisie, having been among the first comptoirs that diggers and négociants went to when they discovered cassiterite there during the Great War, before the Rwandan-supported RCD temporarily occupied Bisie. In contrast, Yves was a pilot, and the rumor in Walikale was that he flew planes for Laurent Kabila during the First Congo War, when the AFDL invaded the east and ousted Mobutu. People speculated that President Laurent Kabila paid his debt to Yves by granting him mining rights and papers to an area not far from, but outside of, Bisie, which turned out to be barren of minerals. According to this theory, a frustrated Yves, now living after the assassination of his friend Laurent Kabila, was continuing to seek compensation for his work during the war by acquiring a substantial mining site, which he would find when he came upon the artisanally fueled movement of Bisie, a mine for which he would eventually try to obtain exploration papers.

By 2004, MPC and Saphir were in fierce competition with one another as comptoirs and, according to people who were involved with both entities, each wanted to push out the other and become a buyer's monopoly. All of this had to do with these companies' efforts to get the best price possible by eliminating middlepersons and other "superfluous" actors and state agents while also controlling and regulating the timing of production with a view to their own purchasing schedule and probably the demands of the foreign smelters to whom they sold. Take, for example, this rather telling reflection by a hole owner who was at Bisie in 2005:

> MPC had been buying minerals in Walikale and Goma for a while, but it was a few years before we hole owners actually saw them face to face at Bisie. They came one day with a helicopter full of cash, and they bought a lot of ore. They were trying to cut out the négociants, and all the diggers were happy because they were able to offer a better price. They did this twice, and everything was OK. Before the third visit, many thousands of people, maybe even tens of thousands, dug in expectation of MPC's arrival. They brought

thousands of tons of cassiterite to MPC—so much that MPC didn't have the money to purchase it all. So the company took a debt [from the *wenye shimo*, or hole owners].

According to this hole owner, MPC left signed papers, a contract, saying that they were leaving with minerals of a certain amount and value and that they would return with the remaining money. Because of this debt, the diggers claimed, MPC would later go to Kinshasa to get exploration papers in an effort to gain ownership rights and so escape their debts and their collaborative relationship with the Congolese hole owners. Their desire to evade debts was combined, it was felt, with jealousy, which, my interlocutors insinuated, may have been informed by their sense of superiority and entitlement as white Europeans: "They said to themselves, 'There's no way these people can keep getting so much from Bisie without us getting something.'" In any event, MPC was upset because they had been excluded from the network of people who were producing and governing there and so took to plotting a way around them from their privileged position on high: "They flew their helicopter over the sites to see all the places that have minerals. MPC had a camera and a computer in that helicopter, so they knew what was in the ground and were able to see the future, to see ahead." In contrast, diggers couldn't discern the future potential beneath them and were too busy "moving" to contemplate it: "Diggers only saw where they were at any given time and even then only what was at the top. They had no idea what else there was."

The memory of MPC's helicopter full of cash is interesting because it provides a window into how the different worldviews and practices of diggers and companies collided in an everyday scenario: the hole owners tried to use "debt" (*deni*) to establish an incremental, ongoing relationship with MPC that would potentially protect both parties from precarious uncertainty, like periodic overproduction (in this particular instance) or a shortfall in real demand (not enough cash in the right place at the right time). In contrast, MPC purposefully created sudden disruptions in the hopes of profiting, but then they found themselves unable to manage the consequences: in the above instance, they caused this rush of supply through their efforts to get the best price possible by showing up suddenly with a helicopter full of cash. It's also clear how MPC's hit-and-run tactics, and their willingness to eliminate "network nodes" that got in the way of their getting the best price possible, would be very frustrating for rival companies (to say nothing of middlepersons and the state officials who sold papers to them) because the sudden manipulation of price had the potential to reorient all activity at Bisie and disrupt any purchasing plans the rival comptoir might have had. In short, companies—MPC and Saphir Inc.—had an interest in trying to reduce the number and types of actors at Bisie while also acquiring more control over the timing of production and purchasing.

Later on, when MPC would come to Bisie with exploration papers from Kinshasa, the hole owners would remember the helicopter event. As one hole owner recalled, "We told them 'This is impossible! You've come after we were already here! There's no prospection left to do. Now you're just pretending this is your place! And you have other people's debts and, because you're holding onto their money, they can't pay or receive loans themselves!'" The hole owners concluded that MPC's sacred Kinshasa papers were based on a lie: the company must have told Kinshasa that there were no people mining at Bisie, since the papers were for prospection that had already been done. "And that's when we said, no, there's no prospection. We discovered this place; we've done the prospection already. And so we stopped selling to MPC because they weren't friends anymore." MPC had scorned the principle of collaboration in favor of the expulsionary promise of papers.

A Funny Thing Happened on the Way to the Cadastre

When, in 2004, the Cadastre in Kinshasa issued a deadline for the registration of new mines, MPC and Saphir Inc. sent representatives there—Mr. Yves from MPC and Pastor Raymond from Saphir Inc.

It is the source of much humor in Walikale that both parties allegedly ended up arriving at the Cadastre office in Kinshasa on the same day at the end of 2004. More accurately, it was Yves and Makabuza's representative, Pastor Raymond, who found themselves together, waiting in line along with some other guy with intentions of applying for rights to a different mine. Most people think that Yves ended up getting the papers because he was first in line, and it's true that the rule of that office is that the first applicant's claims must be evaluated in full before the claims of those that follow can be considered. Some joke that Yves won because he "saw ahead" and, like other white men, "knew time."

But a few insist that it was Raymond who was on time. According to this story, he was first in line at the Cadastre and that made him a shoo-in to get the papers for Bisie, but he had no idea how lucky he was, apparently not knowing anything about the Cadastre's rules. Yves is said to have been third in line and, seeing that Raymond was in front of him, he feared that he wouldn't get his chance. He knew the rule of first come, first served, so he figured out a way to cut in line. People speculate that Yves must have somehow used his cell phone to contact someone in the office because, mysteriously, a Cadastre official came out of the office after some time, approached Raymond, apologized to him that it would be a while, and asked him to wait. The man suggested that he go to a nearby restaurant where he could eat on the office's dime. Perhaps seduced by the prospect of a free meal, Pastor Raymond agreed. He must have been shocked when he came back, sat down, waited some more, looked up, and saw Yves emerge from the office, triumphantly holding papers in his hand—the copies of the forms and receipts he had secured from the Cadastre. According to this account, Raymond was informed that he had delayed for too long and that the office could do nothing for him, at least until they had further researched MPC's claims and MPC had responded to their queries.

Whether or not this story is apocryphal, it does give a sense of how Congolese understand foreign mining companies, and the world order that is pitted against them, to be enacting a long-term plan that is playing itself out incrementally—the whites know what time it is, already have networks in play, and know what they're supposed to be doing, whereas Raymond is depicted as having been influenced by short-term desires, regardless of whether or not this is true. Even if they're late, the whites can still manipulate events, while Congolese often find themselves tricked or manipulated by what looks, on the surface, like reciprocity.

When Yves and Raymond returned to Goma and Makabuza found out that Yves had successfully initiated a claim to Bisie, he was allegedly incensed, but he was not yet beaten. It could be over a year before Kinshasa came through with papers, if they ever did, and MPC still had certain conditions to meet, which included coming to an agreement with the customary authorities—the Bagandula and Bassa—to avoid any future conflicts with them (this is written into Congolese law). Meanwhile, Makabuza could do his best to continue to cement his claims with local and provincial authorities while consolidating his company's relationship to the Bagandula and Bassa. And so began the courting of the indigenous population, a game that Makabuza at first seemed to be winning.

The Bagandula Mining Group versus Mining and Processing Congo

As mentioned, the Bagandula Association, led by Ramazani and Fikiri, was already looking to develop connections with powerful outsiders so that they could strengthen their position against the Office of Mines and the "other governments" that perpetually harassed them about papers. They sought out people with military and bureaucratic connections who might sponsor them to form their own comptoir, which they hoped might later morph into a mining company that would exploit resources at Bisie for their benefit. In addition, they were probably trying to release themselves from the claims of the Bassa clan, whom the government had required them to include as equal members in their association in 2005. Meanwhile, Makabuza needed local alliances that would help him cement a claim to Bisie and build a monopoly while also concealing the history of war and the military connections that came together in and through Saphir Inc.

The two parties, each with their own separate intentions and circumstances, came to each other through Nyanga intermediaries from Walikale—relatively educated and urban folks from a historically agricultural ethnicized group. These intermediaries were the already mentioned Pastor Raymond, who allegedly lost his place in line, and Sheka, the affable coltan trader and locally renowned soccer player who would eventually become the most notorious militiaman and mass rapist in North Kivu. They had each been involved in the coltan trade as merchants and were also recognized civil society leaders, a formal designation in Congo. However it came together, this conglomeration of indigenous and regional actors registered a company with the province, which they called GMB, or the Bagandula Mining Group, financed by Makabuza, with Sheka and Ramazani serving as directors. The name seems to have been intended to conceal any connection to Rwanda and to make GMB appear to be a company owned by the "children of the hill" (*watoto wa kilima*), the Bagandula. In addition, Makabuza is said to have financed GMB to build three community buildings as evidence that, in contrast to MPC, they had "plans for development." As one state official remembered, "This is how GMB entered into conversation with people," by convincing others that only they could bring development to Walikale because of the involvement of Bagandula in the company. In contrast, it was understood that MPC, a company with the "mentality of Wazungu" would understand themselves to be the owners of the hill by virtue of the papers they possessed and would not comprehend the idea of a social obligation to larger groups of people, be they Bagandula or artisanal miners.

Meanwhile, MPC also set about developing the connections it would need to get a title to Bisie. Yves no doubt knew that he would need a contract from the customary authorities to secure rights to Bisie and that there would have to be promises of development similar to what GMB had offered. Once MPC learned that Ramazani, Fikiri, and others in the family had joined forces with Sheka and Makabuza, they might have come to the realization that they had to identify or create divisions within the family of Bagandula while also exploiting the cleavage between Bassa and Bagandula. Fortunately for them, these divisions were already there: the strong, sidelined women in the group had already taken an active role in the making of Bisie, even if they weren't formally recognized as having positions in the association (Marcelin at one point found a lawyer in Walikale to help mete out an agreement between Bassa and Bagandula, for example). She and her niece Vumilia were concerned that their siblings were potentially being taken advantage of by what they assumed was a Rwandan or Rwandan-backed company with military connections (Makabuza's Saphir). There was talk about the danger of "selling the forest to Rwandans." According to many, MPC's representatives capitalized on this fear, and on the history of the Rwandan-backed war, to confirm the anxieties of Marcelin, Vumilia, and others in the lineage, and among the Bassa, that Ramazani and Fikiri were the unknowing victims of a Rwandan plot to take over Bisie, one that would ultimately leave everyone dispossessed. Some insisted that Marcelin sought out MPC first, while others insist that the Bagandula, "unable to see far," were sought out and manipulated by more educated others, who were knowledgeable about bureaucracy and papers. Regardless, some members of the Bagandula clan began to form a pro-MPC alliance with their historic competitors, the Bassa, who were also sure to be left out of any agreement between these Bagandula men and outsiders who were perceived to be Rwandan.

And so, the two companies competed with one another through the family, increasingly driving them apart. After GMB brought Ramazani and Fikiri to Goma to sign a contract with the company, MPC did the same, inviting Marcelin and a Bassa representative to sign a contract with them. But Makabuza found out ahead of time and allegedly paid ANR (the secret police) to arrest Marcelin and her Bassa companion at the airport. When in jail, Makabuza's representatives compelled them to sign a document declaring that they were with GMB and wanted nothing to do with MPC. But, after being released, Marcelin disavowed the contract, playing on the general perception that she was a simple forest woman of low intelligence, unable to see far: she insisted that she didn't speak French. How could she, then, sign a document written in French?

In addition, GMB tried to figure out ways to delegitimize Kinshasa's papers, in case MPC's papers came through. Assuming that, while it was never mined by a company, Bisie nonetheless probably fell within the boundaries of a colonial-era concession, they sought out the government-owned company Sakima, the inheritor of the colonial and postcolonial mining concessions that had mined for gold in Walikale during the colonial period. They created a contract with Sakima and then used this contract to secure papers from the provincial government to buy minerals and, later, papers granting them the right to conduct exploration with a view to eventually carrying out mining operations. In general, Makabuza and GMB were trying to outmaneuver Yves and his national papers by cementing a bureaucratic connection with the past embodied in the company Sakima's historic claims. What they were doing—mobilizing the provincial government against the projected future claims of the central government in Kinshasa—also made sense in the context of that period, roughly 2004–2006. For the preceding several years, the provincial government, run by the Rwandan-backed RCD, had operated independently of the central government, and the Sun City Agreement of 2003 had also affirmed contracts made during that time, so there was some reason to believe that this precedent would continue into the future.

GMB was able to get the local authorities to "believe in their papers" (*kuamini makaratasi yao*) from Sakima so they could exercise a monopoly over Bisie and illegalize MPC, at least locally. But Yves, a professional pilot, used a plane to fly overhead and take photos of the site. Through his technical knowledge and his mastery of optical surveillance and the techniques of visual transparency, he would later be able to show that Sakima never held title to Bisie, which rested outside of the concession. This also meant that Bisie fell within what had some years earlier come to be defined as the ZEA, or legally recognized zone of artisanal extraction.

Despite MPC's countermaneuvers, Makabuza's gambit paid off, at least at first: in early 2006, the chef de territoire called a public assembly in Walikale, in which the army and other state officials were also present, during which he informed everyone, including all the comptoirs of Walikale, that GMB had a monopoly on buying minerals from Bisie. I personally saw a copy of a contract, signed one day before the meeting, in which the army and the chef de territoire agreed to defend GMB and to oversee their security in exchange for 10 percent of the production every month and half of all the money procured from the three main toll stations on the two different paths leading through the forest to Bisie. Soon after, GMB began using the tolls to collect 10 percent of the production of every individual digging at Bisie and 10 percent of the profit of anyone doing business. People say that anyone who resisted this was threatened by the army or the chef de territoire, an appointed administrative official. At one point, an upset Marcelin walked through the forest confronting the toll operators, asking them what percentage of this income would be distributed back to the families of Bagandula or Bassa. Unfortunately, they told her, the tolls had nothing to do with the families, because the company now had papers from the provincial government in Goma, and the Bagandula's claims had been superseded. Marcelin's worst fears seemed to be coming true.

So began a war of letters, which I was able to study copies of: Marcelin wrote letters to GMB and the chef de territoire, accusing them of being "*rançonneurs et extorqueurs*," ransoming and extorting people through their tolls while failing to bring "development" to Bagandula, and insisting that they were harassing those Bagandula that weren't part of their company. Meanwhile, Makabuza wrote a letter to the Goma High Court informing the judge that he had signed a contract with the government mining company Sakima, the actual owner of Bisie mine, allowing them to begin exploitation at Bisie. The letter asked the court for assistance in evicting MPC and competing comptoirs from Bisie, arguing that no other comptoir had the right to buy Bisie minerals. It also asked for the arrest of MPC's Mr. Yves, based on his refusal to respect the law or the injunction of the army and the chef de territoire when they had visited Bisie and declared the mine the property of GMB. In another letter to several offices of the provincial government, GMB accused Mr. Yves of continuing to negotiate with the descent groups at Bisie while buying minerals, despite the fact that GMB had a contract with the government company Sakima, the putative concession owner. They asked ANR, or the "secret" police, to arrest Mr. Yves since he was bringing "insecurity" and "confusion" among the people while preventing GMB from continuing with their work. A few days later, Mr. Yves was arrested by immigration, or DGM, and subsequently released.

Afterward, Mr. Yves and his lawyer wrote their own letter to the court and the government responding to the accusations. They claimed ignorance of GMB's claims regarding its rights to mine at Bisie, declaring that MPC had no intention of abandoning the business of purchasing minerals from miners there. They insisted that the chef de territoire did not have the authority to chase comptoirs from the mine and that Bisie was not the concession of any person or company since the government in Kinshasa had not yet registered the mine, which remained without a registration number. They also insisted that Bisie was not in Sakima's concession, now revealing that Yves had used aerial GPS to determine this.

Not too long after that, MPC's finalized and approved prospection papers came through from Kinshasa.

MPC Triumphant?

When the "government in Kinshasa" came through with papers, MPC prepared to send a delegation to Bisie to prepare the population there for their arrival and to let all the competing interests there know that Bisie had exclusive ownership rights to the hill. In Goma, MPC met with GMB; a lawyer who was present remembered the conversation like this:

> Makabuza was angry, but what could he do? Yves had papers from Kinshasa. So he said, "OK, so you have papers. Fine. But it's a big mountain. If you want to work in Bisie, we're going to have to divide it in half." And Yves refused, because MPC had paid for the whole area, and he wasn't about to give up half of the hill he had paid for because he was afraid of Makabuza. And that's when Alexis said, "OK, but if you don't share it with us you'll be shot at up there." And Yves said, "Fine. Do your best."

A few days later, GMB sent representatives to Bisie who, apparently in the hopes of acquiring support from the population there, were said to have promised to kill everyone at Bisie if GMB were kicked out of the place. People interpreted this as Makabuza publicly threatening genocide against the whole population of Bisie. In doing so, the company was also exploiting people's memories of the Great War, which had technically ended only three years previously and which was not seen as being really over, especially since former members of the RCD and CNDP were still in the army in Walikale.

On October 20, 2006, MPC sent a delegation from Goma to the district capital of Walikale, escorted by their lawyer; a civil society representative from Walikale; two police officers from Walikale; and two representatives from the high court in Goma. The government sent the police and the court representatives into the forest to get rid of GMB and all of the tolls on the way to Bisie and to eliminate all taxes that were not recognized as emanating from "the state." At Walikale, the government held a meeting in the town center in which they explained to the population that MPC had been handed the concession by the government in Kinshasa. MPC's delegation, which included Yves and his colleague Brian, showed their papers and stamps to the assembled crowd and announced their plans for "development" in Walikale: schools, clinics, roads. But even though MPC seemed to be getting the go-ahead, the local government in Walikale remained unconvinced: the chef de territoire, who had supported GMB and banned MPC from buying minerals earlier on, read a letter he had from the court in Goma, which contradicted what the same court had said earlier, about GMB's tolls being illegitimate. The new letter indicated that GMB's tolls were indeed legitimate and that removing them was illegal, since GMB had a contract from Sakima, and it was impossible for the court to prohibit the work of Sakima, a government mining company with a concession claim. This was the first time that the MPC delegation publicly produced papers and photos showing that Bisie was in fact not in Sakima's territory after all. But there was a very awkward silence when civil society leaders asked how they would ensure their own security and that of the Bagandula and Bassa, given that Makabuza had allegedly threatened to kill everyone if GMB were chased out of Bisie.

Afterward, MPC met with local government administrators, Monusco (the military wing of the United Nations, including Major Patrick), and Colonel Sammy, who promised to support MPC with security. They even visited the local Catholic priests, who advised MPC to do everything possible to follow through on their promises to the public. Papers were not enough, they cautioned; collaborating with people was the only way to ensure peace and to guarantee that MPC's work would be successful. A couple of days later, MPC held a football match against Colonel Sammy's army team, suggesting cooperation between these two important forces at Bisie.

The next morning, the preparation team set out for Bisie, along with a contingent of military and police actors, including ANR, to make sure GMB's work was indeed halted and that the tolls that were enriching the company had been removed. MPC's delegation was to follow them in the afternoon, but a large group of Bagandula blocked their way on the road to the town from which they would begin their walk through the forest. Someone, perhaps GMB leaders, had spread a rumor in the family that the rival Bassa lineage had sold Bisie to MPC and had received a great deal of money from the company. The Bagandula had never received a share of these rumored proceeds, and now they were demanding their cut, at one point threatening MPC's delegation with violence. Now Colonel Sammy came to the rescue, arresting Ramazani and another GMB representative from the Bagandula lineage. At this point, Sammy made it clear to the Bagandula that MPC's claims were the only legal and legitimate ones and said that the only reason he had assisted GMB in the first place was because he was operating under orders from his superiors, which he couldn't disobey. Now that he "understood" that GMB had been working with false papers, based on a fake connection to Sakima, and now that MPC had shown up with papers from the "real" government in Kinshasa, his job was to facilitate MPC's work. Colonel Sammy advised Mr. Yves, who was part of the delegation, to "have as much love for Bagandula as he had for Bassa," to treat the two lineages as one, and to respect the contracts that MPC had made with all parties. But some found it significant that, immediately after the delegation resumed its mission, Sammy released Ramazani, perhaps suggesting an actual alliance between the two.

Ramazani had also used this opportunity to ask MPC to help the Bagandula to pay off a debt of $325,000 (or, alternatively, fifteen tons of cassiterite), which Alexis Makabuza was claiming they, or rather GMB, owed him. Makabuza was apparently insisting that, since GMB was not going to realize its claims, Ramazani and Fikiri should pay him the money he had lost trying to incorporate the company—getting papers from the government, paying for corporate personnel, organizing meetings, transporting people to Goma and Kinshasa, and so on. That meant that the whole lineage of Bagandula was being threatened by a dangerous man with Rwandan military connections—someone with whom negotiation would be very difficult. Mr. Yves, forced to turn back to Walikale town, promised that he would talk it over with his colleagues, who were waiting for them in the nearby market town of Mubi.

The next day, MPC set out again for Bisie, and this time they made it through the long trek on foot successfully. When they arrived, they held another meeting with the Bagandula and Bassa families in which they repeated the promises of development that they had made in Walikale town. Afterward, they visited a nearby stream where they made a point of bathing alongside the diggers in an expression of solidarity and shared values, which was also shot through with the idea of mutual cooperation and rebirth through Bisie.

In the morning, MPC received a delegation of hole owners. MPC asked the hole owners to abandon their holes, and in this meeting they told MPC that they would be happy to leave if they could be made whole on their investment—including the money they had invested in "developing" and capitalizing their holes. They also reminded the company of its outstanding debts to the hole owners. For the most part, these entrepreneurs had not been using their own money but were financed by others—banks, politicians, generals, entrepreneurs in Goma or even foreign countries—to whom they were now in debt. If they weren't paid, they would face lawsuits and even imprisonment when they returned home to the cities from which they hailed. While MPC promised to consider these issues, they also made it clear that "one concession cannot have two owners" and reiterated that ultimately it was they who possessed exploration papers from the government: the artisanal miners had to go.

Later in the evening, a delegation led by one of Colonel Sammy's subordinates, a former Mai Mai soldier known only as Rasta, visited MPC's delegation at their camp. Rasta and Yves sat together, and Rasta assured him that all the tolls had been taken down because MPC had arrived. Then Rasta asked Yves what MPC's plans were for feeding all of those soldiers who had been depending on the revenue from these tolls: there were fifty-five soldiers, and MPC would need to give each soldier one hundred dollars a week in perpetuity, the income they anticipated losing now that the pathway tolls were gone. Yves and the others working with MPC of course couldn't agree to this, instead promising to buy each soldier a bag of rice, a bag of cassava flour, a bag of beans, one goat, and twenty liters of cooking fat (again, according to those who were there at the time). Rasta was expecting money, but Yves instead promised food. The party then began to drink together, but Rasta remained silent, quickly drinking what observers remember to have been four large bottles of Primus. Suddenly, he got up and left.

Soon after Rasta left, bullets started flying out of the forest and into MPC's camp.

The Battle of Bisie (Who Shot MPC?)

One of MPC's former (Congolese) lawyers remembers that he had been in the office showing Brian, a colleague of Mr. Yves, something on the computer when the shots went off. "Get down on the ground!" the lawyer commanded, pushing Brian out of his chair and onto the floor. As he dropped to the ground, the bullets blew away the back of Brian's chair. They could hear their Congolese engineer screaming in pain. He had been shot in the leg, and to this day he has a hard time walking.

After a couple of minutes, the shooting stopped. The lawyer looked out the window, only to discover that all of Sammy's soldiers, who were supposed to be there protecting them, were nowhere to be seen. This was surprising, because Rasta's camp was only three meters away, so the lawyer reasoned that it was Sammy's soldiers who had fired on MPC. Meanwhile, my friend Major Patrick, a Congolese soldier working with Monusco and part of the contingent accompanying MPC, had heard the shots and gone running toward the camp, firing into the air. When he got there, he saw that the Congolese engineer, a friend of his from Walikale, had been shot in the leg, and he did what he could to help. After about half an hour, Rasta's soldiers came into the tent and confirmed that it was indeed they who had been firing weapons, but they insisted that they were shooting at Mai Mai who had come to attack Bisie. Nowadays, many people say this was a lie—that in fact they had witnessed FARDC soldiers traveling clandestinely to Bisie, and they had overheard these soldiers talking about a possible attack, which was to take place in the evening.

Major Patrick was concerned and didn't trust the soldiers' story. He decided that he needed to do "whatever he could to save these people." Patrick, not attached to Sammy's battalion, knew he couldn't count on anyone from that battalion for support, so he began quietly recruiting his own force of loyal former Mai Mai soldiers from Bisie, many of whom he had trained to fight against the RCD when he was in the Congolese army (FARDC) during the Great War. That night he slept in the tent alongside Brian, Yves, and his injured friend, the engineer.

The next morning, MPC's personnel went about their business—holding public meetings and promising hospitals and free schools—while struggling to believe that the incident was just a random attack by Mai Mai. But later in the day, Colonel Sammy sent soldiers to MPC's offices, ordering them to leave immediately using the forest route. MPC's personnel refused, having heard a rumor that nine soldiers were lying in wait for them on the road.

"That was the first and only time I ever spoke with President Kabila," Major Patrick recalled. "And it is when I first knew that MPC had powerful protectors in the highest reaches of government." The Belgians were frantically talking to someone on the phone, trying to figure out what to do. Major Patrick remembers, "The person on the other end of the phone wanted to know if there were any soldiers nearby who could assist them, so they handed the phone to me. And there I was at Bisie talking to President Joseph Kabila! He told me that, if anything happens to these guys, he'd know it was my fault." Further galvanized to get MPC safely out of Bisie, Major Patrick tried to get Monusco—the UN military wing—to land a helicopter, but Monusco refused. Following a suggestion from MPC, they instead contacted Banro, a Canadian mining company working in South Kivu. Like MPC, Banro had papers from Kinshasa, but it was facing local resistance in its efforts to expunge artisanal gold miners from Luhwindja in South Kivu and Salamabila in Maniema.

Banro agreed to send a helicopter, and Major Patrick went to Colonel Sammy to inform him of the plan. But, according to Major Patrick, Colonel Sammy refused, saying that the helicopter would create insecurity, since it could be used by Hutu FDLR or Mai Mai to enter Bisie illegally. Sammy insisted that MPC leave Bisie by the main trail, walking the over forty kilometers through the forest at night. And so Major Patrick planned a nighttime escape. When Banro's helicopter landed, Sammy's battalion tried to keep it from taking off, but Major Patrick stood in the helicopter door, facing them down. When Sammy's battalion pointed their weapons at him, he fired into the air, yelling to Sammy's soldiers that his orders came from a higher authority—implying the president and Kinshasa. Patrick got the white men on the helicopter without a shot being fired and left a few hours later on a Monusco plane, feeling he had saved the day for the company, a company that later would dispossess him of his hole at Bisie. Nowadays, the man who saved the "founders" of the world's second-largest tin mine ekes out a living as a very low-paid soldier living in a rat-infested wooden rental home with no toilet in Walikale. He waits for a promotion while others with far less military experience, some of whom even fought for "the enemy" during the war, surpass him in rank.

Before proceeding any further, it's important to address the question, "Who in fact fired on MPC?" The story that has come to us through NGO reports and journalistic media is that it was Sammy's 85th Infantry Battalion acting as a renegade unit to protect its main source of livelihood in the face of a company with legal rights to the mine. This take fits directly into the "blood minerals" narrative that would shape future events at Bisie and pave the way for the company Alphamin's occupation in a few short years. But this is not how anybody who was around at the time talks about it or remembers that time. Rather, they are adamant that others used the army to "create a problem" for the company—mainly the hole owners (though possibly also the rival company GMB), many of whom faced imprisonment, lives of exile in the forest, or worse. In contrast, those hole owners who were "without a debt," fewer in number, were more ready to collaborate with MPC because they had no one to pay off, and some clearly thought they might be able to obtain work with the company. Others paint even broader a picture, claiming that the revolt was essentially a popular uprising and that the army was only the instrument of the insurrection, the representatives of the people.

Thus it is that the Battle of Bisie brings to mind *Murder on the Orient Express*, in that everyone might have done it. But this is also insufficient, because the issue runs deeper. The real problem was that MPC's "mentality of Wazungu" was in conflict with the reality of Bisie, which was based on conflict-ridden collaboration. As one Ministry of Mines official put it, "They thought their papers were the whole world, but Bisie was more than papers. It was more than a piece of ground that you can buy the way you and I would buy a bottle of beer." The company's commodity and bureaucratic fetishism caused them to see Bisie only in terms of its future exchange value and to understand this potential only in terms of the material resources that were in the ground. They could not engage seriously with the whole social and ontological history of Bisie—the multiple kinds of work that had "opened" Bisie to the world. Moreover, they had failed to recognize hard-won local principles of peacebuilding forged through years of war: mainly, that peace comes from incorporating potential opponents into networks and allowing for the flow of value across separated groups to create assemblages predicated on cooperation (see also Ring 2006). By ignoring this, MPC had helped create a monster and seemingly closed themselves out of Bisie for good.

Blood Minerals

MPC was finally figuring out that its papers were just papers: fewer and fewer people were coming to them to sell their minerals (they had gone from getting five hundred tons a month to fifty tons a month), and MPC personnel were growing to feel insecure at their branch office in Walikale because of widespread discontent about their presence and provocations from GMB.

When they returned to Goma in apparent defeat, MPC began inviting humanitarian NGOs—including the globally influential Enough Project and Global Witness—and journalists from around the world to talk about the "blood minerals" of Bisie and how MPC had had its rightful property stolen by an army that was illegally involved in mining. People in Walikale said that the place was quickly flooded with journalists and, later, NGO representatives, their airfare paid for by the company. Some of these people showed up at Bisie and took photos of soldiers at the mines, and women and children at the river, washing minerals. They declared that this was proof of child labor, although, in most cases, the children were simply accompanying their mothers to the river. Sammy's army responded by banning women from the river, thus making it impossible for them to participate in anything directly related to mining. Kevin Bales, who said that everyone at Bisie was either a slave or an abettor of slavery, came to Walikale around this time—he apparently never made it to Bisie. So did *Vice* and the Danish documentarians, who arrived at Bisie just before I did, in 2009. Owing to all this publicity, Bisie soon came to be recognized as a hotbed of conflict minerals and slavery—ironically, several years after it had been occupied by an invading militia that funneled resources from there to Rwanda.

In addition to working with humanitarian organizations and journalists, MPC's personnel began seeking out higher-ups in the army to help them rein in or discipline Colonel Sammy. But every time they looked for a military higher-up for assistance, Sammy would end up cutting that general into the profits by allowing his men to collect revenue from tolls (Tango 4, the Mai Mai Colonel Bindu, and the Inspector General of the entire army were said to have been incorporated into Bisie in this way). As a result, Bisie became more militarized than it had been previously.

For a time, GMB was back in full force. Meanwhile, at Bisie, Sammy's army singled out and persecuted négociants and diggers who they believed were selling to MPC, treating them like wartime enemies: they followed diggers and sometimes beat them or threw them in prison after accusing them of selling to Mai Mai groups or the Hutu FDLR. When they did that, they drew upon the wartime practice of accusing ordinary people of work-ing for the enemy, when in reality they were trying to sabotage MPC. To secure their release, these prisoners were compelled to give money to the army after promising to abandon their "friendship" with MPC. While the army was cracking down on sellers and diggers in this way, they were at the same time responding to the emergent "conflict minerals" narrative by distancing themselves physically from mining. They even formed their own comptoir, apparently in an effort to eschew the accusation that they were entering mines or were directly engaging in mineral extraction.

The military tribunal was overloaded with complaints and accusations regarding Colonel Sammy, and the international NGO community put a great deal of pressure on the government to relocate him. Eventually, in late 2007, the military leaders in Goma called Sammy for a meeting; he waited around in Goma for a few months for his orders to return to Wali-kale only to discover, much to his surprise, that he was being dispatched somewhere else. This was also the end of GMB and its toll system as well as its collaboration with the army; all of this was widely understood as an effort to give MPC an entry into the community by eliminating GMB and the abusive toll system.[1]

However, MPC's directors still faced an uphill battle: they knew that, whatever their papers said, in practice they would have to pay off some of the debts of the PDGs, lest they incite an insurrection, and they needed to find a way to break the connection between the military and the hole owners once and for all. One of MPC's Congolese lawyers, whose uncle was at one point an important advisor to the president, informed me that it had been his idea to use the concept of blood minerals to close the mine at Bisie. As he put it, "[My uncle] asked me what they could do to bring in MPC, and I told him that the international community cares about blood minerals and that's a good reason to shut Bisie down. And so Kabila closed mining in the east in 2010."

Most people in Walikale after 2010 didn't need insider testimony to know that the events at Bisie were the major factor informing President Kabila's 2010 shutdown of artisanal mining in the east and the de facto international embargo of Congolese minerals that followed it. "Bisie was closed in order to open the way for MPC," after the people—especially the hole owners—had shut the company out. The company had "seen far," obtaining "Kinshasa papers," mapping and photographing the terrain from the sky, and situating their work and their struggle against the artisanals within a global narrative that spoke to the "international community." They had used the idiom of "blood minerals" against the armed actors and forms of labor that had formerly "blocked their ways," preventing the company from stealing what the diggers now cast as their "traditional home" and "customary place."

This popular insight reverses entrenched wisdom about conflict minerals legislation: while the standard story is that the passage of the Dodd-Frank Act in the United States provoked Kabila to illegalize mining, in Walikale people make a very compelling case for the idea that the whole narrative of "blood minerals" was largely rooted in the conflict surrounding Bisie and that Kabila's decision to stop the mining came from his desire to use this discourse to bring in, and potentially profit from, industrial mining. After all, Global Witness and the Enough Project, who were at Bisie covering stories of blood minerals after being sought out by MPC, were the ones who lobbied for the Dodd-Frank Act in the first place. According to this popular view, the Dodd-Frank Act emerged from the bottom up rather than the top down—mainly from Bisie, because the events there culminated in a confrontation between the Congolese army and the Congolese president himself.

There are a number of lessons to be gleaned from the case of Bisie: For one, the postwar violence at Bisie did not emerge from a Hobbesian war of all against all surrounding resources, as most people have assumed. Rather, after the war, violence emerged from the efforts of companies trying to consolidate their control with a view to expelling others, and the support that various Congolese actors gave to the companies depended on their perception of which company they thought most likely to expel fewer people. Exclusion was the basis for violence at all scales. For example, the conflicts between family members emerged in the context of exclusion— when companies competed with one another, and the family members got singled out by these companies to speak for larger entities; or when an association, acting like a company, was formed from a descent group (in order to escape harassment from state officials) potentially to the exclusion of certain members. Even the older conflicts between Bassa and Bagandula were not primordial conflicts between clans in a war of all against all, but emerged from the fact that the colonial and postcolonial regimes recognized only one group as having certain kinds of rights to certain territories. The inclinations of people on the ground were to overcome these differences, which had been imposed through colonialism, through such practices as marriage and childbirth (recall the child of the union of these descent groups, who was given the name Essa Yongo, the peacemaker). Too, it was the companies with their wartime military connections that threatened genocide on the population, while the army violated human rights when it became the agent of one company against another. None of this violence had anything to do with the artisanal miners and their practices of multidimensional collaborative inclusion, but with the exclusionary practices of companies working with state figures.

In addition, the case highlights the fact that the ideology of transparency was not something that only recently emerged with digital technology or neoliberalism, but that it goes back to the colonial gaze; transparency was employed as a technique of enclosure and expulsion by colonial and postcolonial authorities and later by MPC, in conjunction with "transparent" paper bureaucracy. Despite their mastery of the instruments and currencies of transparency, the hegemony of these powerful companies was not given or absolute—rather, they had to manufacture their ownership on the ground in real time by interacting and negotiating with others, and failure was a definite possibility. To be successful, MPC found that they had to leave behind the "mentality of Wazungu" and participate in eastern Congolese collaborative practices of inclusion to some degree. In particular, they sought out allies who could help them in the humanitarian industry, who in turn sought out allies of their own, reaching all the way to the halls of government in the United States. In trying to fight Congolese actors, the company ended up creating the kind of global alliance of humanitarian and nonhumanitarian actors that Congolese already imagined to be lined up against them, and which they had long depicted in their war stories (recall chapter 2).

EIGHT

Closure

There are a very few number of smelters globally, and those smelters produce a product that goes into things like an Apple telephone or computer. The onus of custody and source of that tin is placed on Apple itself, so they push that down to the smelter. So the smelter needs to know where does that come from, in the event that Apple is ordered, or anyone on the chain is ordered by one of the conflict free smelt initiatives. . . . So what that does is it sterilizes our mine or our product economically for anyone else. So that helps hugely for us.

—Boris Kamstra, Alphamin CEO, talking about the impact of "conflict-free legislation" and the conflict-free status of the mine at Bisie on the company's value potential on the company's website (www.alphaminresources.com)[1]

If it wasn't for Dodd-Frank, we wouldn't be on the hill [where Bisie is located].

—Trevor Faber, Alphamin COO, in an interview with Bloomberg News

Boris Kamstra, the CEO of Alphamin, comes across as calm and confident, which makes sense given that his company owns the world's most valuable tin mine—as of now, the second largest in the world, with a grade ten times higher than any other (Wilson 2016). The CEO is featured in a series of short video segments on the Alphamin website, which is almost entirely devoted to selling the future potential of this forest-turned-commodity to investors (the Canadian company Alphamin purchased rights to Bisie from MPC around 2012). For most of the video, Kamstra focuses on two main things: the astonishing purity of the ore at Bisie and the technical acumen of Alphamin's geologists and engineers. One of the few times he steps out of character and betrays actual excitement is when he's talking about the accurate calculations of skilled men blindly drilling holes into the ground at awkward angles in the middle of the rainforest. It may not be entirely fair, but I can't help but think of the soldier-colonizer in Ursula K. Le Guin's science fiction novella, *The Word for World Is Forest*, for whom masculinity, violent conquest, and aggressive deforestation for profit are all tied together in a single worldview and in a practical orientation aimed ultimately at closing other people out of the forest (1976).

The CEO's excitement about extraction doesn't prevent him from also romanticizing "nature"—in fact, these two attitudes go hand in hand. He waxes about this "sea of green . . . the most beautiful area you could ever go to." At one point he refers to it as the "Garden of Eden"—suggesting that, while it holds everything that humans could ever need, the devil also lurks there, tempting its indigenes to step beyond their limits and know the world at great risk to their own innocence. I suspect the Garden of Eden comment is a reference to the conflict-ridden history of Walikale, which he insinuates has something to do with conflict minerals, the Apple (pun intended) in this particular story. At various points, he alludes to the fact that the area that Alphamin has rights to is actually much bigger than the mine itself, and that the company plans to use the profits from mining its current site to develop more mines, which would of course mean cutting down more forest. The company's website says that it is part of an ecological initiative, but it also makes it clear that this is entirely voluntary, and that Alphamin has no legal obligation to safeguard ecology, because its rights over the area—its papers—are secure, inviolable, and total.

The temporality of the interview is future oriented, and the future is bright. Because the site that Alphamin owns is now certified conflict-free, the company is going to lead the world in conflict-free tin. Alphamin is going to bring untold profits for others, it is going to saturate the world with clean tin, and it is going to "develop" an area with a "tricky past" into a major tin-producing region for the world. Kamstra not so subtly presents Walikale as a timeless Hobbesian nightmare, a war of all against all: he makes vague references to "intertribal factions and all the rest," different armed groups "running over each other" in the forest. If you didn't know any better, when he says that Walikale was "originally a cashless economy," you would think that he was talking about an eternal past, and you might be persuaded that Alphamin's entrance into this society is indeed world historic. All the more so because, according to the CEO, the "traditional leaders" say that Alphamin is bringing hope back to the region, that "the lights are back in people's eyes," and "the thousand-mile stares" of a few years ago are gone. But when he says "originally," he's actually referring to

2012, when Alphamin bought the company from the comptoir MPC, at which point the society was cashless because artisanal mining had been shut down in the east two years earlier. And it is very likely that artisanal mining was shut down in the east, at least in part, to enable the company Alphamin's predecessor MPC to occupy its concession during a time when it was meeting with violent resistance from the local population, including the army. Because this is Congo, and people already assume it's the heart of darkness, one can persuade people that a situation that is quite contemporary, and that a company had something to do with, is actually a chronic condition—timeless and intrinsic.

While the CEO acknowledges that Bisie was home to artisanal miners before the company arrived (he even calls it El Dorado), he goes to great lengths to stress that artisanal mining was "terribly inefficient" and dangerous—that the diggers only had a 30 percent product recuperation rate and their thirty-meter holes in the ground were dangerously low in oxygen. He strongly suggests that those who were organizing and managing artisanal labor were cruel and villainous. He asserts that diggers were not allowed to alight from their holes until they had yielded a certain volume of minerals (I have never heard this to be true, although I cannot say for certain that something like this never happened to anyone). In discussing a riot by artisanal miners on International Women's Day in 2015, he claims that the diggers' cooperatives lined up women in front of them like a "human shield" and marched on Alphamin's base. Then the men, who were hiding in the background, fired on Alphamin's guards, hoping to incite them to in turn fire on the women and provoke an international incident. I have come across no real evidence that this ever happened.

Rather, what I understand to have happened on that day is that there were two separate protests, one organized by women down in the market town to which they were formally restricted, and one that took place high on the steep hill where the base is, in the evening, at which no women were present. Though it may have looked like something extraordinary to the expatriates at Bisie, events like the one that took place in the market are ordinary on International Women's Day, during which women throughout the country gather together, hold conferences among themselves, and call on powerful men to listen to their viewpoints, concerns, and grievances. During the protest that followed in the evening, male protesters, some of whom had hunting rifles, marched on Alphamin's camp, and the guards allegedly fired on them from on high, injuring several, two of whom later died in hospital. When the guards realized how large the growing crowd was and that a massacre would result if they continued firing into the mass, they abandoned their post, and the protesters rioted and looted the base and its store.

In the video, Kamstra presents Alphamin as a source of peace and development, at times equating the artisanal miners with violence and death. But, for people in Walikale—all of whom depended on mining in some way, even if they weren't themselves mining—the closure of Bisie in 2010 brought a total collapse and a severing of social and economic ties that threatened their very existence. The government, people said, had "cracked the whip" abruptly, taking people by surprise and plunging them into chaos; those with minerals were unable to move their minerals (there were no buyers) and accumulated debts daily while remaining unable to return "home" to face the music. As one person remembering that time put it, "The government demonstrated that it was still a colonial regime by abruptly deciding to implement closure. . . . They cracked the whip. One day they just woke up and said that minerals are outlawed and you can't buy, sell, or dig." Instead, they should have brought salt, meaning they should have compensated the diggers in some way, because everyone knows that, "When you go to Mbuti ["pygmy"] land, you bring salt." That is, if you go to a Mbuti area with a stick to take elephants or ivory by force, you might succeed the first time, simply because the Mbuti want to avoid conflict, but the second time will be different.

This tempo-politics, or the use of timing as a weapon, persisted between the company and the *wenye shimo*, or hole owners: Alphamin and the state wanted the hole owners to evacuate their minerals as quickly as possible, while the hole owners wanted to wait until they could get the best price possible for their minerals. The diggers responded by buying time, hoping that perhaps the mine would be registered "conflict-free" by the government, which would protect them from having to leave because they trafficked in "blood minerals." But the diggers were running out of time as they dug themselves deeper into debt.

Meanwhile, local politicians told them that they would "drink tea with hot pepper" (*mtakunywa chai na pili pili*) if they continued to oppose the state and the company. In this context, many miners and traders imagined a few possible futures for themselves: one involving violent struggle against the company; the second involving relocation to other less-productive sites; and a third entailing a sharing of rights to Bisie, in which artisanal miners and the company would collaborate to extract value from the ground. In 2012, after various visits to Kinshasa, the diggers claimed to have come to an agreement with Alphamin that each party would stay in separate zones and that artisanal mining would continue indefinitely. But a couple of years later the agreement was apparently abrogated, and the company began extending into the sites of the artisanal miners, provoking protests and even riots. Many continued to dig, and the hole owners continued to accumulate minerals and debt because they were prohibited from moving their minerals to market. Indeed, the mine actually expanded dramatically after the closure (by 2015, the mine measured twenty square kilometers and a thousand meters deep, too deep to be a legal artisanal mine according to the Congolese Mining Code).

But all of these minerals were now considered "bloody" according to the government and, apparently, the international law of Dodd-Frank, or what was locally referred to as "Obama's Law." A few hole owners used military or militia connections to move their minerals from Bisie, while the rest awaited a special digitized tag from ITRI (of which Alphamin was a member) that would temporarily render their minerals blood-free, as a prelude to evacuating them for good. By 2015, a small fraction of former miners (approximately fifteen hundred) worked for the Canadian company building a road for a pittance. As one person put it,

> Our diggers, they have been through so much. Probably between 80 percent and 90 percent have debts and can't go to town or anywhere else—they're stuck here. Some sold their homes in towns and can't go back to their families out of shame, so they've decided to stay in this place, their traditional place. They'd rather bring on a war than leave. And since neither group (the company or the miners) is willing to leave Bisie, there will be a war.

Captured Owls in Limbo

Having once been "hunters" (*wawindaji*) who search here and there for their livelihoods, after 2010 diggers likened themselves to owls (*ma hibou*), which flourish at night but are easily captured in the light of day. The idea was that the "light" of NGOs and media had been cast on Bisie, even if the information spread was false. The closure that resulted was a total experience encompassing everything from the workings of the state to the movements of the body. At best, it was a prison sentence, at worst a state of limbo akin to death. No wonder "everyone look[ed] like they're carrying hoes to their own funeral"; they were, "living for example" (*kuishi kwa mfano*), forced to play out a virtual simulacrum of actual life when really they were "dead while still alive," immobilized while watching others move around. As one PDG put it, "Those still at Bisie are *'prisonnièr libre'* [free prisoners]. If they see a motorcycle or car, it's on television [note the reference to watching other people's movement while one remains im-mobilized]. They've been closed/tied (*kufungwa*) there, like they're dead, but they're still alive." Others echoed the Bagandula in describing their situation as akin to "being at the Hague," somewhat free but actually in prison—this joke also suggested that they understood their work and lives to be crimes against international law.

Closure was experienced in registers at once economically and bodily. Local hospital records bore out miners' macabre renderings of their plight: from them, we were able to determine a tenfold (meaning about 1,000 percent) increase in disease rates, mostly from malnutrition, despite a steep decline in the overall population. Everyone had a story about some-one, often a hole owner, who became sick because of "troubles" and was unable to get treatment. In these cases, the worries related to the accu-mulating debt brought *presha* (pressure), a state characterized by an over-worked and overheated body in which processes were moving too quickly. Invariably, the indebted would collapse because of the weight of all his or her thoughts and responsibilities, specifically concerning the loss of house, family, and children. Sometimes these narratives involved people falling into permanent comas, a disease that mimicked the social condition they found themselves in—forever in a state of limbo, or waiting. Meanwhile the price for sex work had declined dramatically in pace with the collapsed mineral prices: sex workers who had earned anywhere from 20,000 to 50,000 francs per job at the height of Bisie complained that they could only expect 3,000 francs, or about three dollars, in 2015.

Meanwhile, "the state," or its officers, had at once withered and grown increasingly pernicious and exacting. Many government offices in Walikale had effectively closed shop, their buildings deserted. There was no mys-tery as to why they had vanished: During the height of Bisie, an estimated four thousand people per day had passed through the town of Njingala on foot, coming and going to Bisie, and the government received $2,000 from taxes by "doing control" (*kufanya control*) of the road, in addition to the uncountable money brought in by individual state actors representing different offices. A small percentage, perhaps a quarter, of shops were open in Njingala in 2015. Whereas during the "time of Bisie" trucks came ten times a day, now there was but one single truck driver that had been wait-ing for a month for all of the sites other than Bisie to be validated green, or "conflict-free," by the government so the driver could go to Goma. Unable to rely on mining, state officials went after nondiggers engaged in various activities, most of whom were women. They waited in the forest for people who had gone to get wood for charcoal, hoping to catch them for not having papers. As one woman complained, "When we sell wood now, we might have to buy ten different papers, and you have no idea what they're for, but suddenly everyone is coming to sell you papers!"

The closure of Bisie led to a closure of the social worlds that depended on Bisie. People ruminated bitterly on how gifts quickly turned into debts as people turned against or ran away from each at a variety of scales. After all, "No one would be in Bisie if it wasn't for debt," meaning that people wouldn't be able to work without regular cash advances from buyers and other financiers. I came to know two women who spent over a year in prison because their husbands had escaped their debt after the closure, and creditors came looking for the spouses to make good on it. Some people became refugees in neighboring countries or entered militias like M23 or the FDLR.

One digger explained, "We had a history of helping each other. If one group of diggers had found something, they'd give money to a group who was still cleaning their dirt, because we know they'll get something later. The hole will pay." But after closure, friends became debt collectors: "Hey! OK, you. Bring me the money (*we, leta pesa!*)." People started to run (*walikimbia*), causing divorce and rootless children vulnerable to joining militias. One man explained how debt entered and imploded marriages, thereby also thwarting the creation of futures through children and the incremental temporal process through which adults become ancestors. The sudden shock of being closed out made it so that relationships and networks predicated on enduring friendship or love came to be perceived retroactively as commodities with specific exchange values:

> A woman might come to Bisie to sell rice. So say she has two bags of rice. The husband might ask for this rice to sell it so he can eat and buy tools to work, which is fine, good. But now the mine is closed, and now he has a debt to her. If he can't pay the debt, he'll leave. At least 90 percent of people who met each other out there at the mine, who married there, divorced when the mine closed. Most of the time this was because of debt, usually because the husband had a debt to someone else, often the wife.

On a larger scale, the transformation of loans into debts fed local and regional violence, as certain individuals with exceedingly large debts became warlords, using violence to either avoid their creditors or pay back their debts. It had become a truism, if not exactly a proverb, that "if you have a gun, no one will come collecting a debt from you." No one epito-mized this more than Sheka, the coltan middleman and former friend of the Bagandula who became the most notorious and brutal warlord in the region. As of this writing, Sheka is accused of organizing the rape of more than 350 women in mid-2010, just after the closure of Bisie, and of abduct-ing thousands of children, who became soldiers. Eastern Congolese often use Sheka as one of several examples of a militia leader who employed acts of rape and violence to bring the international humanitarian community to him in an effort to get a position, or *cheo*, in the army—a strategy that never actually paid off for him. Despite having committed manifold atrocities to Walikale's people, he even ran for the position of Member of Parliament from Walikale in the elections of 2011, campaigning with his child soldiers in the middle of Walikale town while there was a warrant out for his arrest.

Sheka's alleged fast descent into madness became a kind of allegory for Bisie and its male workers, told in ways that were intended to persuade the listener of his absolute milquetoast normalcy before the pressures placed on him by debt. The idea was always that if Sheka could be transformed in this way, it could happen to anybody. Back when he was hired to represent the Bagandula Association, which later morphed into GMB, Sheka was a "man of development" (*mtu wa maendeleo*), a respected leader of civil society with no known history of violence. He studied English in Goma and loved soccer—some said he even taught English at an "institute," though others disputed this point. Many, though, referred to him as a teacher (*mwalimu*). Like any good coltan middleman, he "took money from the rich," who trusted that he "knew the secret of minerals." At some point after the time of closure, though, Sheka "got a problem" (*alipata shida*) and lost all the money he had received. And the actions he ended up taking in response in turn "changed his mind" (literally, his mind/intelligence changed: *akili yake ilibadilika*), dissolving the connection between his past and present selves.

According to a close friend of his I came to know, Sheka went to the for-est with a debt of $70,000 from various sources (some people cited other figures and other creditors) because he "worried for his life" in Walikale. The debt was not his fault, but the natural outcome of his success in the "America" of all mines:

> Yes, he had debt! If you have a hole, and it's bringing you $20,000 a month, who's going to deny you money? Who's going to think you can't get money? Money was everywhere. Bisie is money. Everything in Bisie is money—every rock, everything! Bisie was America. No one would deny a loan to someone who was working at Bisie or had a hole at Bisie.

But after closure, debt turned Sheka into an "animal of the forest": "he had to transform himself into a savage, into a cannibal, a thing of the forest, because of debt."

Many held that Sheka was hired by other middlepersons to scare off Alphamin. In the years after conflict minerals legislation and the government's validation of "conflict-free" mines, many also believed that Sheka was hired by Alphamin to prove that the artisanal miners and their mine was violent. According to one theory, Alphamin needed to make a case for the prevalence of violence at Bisie so the mine would fail to receive a "green" validation from the Office of Mines and the miners would be forcibly relocated to other newly validated sites. The turning point was said to be Sheka's invasion of Bisie in 2014, during which time Alphamin "ran away" and their store was sacked. Some claimed to have seen Sheka visiting the company afterward and reasoned that, at this point, Alphamin and Sheka began to talk with each other, and the company flipped Sheka over to their side. As one miner put it, "whenever they [Alphamin] want us to leave, whenever they want to start mining, suddenly Sheka shows up and kills people, or has his men rob them. Behind him, Alphamin is doing this so they can say that artisanal mining brings blood. That where there are artisanals, there is also blood."

Many people held that Sheka's indebtedness led him to seek out work from one or more of the factions at Bisie (the diggers or the company) at different times. One high-ranking soldier even related a meeting he had allegedly held with Sheka in his house in Walikale, just before the latter attacked people at Bisie. According to him, Sheka wanted the soldier's help to, among other things, invade and kill the Bagandula, the indigenous residents of Bisie, thus paving the way for MPC, which was at the time having a hard time bringing all the Bagandula over to their side: "Sheka was working for MPC at the time. He said he wanted to invade Bisie to make way for MPC." The plan, the soldier claimed, was to ultimately cast blame for the attack on the artisanal miners. The soldier's wife kicked Sheka out of the house "for wanting to pour blood over money. . . . And now here he is doing the things he said he would do."

Reopening? Cleaning Blood Minerals with Tags

After a few years of closure, people in Walikale were ready for any kind of opening, and it finally came in the form of the International Tin Research Institute's bag-and-tag scheme (ITSCI), discussed in more detail in the following chapter. Traceability schemes started at places nearer the road, closer to the cities of Goma and Bukavu, and came to Walikale a little bit later, always promoted with a view to alleviating the tensions between artisanal miners and the company. In Walikale, traceability was promoted in three different ways, with different objects: 1) in the form of a probably false promise to validate Bisie as "conflict-free" so that diggers and the company could dig there together, with the minerals from Bisie finally recognized as "clean" by the whole world, 2) as a way of relocating diggers to other validated sites that were less productive, and 3) as a mechanism for moving the technically bloody minerals of Bisie's hole owners so they could sell them, thus freeing them up while removing this thorn in the side of the company. When the plan was introduced to them by NGO and state officials, the diggers believed that Alphamin could be able to help them in this initiative, since they were connected with the white people (*Wazungu*) who had imposed the international conflict minerals laws and who also held this concept of blood minerals to be important, even sacred. As several put it, American buyers in high-tech firms like Intel might not trust African artisanal miners, but they would certainly believe in a company like MPC/Alphamin, their "brothers" in the same network. As one hole owner remembered,

> After 2010, minerals had been closed (*madini yamefungwa*), there's no market (*hakuna soko*), people are just living up there, but their money is in the hole. The *wenye shimo* and the négociants have sold their homes, their houses and the education of their kids, the whole future—it's all in the hole. People don't know what to do. And then finally in 2014 MPC [Alphamin] sent the *chef de poste*, and they talked to us at the church, and the cooperatives and Bagandula were there. And MPC [Alphamin] said that, in order to sell these minerals to the international community, the minerals can't have blood on them. So to show that there is no blood, they needed to have five holes for validation. They wanted to take samples from five holes to take to Europe, and after that the government would do validation and mining could continue for both us artisanals and Alphamin, working in separate zones. We were desperate, but we also believed that surely their brothers would listen to them more than to us.

In retrospect, diggers now believed that all of this had been a ruse—the company was trying to get the diggers' minerals certified for themselves, because it already had legally recognized rights to Bisie but needed the samples from the artisanal miners to finalize the process. According to the diggers, after these meetings, Alphamin started "doing prospection" and entering into other peoples' holes to secure samples. But they stayed longer than they said they would and extended their prospection into places that were not planned. When the miners protested, Alphamin brought in police—"an army that looked like it was going to war." These people allegedly shot at the miners in 2015, destroying their water pumps and shops while also allegedly stealing gas and goods from their stores. A group of diggers was fuming with anger when they narrated this story. One exclaimed, "I've never seen a country where police can just shoot at people! They shot at people for two days, and no one from the government has entered, and no one has asked who the enemy is and who they're following!"

Through ITRI's new bag-and-tag scheme, diggers and hole owners were also relocated to validated mines, which were in turn used to launder Bisie's ambivalently bloody minerals—ambivalently bloody because they were bloody for artisanal miners but not for the company. Most of these sites were not actually productive sites, and most of them were in conflict zones, but by bringing their Bisie minerals to these sites, miners could at least have them "cleansed of blood" so that they could be bought by foreign electronics companies, for whom blood minerals were taboo. As one Office of Mines worker remembered, "the tagged minerals are actually blood minerals [because they are in places where there is conflict or armed actors are controlling], while Bisie's minerals are blood minerals only technically, because they haven't been validated."

Some diggers were able to go to a 3Ts mine owned by an MP, who was able to use his political connections to get his mine at Busisi registered green, or conflict-free, by the government. Once in the scheme, these diggers also had to sell to his comptoir—if they sold elsewhere, they were dealing in blood minerals. One major irony of all of this, certainly not lost on any of the workers involved in the trade, was that Busisi was not actually "free of blood" according to the rules laid down by ITRI and the Congolese Mining Code—the army soldiers protecting the politician's mine were still there and receiving 10 percent of all profits from the diggers and traders.[2]

Alternatively, the diggers could leave the 3Ts altogether and instead mine gold, but the gold sites in Walikale were run by soldiers, and almost all of the diggers at these places were displaced diggers from Bisie. For example, in 2015, there were about a thousand diggers at a gold mine called Morocco, and they had "a bad Mai Mai problem," as one digger put it to me. They paid "taxes" to the Mai Mai who ran the mine in Morocco and, when they came to town they had the Congolese state officials following them, accusing them of helping the Mai Mai because of the money they were compelled to give them while there. This meant they had to "work with two governments" and spent most of their time in hiding in an effort to avoid the state authorities. Another gold mine was run by a FARDC general who held a monopoly on all commerce at the mine—only he could sell food, cigarettes, and water to the diggers.

Back at Bisie, the hole owners had been holding out for the company to pay off the debts they had accumulated over the years, some of them claiming amounts in the tens and even hundreds of thousands of dollars. But, after a while, the hole owners grew tired and frustrated from growing expenses and debts, and eventually each took payments of what they claimed to have been around $250 from the company and sold their materials for whatever price they could. Eventually, the diggers did receive special tags that would allow them to sell their minerals to a specific comptoir at a deflated price. Reflecting on the system of traceability at Bisie, one miner remembered,

> They tried to convince us that they were bringing us an opening, but it was a final attempt to move us when we were at our most desperate. Traceability as it is now is a contradiction: they give you an opening and then they close you out. It's like if I were to give you this beer and then tell you that you can't drink it. That's what traceability is now—it is not free exchange [echange libre]. Traceability would be good if everything was open and we could know where everything is, and where everything was going, but not if they're saying you can touch this, but you can't touch that.

Concluding Thoughts

In the early days of Bisie, some people, like the Bagandula, Sheka, and Colonel Sammy, tried to set themselves up as mediators with powerful others and to use their position as power brokers to extract rents. Through a practical orientation that Francois Bayart refers to generally as extraversion, Bisie helped people to turn the chaos and unpredictability of their lives into friendships, social networks, and incremental time—even as Bisie also unleashed unprecedented acceleration of events and prices. But some of these people tried to do something qualitatively different than simply practice extraversion: they tried to produce a direct and unmediated relationship between locals and foreign capital that would totally transform them, their situation, and even space-time itself, overcoming a violent recent history while connecting with alternate dimensions (the ancestors and the forest on the one hand and foreign capital on the other). What the artisanal miners and the Bagandula seemed to want was to use minerals to create an opening to another dimension where foreign capital existed in abundance so that they could position themselves in front of the opening in order to control movement. To carry through with the metaphor, we could say that, in order to trigger the opening, they needed to harness the power of the forest, ancestors, and diggers, which came to life through certain substances that also had particular properties and potentials (the social value potential of the black minerals being different than the potential of gold).

The tactic that emerged on the ground consisted not so much in trying to compete in the politics of extraversion, but to use the heavy and socially generative substances in the earth to generate sustainable social relationships that reflected the materiality of the minerals themselves (dense and heavy, requiring interconnected multitudes for their extraction). These minerals and the process of their extraction in turn seemed to take on the qualities that people wanted to see in their own lives and in the larger social world (consistency, incremental temporality, cooperation, transparency). Nonetheless, these owls, flourishing in the dark but easily trapped when exposed to the light, knew that they were at a disadvantage, and they saw their deficiency to lie in three interconnected domains: networks, space-time, and knowledge. These domains took on concrete forms like planning (being able to know when the registration is happening), visualization (mapping), and bureaucracy (having the right papers and knowing how not to be conned by unscrupulous men trying to sell you the wrong papers).

In other words, the tactic of unmediated extraversion (the wormhole) ended up being difficult to control in part because, to invoke a distinction made years ago by Michel de Certeau (1986), it is tactical rather than strategic. The company, in contrast, has the right networks (connections with the president) and knowledge (the ability to use aerial GPS to determine the exact location of Bisie on a map), which enable it to get the right papers from the right, more distant, place (Kinshasa rather than Goma), which in turn helps the company to secure its rights. This network's successful spatiotemporal expansion culminates in the incarceration of everyone else, who are either "imprisoned in the Hague," "living for example," or literally languishing in debtors' prison. In comparison, the knowledge and networks of ordinary workers at Bisie are ad hoc and emerge in the moment, evaporating (like many mines themselves) as quickly as they emerge: a stolen map, an "agreement with some Kumu guy," a fabricated stamp, a sacrifice correctly offered. To those with access to the right networks and bureaucracy, the people engaged in these contingent, bottom-up tactics appear "dirty" and "without *akili*" (intelligence, the ability to plan). Their intelligence was "in a hole" just like them, and their entire consciousness was reduced to the darkness of the hole.

Nonetheless, for those on the ground, Bisie became a model for a certain kind of peacebuilding, where everyone got to "touch money," whereas the model of peacebuilding that the company helped to devise and benefited from threatened violence by closing people out. In the years following the closure of Bisie, a similarly exclusionary peacebuilding strategy, materialized in the system of tagging, was implemented throughout the region by agents who came to be seen as secular priests touting a strange new ritual.

NINE

Game of Tags: Supply Chain Auditing
as Purification Project

In the aftermath of the Second Congo War, foreign companies entered concessions where artisanal miners had been digging for many years, often deploying the papers they had purchased from the central government in Kinshasa decades before. The Canadian company Banro, for example, asserted its rights over properties they had bought from Sominki before the First Congo War. By 2016, it was clear that I was witnessing a related major shift in the nature of conflict around minerals and in the governance of mineral extraction. Six years earlier it had seemed that soldiers dominated most mines and that the major violent conflicts were among different armed actors, but now they were between foreign companies and artisanal miners. In early 2016, protesters in the Maniema mine of Namoya, near the town of Salamabila, demonstrated against their expulsion, by Banro, from the gold fields they had been mining since the liberalization of artisanal gold mining under Mobutu in the 1980s. The commander of the police de mine, allegedly a former soldier in Jean-Pierre Bemba's militia, the Movement for the Liberation of Congo, fired upon the peaceful protesters, killing one. A year later, in 2017, the miners retaliated by forming a militia, which others called "Mai Mai," to resist the company's incursions, attacking Banro and kidnapping its foreign personnel. In Luhwindja, South Kivu, a similar story unfolded as the army prepared to eject more than six thousand artisanal miners from Banro's concessions after the company had worked with customary authorities (the woman king, or *Mwami Kazi*) to remove people who had been living there for generations (and something similar happened in Kamituga, South Kivu). We have already seen how, in Walikale, Alphamin asserted its rights over what it hailed as the largest untapped tin mine in the world. The famous blood minerals of

Bisie were not bloody at all when mined by a company: indeed, Alphamin presented itself as the progressive future of "conflict-free" tin, thanks to its papers from Kinshasa. To be as pointed as can be, I never heard any state or NGO actors argue that companies were dealing in blood minerals—this was a term reserved for the minerals unearthed by artisanal miners.

As police were shooting at peaceful protesters in Salamabila, many diggers were going through an auditing process to ensure that their minerals were not tinged with blood—in some cases (e.g., Bisie) after being expunged from the erstwhile "blood mineral" sites that were being appropriated by the "conflict-free" mining companies who were also partnered with the institutions carrying out the auditing intervention. Prior to this, diggers had been closed out more or less completely by the de facto embargo on minerals, which had led to a dramatic decline in buyers. As Congolese involved in the trade often put it, American companies knew their customers did not want to "touch Congolese blood" every time they picked up a cell phone or computer. The bag-and-tag tracking schemes were intended, in part, to end the starvation and violence that had been brought about by closure, and most people agreed that anything was better than closure. But the spatiotemporal restrictions that the scheme imposed, and the arbitrariness of the cultural and ethical values that were packed into it, made it seem to those involved in the trade like a foreign purification ritual that had to be acknowledged and repurposed to reverse the excommunication brought about by the ban. These concepts—church, priests, sacrament, ritual—often came up when eastern Congolese at the receiving end of the auditing scheme reflected at length on its meaning and purpose (in this context, the terms were synonymous with ritualized separation from the profane, or dirty).

Although there was initially competition among at least two different tracking initiatives, the most widely implemented new tracking scheme in the Kivus and Maniema, called the ITSCI project, was devised and implemented by ITRI, the International Tin Research Institute, a consortium of smelters and mining companies; at the time, ITRI included Alphamin, the company that was concurrently evacuating miners from the erstwhile "blood mineral" site of Bisie, which they now advertised as a source of "conflict-free" tin. ITRI worked in conjunction with a US NGO named Pact, which implemented the scheme in collaboration with Congolese state actors. Those "on the ground," the targets of the intervention, tended to have no idea of the difference between these two entities, ITRI and Pact, often using the terms interchangeably when discussing the system of "traceability" (a term they also used), or simply calling everything associated with it

"ITRI."[1] Though not state actors, these ITRI/Pact representatives were secretive and difficult to talk to and sometimes acted as if they were sovereign over all aspects of digging in the areas in which they were present, even though state actors did not see it that way (they may have also been wary of competitors or bad press).[2] Its promoters—primarily the NGO representatives on the ground—publicly extolled the new "conflict-free" bag-and-tag scheme for having the capacity to at once open peoples' ways and offer diggers a fair price fixed to the London Market Exchange (in fact, it ended up doing the opposite, by restricting mobility and enabling comptoirs who were registered as part of the scheme to set the price for those below them).

Grasping at Tags: A Brief History of the Mystique of Traceability

After the closure of artisanal mining in 2010 and the effective embargo that followed, a host of Congolese lobbying groups intervened in an effort to turn around the situation that had been put in place by the humanitarian apparatus. They went to Washington, DC, where they put pressure on political and corporate interests; one Congolese NGO director, whose Bukavu-based organization worked directly on issues related to artisanal mining, recalled, "We told them that this is not possible, that you still have blood on your hands, because people are still depending on minerals, and if you refuse to buy minerals you're killing people. So the companies said, 'OK, but let's make sure they come from a good place without blood.'" Ultimately, a conference was held in Kinshasa between members of civil society and the smelters, who belonged to the ITRI. ITRI drew upon its knowledge of and experience with the Kimberly Process for tracking conflict-free diamonds and proposed the bag-and-tag tracking scheme now known as ITSCI, or *ma tag*, among Congolese, which was first implemented in Katanga and in different regions in the east (North and South Kivu and Maniema) beginning between 2013 and 2015.[3]

ITRI's bag-and-tag system was aimed at monitoring the supply chain in such a way as to ensure that it was not "contaminated," as more than one representative put it to me, by "blood." In doing so, they pictured the chain from the point of view of the movement of the commodity and filtered out most state actors—notably soldiers but also other state actors who have historically profited from mining, including the DGM (Direction General de Migration), ANR (Agence Nationale de Renseignements), the chef de territoire, police de mine, the Office of Mines, and others. In fact, if one listens to the way ITRI personnel and Congolese closely involved in the project talk about the idea of the plan, it is clear that they conceptualize it as a system for tagging moving things in spite of people. In these accounts, the main actors, or rather actants, are places: the validated mine, where minerals are supposed to receive one type of tag (*tag mine*), the *centre de négoce* (center for exchange), where the minerals are supposed to receive a new, second type of tag (*tag négociant*), and the comptoir (buying house), where the tags are removed, and the minerals are placed in a container. From there, the minerals, along with the two sets of tags, go off to the smelters (usually in Malaysia), where the tags are said to be stored in a giant room-sized vault overflowing with countless tags in case the smelters' tech clients at Apple or Intel do an inspection. This, then, was a commodity-centered approach to the respatialization of the supply chain in that it placed the commodified substance at the center while selecting which people would be allowed to orbit around it. This commodity-centered orientation harmonized with the plan's neoliberal, market-oriented approach to peacebuilding; as one tech publication put it, explaining the project in terms of its understanding of the point of view of tech companies, "The idea is that if audits are even mostly accurate, and companies do stop buying from any smelter that drops out of the program's requirements, then local warlords won't be able to monetize control over mines, and attacks will cease" (Templeton 2016). But from the perspective of those involved in the trade in Congo, tagging was an especially flawed peacebuilding strategy—everyone knows that peace comes from many hands touching money, and a system that nullified people from the outset and only added them in as necessary evils at particular points was bound to create hunger.

At the heart of this asocial vision is the bar-coded tag. The tag is a gift in the sense that it was offered to diggers and communities to save them from the horrible situation of closure, but, as Congolese state actors sell the tags to diggers and miners to make money while also opening people's ways, the gift has become a commodity. Like the eucharist to which it is sometimes likened (as in, "if you're kicked out of the church of ITRI, you won't receive your sacrament"), this commodity has a dual nature: on the one hand, the tag is a material thing but, because it has a barcode, it is also half virtual, as a scanner is intended to be used to upload the information into a database where the geographical history of the tag can be read. It is of no doubt that the digital dimension of the tag is appealing to end-of-chain buyers; it promises freedom from the messiness and corruption of the world, all of which are amplified by the fact that the referent of all this work is Congo, the proverbial heart of darkness and Hobbesian state of nature for the humanitarian industry and just about everyone in the Global North. The project also mobilized a larger, ineffable set of hopes and associations related to the digital: as a well-known California academic who works on cultures of the digital age in the Global North put it to me, self-effacingly, when I showed a slide of an ITSCI conflict-free mining tag during a lecture at a major research university, "I know it's silly, but if I'm being honest I have to admit, when I see that tag in the slide, I feel warm and comfortable, as if everything is going to be OK."

The mystique of the project has depended on a concept of digitization that was largely false. If the tag is in fact purchased "illegally" from Office of Mines officials at the Goma airport (which it could be as of 2018) and not given away at the conflict-free mine, then the information that the scanner later reads on the tag is a lie, just like the photographs uploaded to the websites of humanitarian NGOs, as mentioned in chapter 1. The same is the case if there's no centre de négoce (which there almost never is), and if négociants actually have in their home sitting rooms bags full of tags from sites marked as conflict-free (which they often do—they can have tags mine and tags négociant that they've purchased). This explains why ITRI/Pact representatives were always so nervous when I asked them why there were all these tags but no scanners to upload the information anywhere. They usually stopped the discussion short, and a couple of them said words to the effect of, "You said you were just going to ask me some simple questions about how traceability works. This is something else now." They knew that, without a scanner on-site, it was impossible to ensure that the mining tag had originated at the site for which it was coded. Moreover, the viability of tags for end users rested on this mythology of pure information in a pure database unsullied by the social and, of course, by Congo—without scanners, that technological mystique was threatened, and at least some of those working on this project on the ground recognized this.

While, in practice, people are involved in tagging, the system is set up in such a way that the people are supposed to be determined by the simple, vertically tiered process, not the other way around. And the other way around—the different categories of people who are around determine the nature of a process that is more "sloping upward" than vertical—is the way supply chains in Congo have actually worked. Under the idealized system of traceability (that is, the way it's supposed to work), diggers with the proper things (diggers' cards) are involved, in their proper place (the mine), as are négociants and comptoirs, who also have discrete locations (centre de négoce and comptoir) and appropriate object identifiers (négociants' cards and papers) marking them as valid persons relative to the tracking process. Only two types of state officials are supposed to be involved in the process, and their intervention is intended to be surgical, specific, and spatially bounded: SAESSCAM is supposed to be involved with tagging at the mine, and the Office of Mines with tagging at the centre de négoce. Because this setup excluded a multitude of state actors, it attacked the socially thick and complexly variegated system of rights and regulations that constituted the mining sector, and from very early on many state authorities were very clear with me about the fact that, if they couldn't find a way around this, those excluded from the system would have to find a way to profit from it by opening up smuggling opportunities for others.

Despite the fact that the tag is half virtual and depends on an idea of digitization, the digital infrastructure of the tag relies on very material infrastructure (in particular, roads) and, in the absence of roads, very embodied practices: namely, walking in the woods. This is because, before anything happens, the government must first validate the mine as "green," or without blood. For this to happen, people need to go there and do what's called a "mission de inspection" to investigate the site. This mission includes representatives of the government (the Office of Mines), civil society, and Pact/ITRI (importantly, not diggers and local people, who are the only people who can really supply viable information about any site in the first place). If the site is "without blood"—meaning there are no armed groups, pregnant women, or children at the mine—then the Office of Mines writes a report saying that this particular site is green, and the International Tin Research Institute/Pact officializes the validation, after which the site can receive the sacred tags, which some miners jokingly refer to as a sacrament. The problem with this is that international NGOs (such as the German BGR and IMO) finance all the traceability projects in Congo, not the government, and these NGOs are unwilling to go deep into the forest (I was told any walk longer than six hours) to validate the sites—as one NGO representative put it, "they are afraid." And these very NGOs refuse to subsidize others to go into the forest without them, apparently because of a lack of trust. If no one goes to the site, the "inspection mission" can't happen, and the government won't validate the mine. In South Kivu, as of late 2018, there were approximately nine hundred 3T sites, but only ninety-two had been validated, and all of those that had been validated were easily accessible from the road. In North Kivu, there were a little over six hundred, and seventy-seven had been validated. In general, the most productive sites, those containing the richest and most voluminous reserves of minerals, were among these nonvalidated sites, and the vast majority of them had no "conflict," or armed actors. Since any nonvalidated site was technically bloody, whether or not there was conflict, it meant everyone working in them, or transporting minerals from them, was dealing in blood minerals.

Instead of completely closing these unvalidated mines, one official explained how, in an effort to "open people's ways," state actors ultimately agreed to "accept the shame" of allowing unvalidated (technically, but not really, bloody) minerals to penetrate and "contaminate" ITRI's purified chain:

> ITRI is one, but the work is many, and those who are bringing money [to carry out validation] are also few, and there are many places. Even those places that aren't validated, they're still not really closed. Diggers and négociants bring the minerals from these "bloody" mines to validated mines, which now become contaminated with "blood." Before, these minerals went to Rwanda, but when the government saw this was happening, they said that rather than allow all this money to go elsewhere, let's agree to this shame (*kuitika aibu*) of contamination and allow the minerals to get a tag. Often you can see tags here and there, and you can get them where there's no validation. Even at the South Kivu airport, you can get tags from [the validated mine of] Nyabibwe, and you can't say whether it's SAESSCAM or ITRI doing it [selling the tags]. It's impossible to know—SAESSCAM blames ITRI, and ITRI blames SAESSCAM.

While, in practice, state authorities had "agreed to the shame" of illicitly transferring tags, the in-practice enforcement of the rule placed authority in the hands of on-the-ground state actors who were in a position to target people in the trade in possession of technically bloody ore.

If the virtuality of the tag was appealing to outsiders, it is the tag's materiality that has made it attractive to so many Congolese actors—especially mining police, who suddenly had something to do (they could follow these tags and penalize people for not having them). In particular, tags allowed for the control of people's movement through space—these people became visible as "legal" or "illegal" actors through the tags, thus showing how the ideology of transparency that undergirded the tags was a form of control (I place the words *legal* and *illegal* in quotes because, while the rules of traceability were being violated by people without tags, it wasn't clear that any laws were being violated, because artisanal mining was still legal according to the Mining Code, even in non-green sites). After all, artisanal mining is not illegal in Congo, but not having a tag effectively is, so the new system, superimposed onto the Congolese Mining Code, essentially usurps it, as ITSCI becomes law and the International Tin Research Institute and its partner NGO become sovereign agents, allowing state actors to assert themselves in new ways.[4]

For diggers, one of the most frustrating things about traceability was how their movement through space was being micromanaged through the tags in a way that made them vulnerable to state authorities and those above them in the chain (a theme the remainder of this chapter continues to explore more fully). For example, before traceability, comptoirs were located in the border towns like Goma and Bukavu, where they could more easily evacuate their minerals through neighboring countries, mainly Rwanda. But traceability compelled the comptoirs to relocate to the provincial capitals so the movement of minerals could be monitored within the borders of Congo while also generating formal tax, a portion of which (the retrocession) was supposed to return to the province. What this meant for the supply chain was that the négociant had no real purpose: before, négociants performed a necessary task because they moved minerals from far-flung places to the border towns where the comptoirs were located. Now they existed only because they were legally mandated by the Mining Code—diggers did not have the cards needed to sell to comptoirs, nor were they allowed, according to the rules of traceability, to receive tags négociant. In the end, the diggers suffered at the hands of the négociants, to whom they were compelled to sell, despite the fact that the comptoir's office was often right next to where they worked. For example, in Kalima, Maniema, a comptoir rented a house that belonged to the old colonial mining company in the center of the concession owned by Sakima, and the diggers worked nearby; the négociants, on the other hand, entered the concession from places far afield to buy from the diggers and sell to the comptoir that was right there.

One of the biggest winners in this scheme turned out to be the comptoirs participating in tagging, in that négociants were required to sell to them, and so they had a degree of control over pricing (a monopsony) that they never had before (Radley and Vogel 2015; Diemel and Cuvelier 2015; Diemel and Hilhorst 2018). But traceability was attractive, at least in principle, to a number of different state and nonstate actors, some of whom were able to use it to make mining broadly generative in a way they had wanted for a long time. For example, NGOs that had been working in the mining sector used the bag-and-tag scheme to resuscitate the old idea of a "basket fund" for mining communities—where the fund was the concrete materialization of a possible future born of miners' work, benefiting families. This idea could be attached to tracking, as tracking would allow people to know how much was being produced in a particular place from which the comptoir involved in the scheme was buying. As one Congolese NGO representative remembered, "ITRI just wanted the chain to be clean of blood. They didn't care about the basket fund. That was the NGOs and [Congolese] civil society." Other NGOs working for women's rights used the new system to argue for gender equality in the mining sector, and women began to assert themselves in mining through "the manual of traceability." As the chair of the cooperative of women diggers in Nyabibwe put it, "before traceability, diggers and state officials harassed us and tried to keep us from digging, but with the new rule of sex equality, which is in the traceability manual, we were able to write the governor [of South Kivu] and convince him to defend our rights to dig." Moreover, provincial and national governments were able to use traceability to establish a centralized system of taxation for artisanal mining—a percentage of the value of minerals collected by the comptoir would go to the government in Kinshasa, a portion of which (the "retrocession") would be sent back to the provinces. The debate about what has happened to this money has become a point of dialogue and tension between politicians and communities because the value is widely known and relatively transparent. Finally, at least in the beginning, diggers and négociants also expected to profit directly from this "law," not only because their ways would be opened, but because the négociants claimed to have been promised a price set to that of the world market (*soko ya dunia*) or the London Market Exchange; for them, this was not only about getting more money but about achieving a kind of recognition for their work and work product from those "above" them.

In what remains of this chapter, I delve more deeply into the specificities of how this system played out on the ground as it was implemented in very different places—some geographically remote, some very accessible, some "peaceful," and some conflict-ridden. In these ethnographic cases, social actors deploy the concept of "blood minerals" (*minéraux de sang*) and the system of traceability to surprising and at times maddening effects. The first section, "Blood Minerals That Have No Blood," considers how traceability played out in Maniema province in 2016, a place and time that was free of violent conflict but overwhelmed with "blood minerals." As already discussed, most people in Maniema considered artisanal mining to be their "traditional work" and shifted between mining and agriculture based on seasonal variations in weather. The descendants of company employees tended to work Sakima's concessions, mostly because of proximity, while others either worked in the concession or in spaces further afield. Because of the absence of armed actors, the concept of blood minerals made no sense at all in these Maniema towns, but it was at the foundation of the tracking system, in that unregulated minerals were by definition bloody, and people were policed, punished, and taxed through this system. The second section, "Conflict-Free Minerals That Are Bloody," considers a couple of locations in the Kivus that were actual conflict zones, even though the minerals from there were considered conflict-free, or bloodless. The final section examines Congolese reflections on traceability as a foreign religion, or church, centered around a specific ritual whose dominant symbol was the sacrament of the tag.

Blood Minerals That Have No Blood

It's hard for us not to laugh as we watch Victor the mining policeman pushing his motorbike up the driveway to the old colonial mining company director's house that we're renting from the new company (Sakima) in the mining company ghost town in Maniema province. As discussed in chapter 4, Sakima operates the concession, renting out space to comptoirs while charging fees to the négociants who come to buy minerals from the artisanal miners, who dig among the ruins of the old industrial site. Victor's two jobs, as he understands them, are "to protect the expatriate," meaning the company, which is run by a Congolese from Kinshasa, and "to prevent the trafficking in blood minerals." Protecting the expatriate mostly means trying to catch people who try to rob the concession—he caught one breaking into Sakima's warehouse by descending from the roof with a rope just the other night and sent him to prison. Preventing blood minerals is new and came with the "new law" of traceability which, again, isn't actually Congolese mining law but a policy superimposed onto that law, which the government and police treat as de facto law. Victor needs the bike, which belongs to the mining police, to effectively police "blood minerals" by traveling between the many mining sites in the area, but the last several times we've seen him he hasn't been able to afford gas to ride it. Without the gas, the bike is a burden to him, and so we've taken to calling it his "cross" (*croix yake*). For foreigners, the whole system of traceability may be contained in its dominant symbol, the clean white tag with the black barcode, but traceability wouldn't exist without the mining police, who use the tags as instruments of surveillance to see whether minerals have been tagged at the mine. This system has helped Victor by giving him a raison d'être; as he puts it, before it was "craziness" (*fujo sana*), as diggers went here, there, and everywhere to sell minerals. The police didn't follow any of it because artisanal mining is not illegal.

The mining police work with SAESSCAM, the small-scale artisanal mining office, who in turn are supposed to educate diggers about which sites are validated and which are not, and the routes along which they can pass, but in practice they do not do this, and the diggers' lack of information allows the mining police and SAESSCAM to earn profit for themselves and the state. So, if Victor sees someone on the road carrying cassiterite, he'll first stop the person carrying the minerals and ask where they're from. He'll know from the description of the mine and by looking at the minerals whether the site they're from is validated. There is one site not far from here, within Sakima's concession, which diggers think is validated because there are so many people digging there, but it isn't. Since Victor doesn't have gas for his bike, right now he's focusing mainly on the diggers coming from that nearby site to sell to négociants. Without a functioning motorbike, he has all the more interest in ensuring that diggers remain in the dark about which sites are validated, because there are only a few places that he can police. If the minerals come from an unvalidated site, they "have blood" and so become the "profit of the state" (*profit de l'état*): the mining police sell the minerals to Sakima, and the profit is shared between the court and the police.

Diggers

If you're a digger in Maniema and your minerals come from an unvalidated site, whether in Sakima's concession or outside of it (and the odds are they do), this is what happens: first, you move with your materials, hoping to avoid the police de mine. If they find you, they'll take your minerals, or you'll pay a fine. On the other hand, you may get comparatively lucky and run into a SAESSCAM representative, who instead of taking your minerals or making you pay a fine, will sell you a tag de mine, which SAESSCAM is supposed to give out for free at the site. Now you will be safe from the police and protected by the system of traceability. If you don't meet SAESSCAM and manage to avoid the mining police, you can also go to the closest validated site where tagging is taking place and "wash" your minerals there or visit SAESSCAM in town to buy a tag, which is supposed to be free.

Even if your minerals come from a validated site, they may not receive a tag, because SAESSCAM's budget doesn't allow them to employ enough people to go to all of the validated sites, to say nothing of the many unvalidated ones. Therefore, they instruct diggers to meet them at strategic sites between mines. So, for example, all diggers who were mining at locations close to Chamaka, a day's walk through the forest from Punia town, were supposed to go to Chamaka to get a tag from SAESSCAM. But this gave the mining police a chance to trap the diggers, so even when diggers tried to follow SAESSCAM's rules, the mining police would be there to meet them on their way from the far-flung mines to the "center," Chamaka. The diggers were on their way to get tags, but because they didn't have tags at that moment, they were smuggling blood minerals. After we were walking back from one of these episodes, one miner complained to me that "the mining police are stronger than [President] Kabila in the forest." Emphasizing that this power came from ITRI's system of traceability, he continued, "They have guns, and they place the law [*kuweka sheria*—to put/place/impose the law]. *And they get this power from traceability*, which is supposed to make it so there are no guns where there are minerals!"

Another variation of this general theme played out among diggers who collected too few minerals for SAESSCAM to be bothered with tagging— these were largely groups of women and sometimes their children, who worked with them during school vacation, as if in the fields. Money earned in this way helped pay for these children's school fees. For example, in Punia, diggers dug in small mines by the road, where they would gather small quantities that they brought to the négociants. These places were never technically validated, which meant that they were bloody, and the minerals to be found there were blood minerals. Even if they had come from validated sites, though, SAESSCAM imposed minimums on the quantity of minerals that could receive a tag (at the time, it was twenty-five kilograms). While this "small work" (*kazi ndogo*) may have been insignificant for SAESSCAM, it brought a great deal of money to the very impoverished town of Punia and helped make up for the inconsistencies and insufficiencies of women's agricultural work. When these minerals became blood minerals, the mining police started coming to take these women's minerals by force, after which the police sold them to the négociants while making the diggers pay a fine—in this way, the police got paid on both ends. This put an end to the "small work" of women while driving diggers to larger mines where they and the products of their labor could be "taxed" more effectively by police and SAESSCAM. Some of the ma creuseur who had been in these spots went back to farming, while others went into mining gold, which was not being tracked in the same way. Others went deep into the forest where they wouldn't be found and shut down by police, then traveled with their minerals to faraway Kisangani, Walikale, or Shabunda (all in different provinces), where their minerals would also be blood minerals because they were outside the system of traceability, which was geographically bounded within the province.

While diggers had not gotten the price that they felt they were promised (that of the London Market Exchange), many were optimistic, in part
because they knew that they had the final say on what happened to these
regulatory efforts. In Punia, many diggers had already shifted to diamonds
and gold, even though these minerals were not their preference since, with
gold and diamonds, they could go for days without getting anything. One
man drove home the fact that ultimately traceability and state power depended on diggers, not on state officials and directives from above:

> Traceability depends on us, and there is no traceability without us. We know
> [tagging] will end because we've started to go to gold mines. When enough
> of us have done that, the mining police and SAESSCAM will leave where
> they are and follow us to those gold mines. Then traceability will be finished
> because no one will be around to enforce it. Also, no one will force us to go
> to [the administrative capital to sell] because there won't be any comptoir
> there anymore, because they won't be getting enough material. So we'll be
> able to go where we want again, and the police or the army or whoever is
> there at the time will get something small from us, and people will support
> us because we'll be bringing commerce to their towns.

Négociants

If diggers in Maniema found their movement curtailed by SAESSCAM and
the mining police, the négociants in turn railed against the comptoirs and
the high-level political actors (senators, generals, governors) who they insisted were connected to the comptoirs as investors. The main issue was
that traceability gave comptoirs the ability to control price by manipulating space and time. For one, the rules of traceability limited to whom négociants could sell and where: they could not go outside the province, and
they could only sell to comptoirs who had a relationship with the specific
smelters who were part of ITSCI. This spatial incarceration gave comptoirs
a great deal of leeway in setting price, because if négociants tried to go outside of the system, their minerals would be blood minerals by definition.
Even more infuriating to négociants was how traceability allowed comptoirs to manipulate timing. Before the ITRI tagging scheme, if a négociant
went to a comptoir with a sack of black minerals, the comptoir would
test the minerals with a machine and offer a price. If the négociant didn't
like the price or thought they could do better, they could simply leave, taking the minerals elsewhere. But with the new system, the négociant went
with minerals in a tagged bag, and the comptoir removed the tag and kept

it to be sent on to the smelter, along with the original mine tag. Once that happened, the négociant couldn't leave, because as soon as they walked out of the comptoir's office with a tagless bag, they were carrying blood minerals, which could be confiscated. Instead of controlling timing as before (they could delay and withhold their minerals for a period of time), now the négociant was forced to accept whatever price was being offered at that moment in time.

One comptoir became infamous in Maniema when he took this timing game one step further: the comptoir owner hired a well-known smuggler and currency counterfeiter who had allegedly made a lot of money for the RCD during the war to help the comptoir come up with a scheme to get more profit from his exchanges with négociants. The smuggler devised a method for removing and replacing the tags without anyone being any the wiser. When the négociants would come with their minerals to the comptoir, the comptoir would inform them that it was a busy day, and they would do the testing and weighing in the morning. In the evening, the comptoir and the smuggler would open the bags and begin the process of "dirtying" (*kuchafua*) the minerals by extracting a portion of the high purity ore from the total amount and adding rocks and sand, then splitting the extracted volume among themselves. In the morning, the négociant, who often would have already tested the purity with his or her own less sophisticated machines, would be surprised to discover that the minerals were less pure than they had expected. It took a lot of time and espionage for the négociants to figure out the ruse, partly because one of the things about tagging that had originally appealed to them was that it seemed to make "dirtying" (kuchafua) impossible (after all, "dirtying" is an old trick, one négociants are as likely to pull on diggers as comptoirs are on négociants). In this case, when the négociants complained, the smuggler got support from local politicians to create a new association of négociants—friends who were each issued négociants' cards and sold to the comptoirs at whatever price the latter wanted. This culminated in a major brawl between the old, "real" négociants and the new, "fake" négociants, which ended up with a number of people from both groups in jail for a short stint.

In 2016, the price that négociants in one Maniema town were getting from the only comptoir around was so far below the London Market Exchange price that they had originally been promised that they decided to boycott the comptoir. Instead of bringing their minerals to the comptoir, they kept the minerals in their homes—either with the bags neatly tagged or with the tags lying in a bag purchased from the Office of Mines somewhere nearby. This went on for some weeks before the comptoir sent

SAESSCAM officials, the mining police, and ITRI's NGO (Pact) partners to the homes of the négociants. They claimed that holding on to tagged minerals for an extended period violated the rules of traceability and that some of these minerals were not even tagged. Unless the négociants immediately went with these minerals to the comptoir's office, they would be considered blood minerals and confiscated. "Anyway," the contingent of officials reasoned, "where are you going to go with this stuff? If you go anywhere besides the comptoir's office, you'll be trafficking in blood minerals." As one négociant explained, "The problem is that ITRI is in the system now with the government, working alongside the police. ITRI has become a businessman like us. If you're a négociant and they find you with cassiterite in your house, they'll arrest you, and now your minerals become 'blood minerals.' Sure the soldiers might not be at the mine anymore, but they're waiting for us in the town. OK, so they wear suits and not uniforms, they're soldiers with suits, but I ask you what's supposed to be the difference?"

Ultimately, the négociants felt like they had fewer choices than the diggers, who were usually able to pick up and go somewhere else: "You see," one négociant implored, "We're in the middle of a crisis. We're forced to do what the comptoirs and the government say even though we don't want to, because if they close the minerals on us by claiming our minerals are blood minerals, then the crisis will be even worse for us. Our children will join gangs, and crime will increase." Moreover, they continued to be harassed by state officials who levied new taxes and sold them papers: "All this [tax] money that we're paying just goes into the hands of people [meaning it's not recorded]. Where do they go with it? But if you talk about this too much, they'll say that you're the one bringing blood minerals from South Kivu—that your minerals actually come from Shabunda [an actual conflict zone where some mines were controlled by Mai Mai] or someplace." One time the négociants planned to visit the Office of Mines in Kinshasa to protest about how the state officials were harassing them but were warned beforehand that they'd be arrested if they did so. So, instead of taking a direct plane from Kindu to Kinshasa, they took a circuitous route involving different modes of transport (boats and planes), which ended up taking them through three different provinces before they finally got on a plane to Kinshasa. They were quick to point out the irony of the fact that, while they and their minerals' movements were constrained, in Shabunda minerals were moving *ovyo ovyo*, or freely and randomly, without any interference from ITRI/Pact or the state, despite the fact that there was war between the government and Raia Mtomboki (the Mai Mai group, the infuriated citizens). This meant that while Shabunda's minerals were, by any measure, blood minerals, they were not treated as blood minerals, while Maniema's were: "We haven't entered into war here, but we're more closed out [of the market, but also more generally the world] than those who are making war in Shabunda!"

"Conflict-Free" Minerals That Are Bloody

In contrast to the bloodless "blood minerals" of Maniema, many of the "blood-free" mines in North and South Kivu were actually bloody. One mining town in South Kivu was a war zone (red zone, or *zone rouge*) when I conducted fieldwork there in 2015, with Congolese soldiers fighting in a war with a Mai Mai group (Raia Mtomboki), which was at the time allied with the Hutu FDLR. The state officials there made me pay a "fee" just to be in town because if something were to happen to me, they would be held to task and have a lot of paperwork to do. Almost every family in the town had at least one son working with Mai Mai in the forest, where they controlled gold and coltan mines. The profits from these mines went directly to the militia and the politicians that supported them in their struggle against people they held to be of Tutsi descent, including FARDC soldiers. The families would smuggle food to their Mai Mai kin in the forest, who would in turn send them money. The FARDC soldiers went out to fight the Mai Mai during the day and came back to the town, defeated, in the evening, where they would end up drinking and laughing alongside the families and neighbors of their daytime enemies, who were all subsidizing these very enemies.

The mine at that town was validated green and was part of the ITSCI initiative, so none of this violence technically affected the purity of the mine. But minerals from the surrounding sites and from the conflict-ridden territory of Shabunda were brought to the town to be "washed" (*kusafishwa*). This process happened in the open and was visible to anyone who happened to be there, including me. This work supplemented the income of the state officials, who were otherwise excluded from profiting through traceability, while also opening the ways of the diggers at the other sites. A similar situation unfolded in the South Kivu coltan mine of Numbi, set way up in the hills from the main road linking Goma and Bukavu, and only reachable by motorbike or walking. Sometime after 2013, Numbi was validated green, but for a long while it was surrounded by mines held by FDLR and Mai Mai militias. Thus in both of these towns, diverse state and nonstate actors looking for a way around the scheme were able to take advantage of the scheme's formal "object ontology," its strict focus on maintaining the cleanliness of minerals as they moved through predefined sites, without concern for the complexity of the larger context. As one Mines official there put it, "It doesn't matter whether there's war in the village or not. ITRI looks for three things: that there are no children, no pregnant women, and no weapons at the mine. Even if all around them is war and insecurity, it's OK because their validation is just for the site and not for the village."

By far the most violent "conflict-free" site in which I carried out fieldwork was Rubaya, a North Kivu mining town not far from the provincial capital of Goma, in Masisi territory. Masisi is a largely agricultural area originally inhabited by Hunde people. From the 1930s to the 1950s, the Belgian colonial government relocated Hutu and Tutsi from Rwanda to work in Belgian plantations there. Over time, the population of these groups increased and, after the liberalization of land in the 1970s, they purchased land and asserted their independence from the Hunde autochthones, who found themselves marginalized and increasingly dispossessed of land. During the Second Congo War, the RCD incited Tutsi there to join their forces against the Hutu, and later other people who identified or were identified as Tutsi (including Banyamulenge from South Kivu and Rwandans) were rewarded with land for herding livestock, which they acquired at the expense of Hutu and Hunde.

Rubaya, which residents joke means "bad place" (*rubaya* does mean bad in Kinyarwanda), barely existed before the coltan boom. When the coltan price hike and the war coincided, there were suddenly twenty-seven foreign comptoirs in Goma. At that point, six people who had been digging and selling minerals in Rubaya decided that, given the multiple militarized players that were coming in and making claims, it would be a good idea to "look for papers," as they put it. They went to Goma, and they "bought papers" for the entire carrière from the occupying government, the Rwandan-backed RCD. At around the same time, a Congolese Banyamulenge from South Kivu (who other Congolese now recognized as Tutsi and who they averred held American citizenship)—I'll call him Z.—started doing business there. Z. came up with a plan that differed greatly from that of the other PDGs: instead of going to Goma, he flew to Kinshasa to buy his own papers. This was the middle of the Second Congo War, and they say that the Office of Mines folks in Kinshasa, at the time essentially a foreign country, thought Z. was insane because at the time there was "no conversation or understanding" between the RCD government in the east and the Congolese government in Kinshasa. When he returned, Z. had papers from Kinshasa (national papers), but he never showed them to anyone, waiting for the RCD to leave before doing so.

When the coltan price collapsed in 2001, most of the comptoirs left. Those six people with their papers from the RCD in Goma remained as the recognized owners of that place, but life was at an all-time low: middle-persons sold their homes in Goma to pay off the debts they had accrued to their financial backers (some returned to "live in the villages," while others went to refugee camps in Uganda in the hopes of making their way to Europe or the United States; a few joined armed groups). In 2004, after the Sun City Agreement in South Africa, comptoirs reentered in this new context, under a new theoretically unified government constituted by the reaggregation of warring groups into the Congolese army and state bureaucracy. The prices producers were receiving improved modestly during this time, until 2010, when President Kabila closed artisanal mining, and everything collapsed again.

When President Kabila agreed to reopen mining in Rubaya, one condition was that ITRI be involved in the process of traceability. At this point, Z., who was now a senator, began brandishing the papers for his company, now called MHI, claiming that the entire mine belonged to him. But the other six had papers too—only their papers were from the province and, by most interpretations, Z.'s national documents were "higher than" (*juu ya*) theirs. When Z., or MHI, tried to refuse all other PDGs the right to access their holes, conflict ensued. To resist MHI, the other hole owners banded together and used the Mining Code of 2002 to form what they called a cooperative, but what others might call a rival cartel. They called it Cooperamma, echoing the French cooperate and the French-Swahili slang *fanya kope* (make cooperation); its leadership was comprised largely of Hutu landowners. Cooperamma claimed that, in the beginning, Z. had the smallest space of all of them, but that he had performed *maghendo* (under-handed acts, a term also used for smuggling) to get these papers from Kin-shasa. One time, when Z. tried to reach Rubaya, Cooperamma closed the road, and he went three months without being able to show up at his site.

At some point, Z. changed the name of his company to SMB (Société Minière de Bisunzu). The company then entered into an agreement with ITRI/Pact to be the sole comptoir involved in the system of tagged minerals, with ITRI/Pact assuming that Z. was the concession owner since he had these papers from Kinshasa. ITRI/Pact, following the manual of traceability, also seems to have acted as if everyone at Rubaya had to sell to SMB, because at the time that was the only company that had come to an agreement with ITRI. As one member of Cooperamma put it to me, "ITRI *believes* in the papers of Kinshasa more than the papers of the province." Pact, the NGO working with ITRI, held numerous seminars in Rubaya to teach people what ITSCI was, explaining that the scheme would allow middlepersons to sell their minerals without being hassled and taxed by multiple state authorities.

According to the négociants in Rubaya, in this newly empowered position, Z's SMB, now with the backing of ITRI/Pact, entered into a contract agreement with them, stating that if the négociants brought minerals to be tagged by ITRI/Pact and sold by SMB, they would be paid within two weeks. Négociants began bringing their stones, with an average of 730 tons a month passing through an inspection by the Office of Mines, ITRI, and SMB. But things started to change when there was an apparent falling out between Z. and his buyer in the United States. Suddenly it was a month or more before people received money from SMB, which caused production to go down since négociants couldn't forward money and supplies to diggers. Some middlepersons looked for interest-bearing loans from SMB to see them through. As one Hutu trader, who was also in Cooperamma explained, "You're supposed to get your money from the buying house ahead of time, so there can be movement, and you won't accrue debts. . . . But if that doesn't happen, you can get a debt. After three months, you won't be a businessman anymore."

The breakdown of relations between SMB and the négociants started in May of 2015. By September, there was "war" brewing in Rubaya between the diggers and the négociants and, in Goma, between the middlepersons and SMB. SMB's new Chinese buyer allegedly wouldn't buy the minerals without seeing them, and the négociants wouldn't allow the minerals to leave Goma until they had received their money. About 150 négociants camped outside the Goma office in an effort to ensure that SMB couldn't depart with the $5 million worth of coltan that the négociants had "sold" to the comptoir in accordance with the contract saying they would be paid within two weeks (it had been three months), and that debts from the comptoir SMB would never exceed $1 million. It turns out that the comptoir was also using time to his advantage: the world market price had dropped again after previously experiencing a burst, and so the value of what the company possessed had decreased. SMB was waiting for the price to go up. Unlike the middlepersons and the diggers, the comptoir/concession was in a position to use time like a weapon, holding onto the minerals to try and get the best price they could. As one négociant put it, "By waiting, SMB can destroy everyone else, and now people won't have money, and any forwarded loans to cover the costs of production will come from SMB. This makes SMB a money lender as well as a comptoir." And so a relationship based on reciprocal exchange and trust had imploded into debt which threatened to become war—a war of papers in Goma and, in Rubaya, a war of weapons.

In the years that followed, Z., the senator who most people recognized as a Congolese Tutsi from South Kivu collaborating with other Congolese Tutsi from Masisi, developed a dangerous rivalry with a Hutu parliamentarian, the director of Cooperamma. The Hutu parliamentarian and other landowners in Cooperamma leased out land on which other people, mainly Hunde and Hutu, "owned" holes. These hole owners provided access to diggers who worked under them and with whom they split the volume of minerals extracted. By 2015, the Cooperamma mine, which was outside of SMB's concession, had diggers from all over, including many who were displaced from Bisie by Alphamin and the government. Many of the diggers who were from Masisi had been dispossessed of land during the war, and their land had been taken by people who identified or were identified by others as Tutsi, who used it to herd livestock.

SMB used the system of traceability to make a claim on Cooperamma's mine: according to their understanding of "Obama's Law" and the "rule of ITRI" (*sheria ya ITRI*), only sites that were within the system of traceability were green, and only the senator's site was within this system. SMB argued that because it had "brought ITRI" and traceability to Rubaya, providing a way for Cooperamma's cooperative of négociants and landowners to sell their minerals in the first place, Cooperamma's mine should also be considered part of SMB's concession. Cooperamma resisted SMB's claim, but the Cooperamma négociants continued to sell the minerals they acquired from diggers at the Cooperamma site and the SMB site exclusively to SMB, the ITRI-affiliated comptoir.

Conflict ensued when SMB began placing security personnel at Cooperamma's mine to supplement the government's soldiers. The government had placed soldiers at the entrance to both mines, SMB's and Cooperamma's, in an effort to prevent nocturnal attacks between the two groups and to ensure that unregistered miners and traders did not enter the sites to dig at night. This was arguably a violation of the rules of ITRI, which say that for a site to be "blood-free," armed actors cannot enter. By adding their own guards, SMB was also making a claim on the Cooperamma site, owned by the Hutu and Hunde landowners and represented by the Hutu congressman. The situation turned violent when diggers made a deal (*walifanya kope*, or they made cooperation) with the soldiers, giving them a percentage of the minerals dug in exchange for access to the mine at night (precisely what the soldiers were supposed to protect against). But when the hole owners discovered that workers who were not among their own were working in their holes with the help of armed guards who were also profiting, they were angry because they had used their own money to develop these holes. So the hole owners allegedly armed the diggers under their command to shoot at the guards at night, using machine guns they had allegedly purchased from Congolese soldiers with the money from the ore they had dug or the ore itself. When they did this, these mostly Hutu gunmen were said to have made a point of avoiding the Congolese soldiers and firing mainly on SMB's security, who were Tutsi, some of them having served in the Rwandan-backed CNDP and RCD when they were murdering Hutu in Rubaya during the war. All of this violence allegedly happened while SMB's concession was "blood-free" and within the ITRI tracking scheme.

Meanwhile, Cooperamma's négociants were still selling coltan from the Cooperamma mine and the SMB-owned concession to the comptoir SMB, but looking for a way around it, because the comptoir was still holding onto the négociants' minerals and not paying them in accordance with the contract. The rules of traceability stipulated that the négociants had to sell their minerals on-site to the comptoir, and the only comptoir in town was SMB. To get out of this, the Hutu parliamentarian used his connections with state officials to grant the négociants permission to sell their product at SMB's office in Goma rather than the office in Rubaya. In other words, they were no longer locked into the geographic rules of traceability, which stipulated that the "tag mine" had to be affixed to the minerals on-site and traded on-site. Now that the négociants had the product from both sites, SMB's and Cooperamma's, in Goma, it was easier for them to sell to other comptoirs besides SMB: under this system, they estimated that they sold 40 percent to SMB and 60 percent to another comptoir.

When SMB discovered they were losing their monopsony, they were able to use ITRI and traceability to completely shut down mining in Rubaya, claiming that all of Rubaya's minerals were "blood minerals." They were able to argue this because of the past instances of violence, mainly the armed attacks on their security guards. The senator, Z., was said to have used his connections to call "the government," which in turn called in the army to shut down the mine. When the mine was shut down, the diggers and Cooperamma protested, and the Hutu congressman found himself in open conflict with the Tutsi senator. They met and decided to send two delegations—one led by the parliamentarian, the other by the senator—to Kinshasa for a meeting with the Office of Mines. Both groups apparently left the meeting in Kinshasa with the understanding that the cooperative Cooperamma would continue to sell to SMB and that mining would be "opened" again. It was also agreed that the government would register all diggers at these sites in order to reduce smuggling and facilitate the taxation of the diggers and the comptoir. But for the diggers and the cooperative, this was a warning sign because they understood this form of registration to be a first step toward industrialization by the company. Those with recent experience in Bisie knew that when a comptoir wants to "evolve" into an industrial company, as MPC did in Bisie, the company may try to identify workers with a view to compensating them in the future, while also limiting the number of people to be compensated since the population of diggers is always changing.

After the delegations returned, the soldiers continued to guard the mine, closing it to diggers and PDGs. In protest, the diggers shut down the road leading to Masisi, the major food-growing region for Goma, thus blocking the flow of agricultural produce to Goma. In response, the governor came out three separate times, each time promising to open the mine after certain conditions were met. The first condition was that the diggers reopen the road. The second condition was that the government wanted the approximately fifteen hundred workers' houses at the SMB site to be taken down, ostensibly to put an end to nocturnal mining. Relocating the workers away from the site would also make it easier to shut down mining in the future without open resistance. Finally, the governor wanted all diggers to be registered. All these conditions were agreed to, although the last two were highly contentious. Nonetheless, the roads were opened, the houses destroyed, and SMB brought a machine, the kind Americans find at the DMV, to take photographs of every digger and issue cards (again the theme of transparency and visibility, linked to the control of diggers' movement).

But after the diggers had carried out all these demands, the soldiers remained at the mines, and mining continued to be closed. During that time (early to mid-2018), Rubaya was an extremely dangerous place because the soldiers were ensuring that no mining took place, while guns proliferated throughout the town (many told us that guns were the only "inheritance" that mattered because if you didn't have a gun, your property would just be taken away). As few had money and many people had guns, targeted robbery and general theft were very common occurrences. The PDGs, or hole owners, were said to have armed themselves with AR-15s and machine guns in preparation for a war against SMB's occupation. PDGs were said to keep machine guns in their house and walked with AK-47s at night when they had to protect their holes. Those with guns were said to prey on whoever they could and sometimes acted in ways that reminded me of gangster films: we once witnessed a group of PDGs with concealed weapons beat a man with a stick for failing to greet them properly, with a respectful two-armed handshake, when he entered the establishment where they were sitting, while a Congolese soldier looked on, doing nothing. When the soldier was asked why he hadn't intervened in this case, his response was telling: he said he was only there to protect the mine (read: property) not to enter into personal conflicts. He seemed to be communicating a deeper truth about regulatory efforts—although, in his defense, the soldier also offered that the PDGs' violent actions could be an attempt to incite him in an effort to instigate a conflict with the company, hence his unwillingness to respond.

To sum up this far, in Rubaya, the ITRI conflict-free mineral scheme had enabled one comptoir, identified with a specific (Tutsi) ethnic group, to acquire a monopsony over trade in a multiethnic area. That comptoir then tried to use traceability and its connections with ITRI to expropriate land from another ethnicized (Hutu) group, which had been targeted for genocide by Tutsi-dominated militias during the war. During this time, rival groups fought against one another at the mine, which the company that was implementing ITSCI claimed as its "conflict-free" property; this violent conflict had an ethnicized dimension. Finally, the comptoir used ITRI and traceability to shut down mining completely, thus initiating a time of everyday war.

The Church of Commodity Fetishism

At high-profile mines near the city and near the road, it was harder to innovate ways to open people's ways. In these places, those involved in the mining trade felt more completely imprisoned by the (post)colonial gaze of state and international actors. They also often claimed a privileged insight into what the "international community" wanted from traceability and from them. For example, Nyabibwe, a cassiterite mining town where the mine (Kalimbi) was literally on the side of the main road linking the provincial capitals of Goma and Bukavu, was the only place I conducted research where everyone meticulously followed the "rules of traceability." Kalimbi, the first mine in all of eastern Congo to "receive" the ITSCI initiative, was a kind of Potemkin village for traceability, regularly visited by emissaries and dignitaries from foreign countries and companies, including a couple of European princes (allegedly), Martin Kobler (the head of the United Nations Stabilization Mission in Congo), and representatives from Motorola, Nokia, and Intel, among others. Because it was so close to the road, any violation of the rules would easily be noticed by anyone paying attention.

Diggers at Nyabibwe, especially those who worked for the cooperatives that sold to the négociants, were obsessed with being watched. They claimed that, some years ago, Martin Kobler came out, held out an iPhone for everyone to see, and informed them that minerals from Nyabibwe were in that phone. "Martin Kobler told us to please take care of this mine, to protect it well, and to not bring conflict to it. That was an order to us from above! The international community is watching the mine all the time. They see everything we do." Another added, "Kobler told us 'there's a satellite following this place so do everything you can to avoid conflicts because everyone will know.' And he showed us his iPhone, which none of us can even dream of having, and said it's from here, so this is an important place (*nafasi ya juu sana*), and you must protect it." Diggers at Nyabibwe were convinced that these satellites followed them wherever they went and recorded every conflict that went down in the mines. In order to prevent conflict and make sure that the police didn't come to the mines, which could make the minerals bloody, the diggers' cooperative established its own security force and imposed a system of heavy fines for any instances of violence: grab someone by the shirt, pay a fifty-dollar fine to the cooperative. Punch to the body: a hundred bucks. Punch to the face: two hundred bucks.

As one miner put it, "The eyes of the world are on us. The satellites up there are watching what happens here, like a film, and we get a call from the government that there's a conflict in the mine before anybody else knows." They made explicit connections between the minerals in the mines and the gorillas in the nearby park: "If you touch a gorilla or do something to one, they'll catch you on film and get you later, and it's the same with minerals from Nyabibwe."[5] When I expressed my doubts that Nyabibwe was like *The Truman Show*, they challenged me in turn: "How is it that we have a minor conflict between the cooperatives about a water pump or ownership of a tractor, and the next day the Office of Mines is on the cell phone, asking me about this conflict?"

Diggers at Nyabibwe felt that they did everything they could to "follow traceability," but, as one put it, "it has become colonialism under the guise of traceability." If it wasn't for traceability, they insisted they'd all be rich because they'd be able to sell their product everywhere for a good price. The tags were *craca*, handcuffs, preventing them from moving in the ways they wanted and tying them to more powerful comptoirs. Moreover, state actors had used traceability to bring all kinds of tax, including the dreaded tax to the customary authority, or Mwami, who didn't even live near there. If they resisted, the government would close the mine, claiming that there's conflict between the government and the diggers, so now all these minerals are "blood minerals."

What the system of traceability communicated to them most of all was that the mine and its products were sacred, while miners' lives were not. As one digger put it, "They love that mine; they take care of it very well. But they don't care about people; they don't care that we're being tormented— sacrificed." Another added, "We are being sacrificed so the mine can be pure. . . . If someone dies in the mine, it's like a fly is dying, so long as it's not the result of conflict." If the international community cared about diggers, then they and the state would have made good on the initial promise that diggers and négociants would receive a better price for their minerals—getting the price of the world market (the *soko ya dunia*, which many imagined to be an actual market somewhere) or the London Market Exchange meant being recognized as an equal part of the world. Instead, they were compelled by the rules of traceability to sell to the comptoir in the provincial capital of Bukavu, which used its monopsony to "set the price": "When that *craca* [handcuffs, meaning tag] is closed, it's the same as closing the price." Anyway, if the international community actually cared about people's lives, they would have heard more about "blood minerals" during the war, when they were being killed. Claude, the head of one of the diggers' cooperatives, remembers all the terrible things that happened to people here and the millions of dollars' worth of cassiterite that was taken to Rwanda. All those trucks. How the RCD used Hutu informants to flush out other Hutu and others who came to be identified as "enemies." He remembers everyone who was taken to the top of that hill where the Vodacom cell tower is now, above the soccer field, to be shot in the head. One time, Claude was taken up there to be shot for trying to retrieve the radio that the soldiers had taken from his house, but the assassin changed his mind at the last minute because he "liked his face," after which he became the assassin's personal healer because the assassin assumed he was blessed. "No one talked about blood minerals then," Claude once complained to me, shaking his head.

At Nyabibwe, it was especially clear that the mine was clean and the miners unclean and that tags were intended to maintain the former's purity. If, in other locations, it couldn't be ensured that mines were actually pure, people still understood maintaining the illusion of purity to be the goal. Throughout the Kivus and Maniema, themes of conversion, religious taboo, sacrality and purification, and sacrifice were common in discussions of tags. Minerals without tags were impure to outsiders: as a négociant in Maniema put it, "ITRI is a church and traceability is their bible, if it says it is going to be this then it is that. If you bring discussion, they'll kick you out of the church, out of the minerals, and you won't receive your sacrament." Victor, the mining policeman in Maniema, made it clear that the NGO training course on conflict minerals he had taken had educated him in this cosmology of purity and that this conversion had transformed him into a true human being when he said,

> I know you Americans think that we're still living in trees, but when you go home I want you to tell people that we're not. We're human beings. I know the difference between a mineral with blood on it and a mineral that doesn't have blood. If a soldier enters into a mine, then the minerals in that mine have blood on them, and the entire mine is also bloody. The companies won't want those anymore because they are not clean.

He understood that Americans wanted their minerals to transcend Congolese history and to be separated from the image Americans had of Congo and that this was an important aspect of the work traceability was accomplishing. Another put it this way: "If they catch you with minerals without tags, it is like being caught with pork for a Muslim. Like you've murdered someone—how do you say it? Tabu." For diggers, at the core of this ritual was the sacralization of materials at the expense of sacrificed human life. If ITRI was a church, as diggers said, you might call it the church of commodity fetishism—a religion that elevated the minerals to be used in cell phones and electronic devices, and the technologies themselves, to the status of holy objects (and the mines as rights-bearing actants) while treating Congolese people like inconvenient things who potentially made sacred minerals impure.

Selling Indulgences: The Good Cop

Tracking is an element of what has been called audit culture, which we increasingly experience through digital devices, as we use these technologies to track metabolism, track spending habits, track the personality of the whole person, track the number of cars in parking lots (Strathern 2000; Shore 2015). In the tagging schemes, what was being tracked was Congo itself, which was to be ritually removed from the digital devices that seemed to allow the mind to transcend the limits of body and place. Congolese in the mining trade definitely picked up on what is at stake in this ritual: they understand outsiders believe that Congo is dirty and violent and that, if technology is to be clean, it has to be released from Congo, which means tracking and monitoring and removing as many processes as possible from Congolese hands. But this removal is itself violent, because everyone knows that peace comes from many hands touching money. Congolese recognize that miners—many of them dispossessed former farmers and traders pushed into mining during the war because of the global demand for mineral-rich electronic devices—are more powerful than anyone cares to admit—that in these mineral-rich areas, communities touch money only when artisanal miners are extracting value from the earth. No company can replace what they do, as the drama at Bisie demonstrates, and no tracking scheme can truly prohibit people from making money in the way they want. As many people said to me, "One day you'll see that traceability has ended in just an instant, within minutes. Movement will return to us."

Let's return for a moment to Victor, the cross-bearing mining policeman in infrastructurally remote Maniema. Before coming to Kalima, Victor worked for Banro at Salamabila, where he learned an important lesson about closing people out and following the rules too closely. As Victor recalls, Banro wanted to close all the artisanal miners out of the industrial mine, and it was war with diggers every single day: "You've closed us out of here—how do we live?" they lamented. For a while, they were saved by a new police leader "with intelligence," who said to Banro, "We'll protect your place, but let them come and do *maghendo* [smuggle, bend the law] once in a while. After all, these people have been digging here since colonial times, and you can't just close them out because they have nowhere to go." Victor then drew us an analogy of a snake trapped in a house in which the diggers were the snake and the house was the property of the company (again, the digger as snake is an oft-repeated metaphor, with overtones at once vernacular and Biblical, which was sometimes embraced by diggers and sometimes thrust upon them by state authorities or others wishing to insult or demonize them). Victor explained that, if you want to get a snake out of your house, or even to kill it, you have to open the door a bit and give it a chance to leave you in peace in your home; if you close all the snake's ways, it will try to kill you and end up taking over your house. So the police "opened up a way," and it worked—"there were parties all the time"—but in the end Banro got rid of that police commander and installed new police. When the new police came, Victor tried to teach them, drawing on the training sessions he had taken through ITRI/Pact. "But they just scorned me, saying 'oh, you think you're from the Sorbonne?!'" The former MLC militiaman who shot a protester in Salamabila was a prime example: instead of matching protesters' sticks with sticks, he vowed that "if people play here, I'll kill them." As Victor put it, "His mind was still in the war, he didn't realize that that time was over," and now he was paying for his indiscretion with imprisonment. What Banro had failed to realize was that "to close everything brings problems. You must open a small path so that people can live and get something. That's what living with people means." But Banro learned this soon enough when people rioted and stole everything Banro had in Salamabila and, afterward, when they kidnapped Banro's personnel. Riots were expensive, and Banro had to bring food every day to try to prevent people from breaking into the stores.

Victor took the lessons learned from that time, how to come to an agreement with diggers, and put them to work in tracking, even if what he was doing violated the "rules of traceability": "If the diggers get some *mchanga* [sand] rather than blocking them, I say, 'Come, and we wash it together' with the police." The flexibility that some mining police engineered into the system was intended to help not only diggers but government officials as well. For example, SAESSCAM officers were not supposed to have holes under the Mining Code or the new rules of traceability. As Victor put it,

> Does the law say this is OK? No. No, I studied traceability, and no one in the government can have a hole or be allowed in this work. But this is Congo. There at the *chantier* (quarry), SAESSCAM says they need a hole "to receive guests" (*kupokea wageni*). If I want to arrest them, they say, "Go ahead, but how will we receive you in the future? If we have this mine, then we can receive you—you'll get something small, and without us, you'll get nothing."

Moreover, because SAESSCAM was located at the mine, they could become proxy hosts for other government agents. That is, if personnel from the Office of Mines come to do some work at the mine, they show their "mission order" (*ordre de mission*) to the chief, who in turn sends them to the mine so they can be "received" (*kupokelewa*) as guests by SAESSCAM. Thus it is that the SAESSCAM personnel refer to their hole as "the cafeteria"— another reference to the old mining companies and their practice of food provision—that allows them to provide visitors with food and soap when they visit. Of course, in opening people's ways, the police were also trying to save the system for themselves, because if everyone left because they were being harassed, there wouldn't be any system at all anymore, and no state actors would touch money from mining.

In short, Victor understood that he had to allow for what Abdou Maliq Simone calls the "infrastructure of people" to become more significant than the pristine tracking of things, "opening people's ways" even if he was also simultaneously closing certain people out (2004).

Concluding Thoughts

It was never clear to me whether tech companies truly believed tagging would work or whether it was simply that the smelters understood the culture of the digital age in the United States and the related enchantment of a bar-coded tag well enough to be able to sell this idea to them. My fieldwork was limited to Congo, so there were a lot of things I had to infer and which Congolese had to infer as well, and this is one of the main points of this chapter. On the one hand, tagging seemed to be a very specific and ethnographically concrete example of the hubris of technosolutionism, as tech companies and those acting in their interest seemed to imagine that they could transcend Congo, and the "impure" improvisational practices of Congolese, with digitized tags and a database. On a basic level, it was ironic that the corporate response to the recognition that there was potentially violent geologic materiality in digital devices was to propose a technological fix, to double down on digital solutionism. The case of tagging also shows how the fantasies of virtuality and disembodiment of people in the Global North in general, and Silicon Valley in particular, ended up contributing to a viscerally exploitative capitalist system by crucifying those at the bottom on the cross of traceability. For many, traceability seemed like a church, not so much in the sense of being overtly "religious," but in the sense of excluding those deemed unclean and enacting a kind of cosmology of transparency and enclosure. It also seemed to build on enduring colonial ideas of light and darkness, the pure and the impure—concepts which Congolese understood to be part and parcel of their long-term interaction with whites/Europeans going back to the missions.

From the point of view of many involved in the trade, the theory that informed the practice of tagging was exclusionary and separative, and in this it shared something with the violent exclusionary practices of extractive companies that were asserting themselves in Congo and were also sometimes partners in ITRI. And it also had something in common with the entire colonial history of closing people out of the sources of movement, from mining to the forest itself. Tagging, which sought to separate out certain kinds of things from other things, parsed out certain actors from other actors and most actors from the things, people, and places that allowed people to collaboratively generate value. It was almost incidental that ITRI was enacting a specific Western vision of how states and economies should interact with one another (for example, that certain state actors should remain separate from, but also monitor, economic actors from above, while other state actors, mainly those with guns, should remain invisible), or that the intervention also tried to impose a software design logic onto the system of nation-states and national sovereignty, consisting in an attempt to change the geopolitical order into something like what Benjamin Bratton calls the stack—a vertically thick and layered, rather than horizontal, geopolitical cartography (2016). What bothered Congolese more generally was that the atomizing and separative logic of the tagging scheme flew in the face of people's need to move, touch money, and *fanya cope* (make cooperation), concepts that emerged, in part, in response to and in conversation with this intervention and others like it. Traceability sacralized the separateness of commodities, be they places, minerals, or technologies, over people's need to live and move.

And so the oft-mentioned avoidance of blood that was built into the system seemed to be an avoidance of not only violence but also of blood in the sense of life and vitality—flesh and blood. What the "international community" seemed to want to do was to carve out certain mines, from which their stones were derived, and to separate them from the rest of Congo—its violence, blood, and dirt—so to ensure that their minerals were "clean" even if Congo was not. But this also meant separating people from their livelihoods and things from the social relations that made their extraction worthwhile. There was something very familiar and disconcerting about these efforts on the part of outsiders to block ordinary movement in the name of a putatively universal law that was actually highly particular. It evoked long-established racial hierarchies and ideas about the unclean; as Achille Mbembe has put it,

> Life under the sign of race has always been equivalent to life in a zoo. . . . These animals are taken from their natural habitat by humans who, having seized them, do not kill them but instead assign them to a vast subdivided enclosure, if required into several mini-ecosystems. (2019, 167)

To be sure, tracking probably didn't have to work like an antisocial machine. It's possible to imagine a situation where something like traceability might have worked out well for diggers and their families and allowed people living in artisanal mining communities to benefit from the "opening of their ways." They could, for example, have been asked to participate directly in the process of validation, which would have made it easier to validate all those sites (roughly 90 percent, as of 2018) that remained unvalidated. That way, they and their homes wouldn't have been arbitrarily criminalized because some urban and foreign elites were too cash poor, suspicious, or scared to walk in the woods. It's not hard to envision a system of local communities using certain techniques and technologies to partner directly with foreign companies, but that would have meant listening to and taking seriously people and practices that the NGOs and the companies treated as intrinsically unclean and violent.

Those involved in the trade could at least take solace in the fact that the whole edifice relied ultimately on the work of diggers, and if they were harassed or closed out too much, it would collapse under its own weight. Diggers were not abject victims, because the state came to them, and it was they who made the state possible; so if they were to leave and go to the forest, it might look like they were being expelled, but it would also be sabotage on their part, in that it would put an end to the site-specific project of "transparent" traceability, as state figures followed them somewhere else. And movement would begin once again, anew.

Chains, Holes, and Wormholes

There was once a dream in the West that the information age would free us all from past limits to progress and expansion (including ecological limits), even from the history of violence and dispossession that has accompanied capitalism. That dream of final release from the earth and everything associated with it didn't just pop up in the so-called digital age—it has a long history that could be traced back, in some form, to the Enlightenment and probably further back (see, e.g., Rist 2008). At least since the late nineteenth century, the idea of the Congo, the proverbial heart of darkness, has often been the inverse accompaniment to this promise of progress as freedom from earth and its attendant meanings, even if as an absent presence. Of course, the so-called information age did not make good on this promise. Rather, the dependence of "the digital" and the virtual on minerals that come from the earth—including one of the two most significant and threatened rainforests on the planet—makes it abundantly clear that computerization will certainly not allow us to leave the earth behind, no matter how digitally networked we are, or how much time we spend in *Second Life*, or whatever. At the same time, there are new things on the horizon, emerging out of and posing a challenge to the old world order, and among them are the transdimensional worlds produced around artisanal mining, brought about in part by the demand for minerals in the "information economy." In this conclusion, I want to briefly consider some of these potentials; but first, I want to sketch out the picture of contemporary global capitalism that I believe emerges from this material.

As I'm finishing putting this book together, I find myself reading a newspaper article entitled "Congo Mine Deploys Digital Weapons in Fight against Conflict Minerals" (Ross and Lewis 2019); it concerns a new traceability scheme at the mine at Rubaya in North Kivu, not too far from

Goma. A German company (RCS Global) has initiated a new project for the comptoir SMB (mentioned in chapter 9) in which the data from the "conflict-free" tag is claimed to actually be uploaded in situ onto a database using GPS technology and scanners (a technique for disentangling information from Congolese social practices and histories like paper-based record keeping, which I gleaned to be the ambition of ITRI/Pact, but one which had never come to fruition because there were almost never scanners on-site). The article begins with a familiar juxtaposition, inciting the reader to contrast the muddiness of Congolese "nature" with the transcendent ethereality of the futuristic technological cloud, detached from muddy earth:

> In a small shack overlooking muddy pits hewn out of eastern Congo's rolling green hills, a government official puts a barcoded tag on a sack of ore rich in tantalum, a rare metal widely used in smartphones. With a handheld device linked to a server in the cloud, the agent scans the barcode, uploading data including the sealed bag's weight, when it was tagged, and by whom. (Ross and Lewis 2019)

The new intervention is also depicted in terms of an evolution away from the materiality of past interventions, as up until now, "companies mostly rely on a paper-based certification scheme," which makes it easier to "taint" (read dirty) the product with unclean minerals from other sites.

As an article about Congo goes, this one's written in a pretty neutral and prosaic tone, seemingly just delivering up information about a new project. Still, it cannot easily escape certain unmistakable, well-worn themes about the heart of darkness, which are fleshed out by fast comparison to the "Digital," a concept that in turn gains symbolic significance through comparison to Congo, or the idea of Congo. In the above passage alone, "past" and "future" are juxtaposed in a way that's so familiar it's likely to go unnoticed—the movement from the ground to the air, from being-in-nature to mastery of nature, from geographic and cultural distinctiveness to universal global belonging, from materiality to immateriality, from darkness to light, from invisibility to transparency. The temporal dimensions are not explicit but implicit—this is a new project that comes from outside Congo in the form of the most current technology; according to the well-worn narrative, the contemporary, which assumes the form of technology, has to be brought to Congo, a place already understood to represent the primordial past, outside of History. These associations are woven into the text almost imperceptibly, no doubt to the author as well, in the same way as they are interwoven to form the commonsense of Euro-American thought, the conceptual water that Euro-Americans swim in. They are also woven into the company's efforts to regulate the supply chain.

The article goes on to explain that tech companies (Apple is mentioned, among others) have to "prove their supply chains are clean" and that the goal of the intervention is to ensure that the coltan that makes it to them is actually conflict-free and not rendered "unclean" or dirtied by minerals from other mines. The authors continue:

> Maubrey [the managing director of RCS] said the new system had helped prevent tainted ore being mixed in with SMB's products by creating new obstacles. To use stolen tags, for example, a smuggler would also need to steal both the scanner and the laptop.

The company's language, mediated somewhat by the author, implies that the minerals and tags are the private property of the company, so misuse of them is theft. This is not surprising, but it is different from how Congolese state officials and civilians tended to see the misuse of tags under ITRI/Pact, and I imagine that it is different from how they see the tags of RCS. Mainly, those involved in the trade saw the illicit distribution of tags as, at worst, "corruption" or smuggling (*maghendo*) rather than theft, in part because they tended not to see the tags as the property of the companies buying the ore but of the government that oversaw the distribution of the tags. This is because state functions (tax collection and policing of those dealing in "conflict minerals") and NGO aspirations (e.g., community "development" interventions) were, in practice, intermingled with the product decontamination imperatives of companies (cleansing conflict minerals). As discussed in chapter 9, many people involved in the trade looked more or less favorably on state officials' distribution of tags outside of the system because it prevented total closure, "opened the ways" of state and nonstate actors, and minimized the risk of poverty and violence in the long run. This was generally in keeping with broader practices and ethics that emerged through the trade in minerals, which were based on an expansive concept of collaborative movement. But for the company, the tag and the ore are their property, and they are therefore intrinsically clean, remaining so for as long as they are removed from potentially contaminating Congolese social arrangements. For this reason, it doesn't matter that, as discussed in the previous chapter, Rubaya is a place where violence is common, or that comptoirs are imbricated in a history of ethnicized violence, so long as the mine (a commodity) can be shown to be "green" according to the government and the World Bank–drafted Congolese Mining Code.

The hope is that the company's minerals, as well as knowledge about their source, direction, and whereabouts, will be safe in the cloud "above" the earth (though not really), unsullied by Congolese people and their dirty tricks. Tagging, then, is about tracking the movement of things and ensuring that shallow, "frictionless"—to borrow a concept from Anna Tsing—globalization takes places in a manner that doesn't make the commodities of corporations unclean (Tsing 2005); it works, at least in theory, by sequestering valuable, commodified substances from Congolese social and political life. Those who fall outside of the light shined by this visualizing apparatus are likely to remain officially (though not really) in the dark and so subject to predatorial policing by state agents. And those who suffer most from this policing are likely to be the most marginal, the ones who are least able to insinuate themselves into the "legitimate system"— for example, women and porters (as discussed in chapter 9). Better than closing everyone out completely by illegalizing mining, tagging still puts those who are outside of it at risk of immobilization and expropriation while empowering those at the top of the supply chain in relation to those below. This is a project that enforces and encourages fenced-in private property (the property of the company, the tag as property, the clean mine as untainted property of an owner).

The rhetorical work taking place in the article, and in the project that it describes, has been repeated since the time of King Leopold. Mainly, an idea of Congo as violent is conjured up by a combination of media and corporate actors, apparently with a view to protecting Congolese, but the humanitarian language and pretenses are also instruments of enclosure. In this case, a comptoir is acquiring a monopoly over ownership of the mine at Rubaya (pushing out competitors who are not part of the tracking initiative) and over the local trade in minerals, and certain miners, traders, and transporters are being exposed and made vulnerable to state actors operating without a meaningful salary of their own, or perhaps any salary. To be sure, this enclosure is less total than some of the other forms of enclosure happening in Congo, such as the expropriation of land by mining companies and the sometimes-violent expulsion of people from the forest in the name of conservation and biodiversity (Blomley et al. 2013; Longo 2019 and 2020; see also Kelly 2011). But one of the arguments I've made in this book is that tagging can't be easily disentangled from these other contemporaneous expropriations because on a larger level tagging has become part of the process through which diggers are redirected away from company-owned sites. In contrast, the evasion of regulators, and the imperative to stay mobile in the face of regulation, fueled what I have referred to in this book as the ethics of invisibility—studious efforts to avoid indebtedness and capture by state actors, whose language and papers, indeed whose very presence, chased minerals away.

Overall, what is happening in Rubaya, as in other places where tagging was taking place, is an outsourced intervention by tech companies to define—or at least to appear to try to define—which actors and locations are in their supply chain and which are not (after all, the tech companies simply do not know). One can also infer that the tech companies that are being asked to monitor their supply chain see everybody involved in digging these minerals as *potentially* part of their supply chain because they don't know who is in it and who is not. So they try to find ways to define it—ironically, by outsourcing the job to their supply chain. What they are also doing, unintentionally, is relegating those outside of their visible chain to a host of abuses at the hands of state actors. At the same time, and as discussed in chapter 9, they are not actually able to regulate the chain, though this may not matter for them so long as they can create the effective illusion of doing so (Section 1502 of the Dodd-Frank Act says they have to exercise "due diligence"). For the diggers, who are trying to stay mobile, when and where the regulatory efforts do kick in, they feel very much like a chain is being imposed on them—recall that they refer to the tags as *craca*, or handcuffs, an image similar to that of the chain. Those involved in the trade saw minerals as equally resistant to immobilization: because of ancestors' influence over their visibility, minerals were mobile, and not fixed, as geologists would have it.

Clearly, the companies are attempting to create a visible supply chain that they can claim ownership of, but is it not also the case that all people involved in artisanal mining are part of the *total* global supply chain for these substances, including those who fall outside of the traceability initiatives? At the same time—beyond the issue of whether or not companies are ultimately able to control their supply chain—the concept of the supply chain, though useful, is also insufficient to make sense of the totality of work and innovation that those involved in it perform, whether they're technically part of the supply chain or not. In other words, even if the concept of the supply chain is an apt construct for thinking about their relationship to capitalism, does the fact that diggers are supplying minerals for companies mean that what they do and the social arrangements they make are more or less defined by this particular organizing principle, or system, for making minerals available to companies? How so and to what extent? Or, might those at the bottom of the "chain" also be generating something that capitalism can't control, and that perhaps even threatens capitalism? Perhaps this is part of the reason why there is so much hand-wringing about the need to regulate them. In this vein, it is interesting that even the supporters of artisanal mining (and there are far fewer of these than there are detractors) feel the need to make the argument for this form of mining from the point of view of the productivity of national economies and the economic well-being of states. Artisanal mining can also be taxed, they insist, or artisanal miners should be granted permission by states to exist alongside industrial mining because of the artisanals' contribution to the economy (perhaps they should exist in the same location or perhaps in a different zone—a big argument in itself in the "development" literature regarding artisanal mining; see, e.g., Hilson and Owen 2020). In other words, in most academic accounts of artisanal or small-scale mining, the power of artisanal mining, or miners, to produce a qualitatively different kind of situation or world is discursively subsumed within a particular vision of state-centered economic development—almost as if those who are writing these reports are trying to convince people in positions of power that artisanal mining doesn't pose a challenge to the world order as it currently exists.

Let us first consider fully the fact that diggers and traders of coltan are ultimately part of a larger system, which Anna Tsing and others have referred to as "supply chain capitalism" (Tsing 2009). Up to a point, at least, this is incontrovertibly true. Supply chain capitalism is a relatively decentralized mode of capitalist production in which outsourcing to suppliers at lower levels or "links" in the "chain" is a crucial part of capitalist production. This increasingly widespread iteration of contemporary capitalism generates profits for "top of the chain" firms that benefit from cheap labor and resources at the bottom of the chain; they also derive benefits from the relative autonomy of lower-level actors (firms at the top of the chain generally do not have to directly regulate those below them, especially those far below them, who are regulated by other factors and actors; moreover, they are not usually perceived to be directly responsible for what happens in those autonomous spheres). Supply chain capitalism relies on a loose assemblage of autonomous but networked sociocultural, spatiotemporal, and economic arrangements to generate profit, allowing goods to be produced more cheaply than they would through a centralized economic system (in particular, the mid-twentieth-century Fordist model in which large firms controlled most or all aspects of production). Some of the on-the-ground concepts and practices that can operate as elements of the supply chain may not seem to have any necessary relationship to capitalism, such as ideas and practices surrounding masculinity or particular kinship and marriage practices. As Tsing puts it, "Supply chain capitalism makes use of diverse social-economic niches through which goods and services can be produced more cheaply," (171) and "no firm has to personally invent patriarchy, colonialism, war, racism, or imprisonment, yet each of these is privileged in supply chain labor mobilization" (151).

Thinking of capitalism in terms of decentered elements and diversity does not necessarily imply a kind of structural functionalism in which nearly everything happening supports the "needs" of capitalism; rather, diversity and difference in the supply chain exist as potential "resources" that are available for capitalists to draw on and take advantage of. It would not be difficult to read the material presented in this ethnography through the prism of the "chain," emphasizing how the various elements presented here play into this larger global system. First, there is the creation, through war, of a large, displaced population with nowhere to go but mining or perhaps other forms of resource extraction involving wildlife, timber, and charcoal; their uprootedness is exacerbated by the destruction of families, family life, and the incremental time that emerged through these relationships and practices (although here it should be pointed out that mining today also rebuilds those relationships, especially through marriage payments and remittances). Among those who went into mining, these earlier networks and relations have often been at least somewhat eclipsed by more horizontal relationships based on friendship and more vertical ones based on debt (whether "contractual" or decidedly noncontractual), which often compel people to remain in the trade even when they might wish to leave.

A second feature of this supply chain, operating in tandem with displacement and generalized uprootedness, is the decentralized and bottom-up nature of the organization of extraction and distribution, which allows for a cheaper product for actors higher up in the chain. The diggers are generally not paid in money but in minerals, and they then convert these minerals into the goods they need to survive—generally selling cheap and buying high, because the cost of "imported" commodities in and around infrastructurally remote mines is so high. This is one of the forces that can keep them stuck where they are, compelled to work to pay off debts, and it is also one of the factors that encourages them to extract quickly, as minerals are their power, their "pen," and generally their source of independence. Their work thus produces great supply: in every Congolese mining zone with a history of company concessions, one is regaled with stories about how artisanal miners extract more volume more quickly at a lower cost than industrial mining was ever able to do. The managers of companies are in awe of these people that they do not manage, but simply exist alongside of, extracting rents from their work because they have papers; for their part, the Office of Mines officials are ready to tally off statistics about how much less Sominki extracted in thirty years than artisanal miners in a single location in just a few (Bisie was especially remarkable in this regard but people in other concessions, like Kalima, Kailo, and Punia, have similar stories). The transformation of old industrial companies into landlords who collect rents from lower-level middlepersons no doubt cuts down on the overall cost of minerals for those further up, as does the emergence of adaptive artisan-entrepreneurs around the ruins of industrial mining—people who are willing and able to use their *akili*, or intelligence/cleverness, to transform waste into productive machinery to launch a new semi-industrial "phase" of mining with some investment from state actors (including warlords or former warlords).

Add to this the cheap but efficient system of labor organization around holes, which enables seamless extraction, often utilizing expertise culled from earlier generations of centralized mining (think of Lagome's father, the engineer, as well as the redeployed history of food provision and all the drama surrounding old maps in general), and we have a system that is autonomously organized and which relies on local cultural values and histories to make extraction happen on the cheap (e.g., the values of reciprocity and friendship—five men sharing a cigarette in a hole—as well as men's mobilization of women's labor at the hole, in the form of cooking for a husband's friends). This system of cheap extraction is supplemented by a system of porterage that draws from the "resource" of the historical experience of forced porterage during the Swahili slave trade, the colonial period, and the wars. It is also abetted by the lack of information at the bottom of the chain (interesting to think about in tandem with the concept of the information age), which is part of what diggers mean by being in a hole ("What is the true price?" and "Where is the world market located?" being among the main questions they ask).

In general, the contingent and on-the-fly nature of artisanal mining, in which people are highly mobile and relationships are developed anew at the hole and between diggers and négociants in a way that is both fraught and collaborative, makes for a system that is stable in the moment but also open to sudden shifts and transformations, mainly changes in price or demand (as Tsing puts it, "Supply chain diversity needs to be understood in relation to contingency, experimentation, negotiation, and unstable commitments" [150–151]). The emphasis on mobility and adaptability— "being supple"—also has something to do with the material qualities of these minerals, including their relative accessibility, the relative ease with which they can be separated from other rocks, their changeability over time (meaning what is valuable in them can change over time, and they also become harder to extract over time; also, according to many, they can be taken away by ancestors), and the fact that they exist in disparate locations, each one of which is different and each one of which requires different kinds of negotiations (including negotiation with spirits). All of this requires a certain amount of adaptability and rapid response to changing conditions.

One could add to this a number of other factors—an earlier, Althusser-inspired anthropology would have called them ideological—which feed into this system. Some of these are directly related to the way in which diggers and traders conceptualize and identify with their work. One might be the very practice and concept of movement itself, which consists partly in a readiness, on the part of low-level actors, to be available for changes in the demand for bottom-of-the-chain resources (be they minerals or something else) and the related pride with which men, enlivened by a renewed sense of masculinity following wartime experiences they describe as (among other things) emasculating, bend their bodies and move like "machines." Think for example of the Machine, introduced in chapter 1, who boasts of being able to work nonstop; the fact that he embraces the concept of being a nonhuman instrument may suggest that his exploitation has become a matter of personal pride (Tsing calls this "superexploitation"), an identity that can be drawn upon as a resource for capitalism (this is also one possible way—one that I don't support—of thinking about the ways in which diggers turn themselves into moles or snakes—that is, to think of these transformations in terms of dehumanization rather than multispecies/multidimensional collaboration). Another important local concept, or desire, is the promise of becoming a PDG, or hole owner—the idea that it is possible to work oneself up from nothing to become the "owner" of a hole, an apparently (though in practice rarely) independent entrepreneur in a position to mobilize labor and get rich; this dream of upward mobility is definitely part of the appeal of artisanal mining (in reality, many, if not most, hole owners are financed by powerful backers and did not rise from the bottom—although a fair number did). In general, people involved in the trade thought of themselves as independent entrepreneurs, often positively heroic ones (they turned flashlights into houses) rather than workers in a supply chain; but this feeling—should we call it an illusion?—of independence among those at the bottom of the chain is part of what makes it operate as a total, diversified economic system that doesn't appear to be interconnected on the surface (parallels here to how workers in the so-called sharing economy—for example, Uber—think of themselves as independent operators rather than low-paid, non-unionized workers).

To these one could add a number of other concepts and practices that go into the making of this extractive system, such as the concept, or fact, that ancestors and spirits are responsible for extraction. If ancestors are responsible for this work, then are they not also at least part-time workers in a loosely defined supply chain, recruited by their descendants into securing safe labor conditions and easy extraction? One could fairly ask oneself how capitalism can fail to be victorious if the whole forest and all of its dimensions are on its side, but the issue is not that simple, because they are only effective if their descendants have recognized, acknowledged, and conversed with them, which is but one of the reasons why they and the forest are a wild card with respect to the global supply chain. Still, one could interpret the social-temporal arrangements between ancestors and the living in two, not necessarily incompatible, ways: On the one hand, by bringing people together in collaboration and enforcing a spirit of reciprocity, ancestors are empowering diggers and land occupants while resisting enclosure, the usurpation of property by companies, and practices associated with selfishness and deception at the hole (e.g., the practices of négociants and state officials, which chase minerals away). On the other hand, by encouraging reciprocity at the hole they are also facilitating the mobilization of social arrangements, including women's labor (preparing food for diggers, for example), in a way that enables cheaper extraction, benefiting buyers further up the chain.

Finally, we could add the very idea of the heart of darkness itself as a powerful resource for supply chain capitalism—Congo is seen as being so distant from tech firms and the digital age in general (I can tell you from my experience presenting parts of this book at talks and conferences that Congo is frequently held to be "outside of" the digital age) that diggers are rarely seen as bottom-of-the-chain workers. This is why efforts by companies to regulate their supply chain don't even look like acts of regulation but like benign humanitarian interventions carried out from a separate dimension.

The struggles between artisanal miners and companies also acquire a different level of meaning when interpreted in terms of the larger context of supply chains and their regulation. From this perspective, the Battle of Bisie was a fight over turf between differentially positioned actors at the bottom of the chain, in which those participating in and producing a more decentralized mode of extraction struggled for survival and position against those trying to bring into being a more centralized, territorially enclosed, and hierarchical system; the latter were supported by the "state in Kinshasa" while the former were supported by lower-level state actors, like the territorial chief, who had acquired relative power vis-à-vis the central state during the war. Either way, the tech firms at the top got to keep their hands clean—they appear to have nothing to do with the violent struggle happening at the bottom, even though they are ultimately the raison d'être for the whole drama. The way it ended up turning out—Alphamin emerging as a "world leader in conflict-free tin" against the putatively Hobbesian state of nature on the ground, as the artisanal miners submitted to a new regime of regulation in the form of the tags—certainly seemed to harmonize with the "needs" of supply chain management at that moment in time, when new international regulations were being passed related to Congolese "conflict minerals" (interestingly, changes which also *came about*, at least in part, because of the struggles between companies and artisanal miners in the immediate tentatively postwar period, as events following the Battle of Bisie show).

Nothing to Lose but the Chain

The idea of the chain highlights global territorial linkages and is a good metaphor for explaining how this form of capitalism operates as a system in contrast to other forms of capitalism; it also captures how some people experienced regulations like the tags/*craca* (handcuffs), as ineffective as they turned out to be in most places. It also happens to be a good representation of what I imagine to be the corporate view of global capitalism—hence the many management publications that pop up following a quick Google search for "supply chains." While I'm not suggesting that this chain does not exist, the metaphor—one of bondage, limitation, and control—hardly captures the active energy or creativity of artisanal mining, especially in places and times of movement, nor the world-altering transformation, rebirth, innovation, and breaking-through-limitations that accompanied it. Nor does it capture the ways in which artisanal miners and traders resisted and challenged global capitalism—the fact that artisanal mining consisted in collaborative, multidimensional resistance to enclosure. In particular, their practices made it more difficult for companies to privately monopolize land (though it also attracted them in the first place, as in the case of Bisie). Diggers modeled an alternative to state-centered politics that also challenged the state and companies, as well as a whole host of taken-for-granted assumptions and arrangements related to peace, prosperity, and "resources" (in quotation marks here because substances have to first be made into resources through an array of practices, some of them "religious," in which diggers were intimately involved). According to this line of thought, those in the mining trade are better thought of not as exemplars of abjection but as creative pioneers of new modes of movement, providing insight into what an expansive concept of movement would mean, even if they often found themselves "stuck in a hole," and even if expulsion and dispossession have forced them into their situation. This line of thought is developed further below.

Isabelle Stengers, a philosopher of science, has discussed how "the commons," once destroyed by capitalism, might again reemerge from it, as workers resist the enclosure and destruction of practices and knowledge that is common to them: "without wanting to, capitalism would thus contribute to the possibility of a humanity reconciled with itself, a mobile creative multitude" (2015, 84). One of her main protagonists are computer programmers, who she refers to as an "immaterial proletariat" seeking to make their work and knowledge common in opposition to corporations that have sought to enclose and destroy what they had made. The programmers are unlikely heroes, in part because they are not utopian revolutionaries trying to save the planet but workers trying to protect their work product: "[They] resisted what was endeavoring to separate them from what was common to *them*, not the appropriation of the common good of humanity" (2015, 85). Nonetheless, "it was computer programmers, whose work was directly targeted by the patenting of their algorithms, that is to say, their very languages, who named what was threatening them thus, and created a response, the now celebrated GNU public license," allowing for open-source software that can be changed according to the needs of users (2015, 81). Stengers's point here is similar to the arguments David Graeber made about how a new world might emerge from within the old state-centered capitalist one, visible in the ordinary collaborative decision-making practices of anarchist activists and others (2011). As he put it at one point, "Communism already exists in our intimate relations with each other on a million different levels."

It's worth bearing in mind that, for most of Stengers's audience, a good part of the appeal of the programmers is that their labor is immaterial, a prerequisite for their being at the forefront of time's arrow. In truth, the labor of the programmers is not actually immaterial because it is made possible by the work of many others, including those who dig up the minerals used in computers (keeping in mind that an estimated 25 percent of tantalum on the global market comes from artisanal mining; IGF 2018). Their work has as much impact on the earth as that of the miners, even if they are not directly involved in extraction. And artisanal miners, a creative multitude working in holes in the ground, are at least as likely candidates for the revolt of the commons that Stengers describes. At the same time as they and the supply chains of which they are part are useful for capitalism, they also resist and oppose this system, coopting and deploying the state in ways that challenge state power and the allied efforts of the state and foreign capital to expropriate earth.

In other words, it may not be wrong, after all, to associate the digital age with a major shift, or revolution, in capitalism and state-centered politics, as many already have. But it may also be that one of the main revolutionary transformations brought about by the digital age is the turn toward artisanal mining rather than the work of software developers and computer programmers (keeping in mind, of course, that artisanal mining in its current form is enabled by certain digital technologies, especially the cell phone).[1] We should think more deeply about the kinds of social formation and politics that emerge around the hole, because they are directly opposed to the corporate idea of transparency and the idea of links in a chain. Rather than the link in the chain, the image of the vortex, or the wormhole, more closely approximates the open-ended, transformative, and liminal qualities of mines, the energy produced through artisanal mining, and the political potentials that emerge from artisanal collaboration through earth.

Holes and Wormholes

If transparency and disembodied information are one model of the future, based mainly on cultural evolutionist Euro-American understandings of technology, materiality, and disembodiment, there were others that emerged in the forest and in the ground. One prominent figure was the more or less opposite one of the hole, a concept that also proliferates, in a somewhat different form, in Kinshasa. Indeed, Filip De Boeck and Sammy Baloji have framed much of their coauthored work on the city of Kinshasa around the trope and analytic of the hole, or *libulu* in Lingala (De Boeck and Baloji 2016; De Boeck 2019); rather than being synonymous with lack, they describe libulu as "an opening, a possibility, at least for those who know how to read an alternative meaning into its blackness" (2016, 14). In Kinshasa, the term *hole* can refer to literal holes (as in giant potholes and craters that have emerged in the city), the quality of life in the city, and the opacity of the informal economy. But De Boeck and Baloji point out that holes are not just figures of lack but vortexes of social collaboration and gathering where people and activities come together in the act of "suturing the city" (a number of nightclubs have taken on the name, and "dancing in a hole" has become a kind of allegory for life in Kinshasa). As they put it, "Urban living constantly attempts to 'suture' the city, finding ways to stitch gains and losses, or pasts and futures together in the moment of the 'urban now'" (2017).

For De Boeck and Baloji, the informal, or what we could also call the artisanal life of the hole, is opposed to the large urban planning and "cleaning" projects that have been taking place in Kinshasa to the exclusion of ordinary Kinois. In discussing the state-driven practice of "cleaning up" the city, De Boeck shows how it attacks the "capacity for insertion" and the "crucial creative capacity" that are central to getting along in the city:

> For some years now, a successive series of city governors has been engaged in "cleaning up" the city. This cleansing basically boils down to a hard-handed politics of erasure, destroying "irregular," "anarchic" and unruly housing constructions, bulldozing bars and terraces considered to be too close to the roadside, and banning containers, which Kinois commonly convert into little shops, from the street. . . . In Kinshasa, every singular life is embedded in a multiplicity of relationships. Many of these relationships are defined by family and kinship ties, but many others have to do with the specific ways in which one inserts oneself—has to insert oneself—in multiple complex, often overlapping, networks that include friends, neighbors, colleagues, acquaintances, members of one's church congregation, professional relations and so on. The capacity at insertion constitutes the prerequisite for a life worth living in this kind of urban environment, in economic as well as social terms. The state's brutal destruction of citizens' material and social environments under the guise of an urban reform that once again seems to be inspired by the earlier moral models of colonialist modernity, therefore forms a violent attack on precisely that crucial creative capacity that is a sine qua non to belong, and to belong together, in the city. (De Boeck 2011)

In the rural and sylvan worlds of coltan mining, far from the manifest cosmopolitanism of megalopolises like Kinshasa (or even the much more modest cities of Goma, Bukavu, and Kisangani that I am familiar with), this conflict between cleaning/purification and what I have referred to simply as movement (similar, though, to De Boeck and Baloji's "suturing") is written on an even larger and more global scale, as displaced Congolese try to make lives and futures for themselves in the wake of war by collaborating with each other and weaving together different elements to try to make connections to the world outside of Congo. Here it is not just the state but the whole world that seems aimed at cordoning them off, removing them from the source of life, and parceling out the forests and rural areas for their own extraction. These holes in the forest are, despite their geographic remoteness, even more dramatic and bald-facedly global vortexes than the urban giant of Kinshasa, in that they attract global, local, and regional forces in a maelstrom where each element is thrown together with others, their normative hierarchy at least potentially upended.

Mining holes are, first and foremost, openings in the earth that allow for new possibilities based simultaneously on new innovations and relationships and on escape from dominant—and dominating—people, institutions, and ways of perceiving and organizing the world. For diggers, their holes (*mashimo* in Swahili) stand in opposition to the surveillance that comes with transparency, or being in the light. The term *hole* can also reference indebtedness and the condition of avoiding state authorities— being in a hole or escaping into a hole. But holes are also the source of wealth and can keep giving for as long as one's work and social relationships (which, through the extension of credit, can enable the creation of deeper holes through the use of pumps and machines) last. Holes are also sites of generative collaboration and exchange, as diggers can turn even the most aggressive and well-armed enemy into a friend by working in a hole (although holes can also be sites of exclusion, allowing those in the hole to conceal the value of their work from others, including women). At a broad level, the hole is a form and outcome of movement, as well as a symbol and mode of action and belonging that stands in contrast to (post)colonialist and capitalist enclosure.

Holes are metaphors and engines of transformation. Far from being empty, they are filled with all kinds of things: they are multispecies, multidimensional, and multitemporal vortexes that allow access to many times, dimensions, and places. Mining holes are places of (outwardly masculine) camaraderie and rehabilitation that can feel like home. At the same time, these holes are challenging, forcing you to adapt mentally, physically, and spiritually. It's impossible to see far (*kuona mbele*) when you're in a hole, but other ways of perceiving, including hearing and touching, also present themselves. Holes are also feats of engineering, and the deeper the hole the greater the feat. Because they are almost always made possible by the financing work of others, including women, they are also physical manifestations of trust. Holes can become famous, acquiring great, universalizing names that are also specific (e.g., Modern Times), and their fame can attract more people. At the same time, hole creation is guided by forces that no one controls, such as price fluctuations, so they do not spring wholly from volition. When they are working, holes can bring other worlds into being and bring the world to the level of the hole so that it appears in and through the hole, remade by the hole. At the same time as they open up new worlds, holes can also be destructive to that which surround them— they can ruin preexisting worlds and, while promising salvation, can also bring chaos and loss. Holes are a gamble, and they are fragile creations. But under the right circumstances, they have the ability to create a kind of hypermobility that is at once spatial and temporal.

Through the hole, diggers' movement takes them to other dimensions, as they consort with entities that others would prefer to avoid; the digger's life is tied up in that of other species and beings. This is why so many insist (though often with a great deal of ambivalence) that dirt is not shameful but a valuable badge—even a magnet—that brings wealth, minerals, and friends (in contrast to those consigned to offices, or consigned to dream of offices, for whom dirt brings only scorn, poverty, and humiliation). At home almost everywhere, as long as there is a hole to crawl into, diggers can build relationships on the fly and turn the tables on those who are above them, transforming despised things (themselves included) into things of value—and their work circulates out to others in the form of money, even if they themselves don't always get to hold onto it or even see it. While others work to invest their futures in durable houses, they find themselves less able to do so, but this is not a failing so much as a sacrifice for others.

Diggers are not only exemplary because of their desire for spatiotemporal mobility but because of how they exist in the in-between spaces, moving between "worlds" and ways of being, adapting to them when appropriate without necessarily being committed to any single one. In this way, they are also theorists of movement whose practices are guided by certain understandings that are also spelled out conceptually and that are worth taking seriously, in the sense of considering ways to echo them as a way of being in or relating to the world. For one, there is no single kind of digger, with a single understanding or approach to earth; rather, diggers adapt to the places and to the earth in which they find themselves—they will surely have to dig a hole, but how deep is an open question (the answer, of course, is "it depends"). In a similar vein, there was no single understanding of the forest, or of the ancestors and spirits that lived in or comprised it, but all these various understandings came to bear through mining. And each hole will necessitate a different way of being—different senses will come to dominate in a deep hole, as opposed to a shallow one, although diggers must be open to using all the senses available to them and even ones that are not or "should" not be (this is part of what makes it physically exhausting to be a digger). Hence their tendency to liken themselves to burrowing animals that perceive and move in ways that are fundamentally different than humans, conditioned by the qualities of the earth in different locations. Diggers are also open to different ways of understanding earth, and their relationship to these understandings is often noncommittal, or at least open, because, after all, they are on the move, traveling in strange lands (though it is true that some diggers stay close to home, but even then they are travelers). As a rule, diggers do not nitpick about what forces are in the ground or what they signify (they have no time to try to classify the X number of supposedly mutually exclusive ontologies that might exist in the forest, but they are happy to move between them). Diggers are not theologians, as some say, though they can be scientists, as they also remind us (*geologues* is the term many prefer).

For many, maybe most, Congolese people, especially for those operating from the vantage of postcolonial status and authority, digging in a hole is degrading and dirty work. It is opposed to office work or working with papers, which diggers of course see as genuinely dirty, deceptive, and abusive, for reasons that should be abundantly clear from this book. Interestingly, while many see it as disorderly, generative trust-based social relations and collective well-being are created through this "dirty" but actually orderly work rather than through the exclusion of "dirt" (*chafu*) from clean work, as regulators and normative anthropological thought tend to have it. This is interesting for a lot of reasons, practical and theoretical: for one, for most anthropological thought, stemming in part from the work of Mary Douglas, the exclusion of dirt and disorder is how social norms and society are made. According to this model, society is like a giant filtration device excluding the unclean from entering its gates, and the making of society is, in turn, an ongoing purification project. In this, anthropologists, companies, and state and NGO regulators are aligned, sharing in the same basic assumptions about social norms and order being generated through the exclusion of the unclean (for modernist anthropology, the repulsion created by the unclean is one of the main reasons why society is able to exist). But for those involved in mining, the making of social worlds, which included ancestors and other spirits and forces, happened by engaging with and through the multiplicity of dirt head-on and rejecting the colonial history of associating cleanliness, and whiteness, with civilization, development, and the good. So the hole, and the liminal work of digging in holes, suggests the emergence of a kind of sociality that isn't based on a geographically bounded organismic or mechanistic model, but which instead emerges out of collaborative movement in the broadest sense of the term. It is also not based on ownership (in contrast, the company's minerals are clean if they can prove total ownership of them from beginning to end). In a similar vein, artisanal mining offered opportunities to break out of spatial limitations that had been imposed since colonial times. Indeed, mining often offered more opportunity for transcendence and empowerment than "the digital" (although, again, the digital was tied up in mining, especially through the cell phone and smartphone), creating wormholes that tore through normative spatial and temporal barriers to connect people to worlds they had long been prohibited from entering.

It is for all of the above reasons that I have drawn upon the metaphor of the wormhole (unlike movement and the hole, not a local concept, but my own attempt to translate what is happening) to describe particularly spectacular holes, like the one at Bisie (a hole composed of many holes with many holes). Speculated to be brought into being by "exotic matter" that violates the rules of physics, or by a great deal of energy, wormholes are possible structures ("bridges," they're often called) connecting vast distances in space-time. In some physics theories, and in most science fiction, they are traversable, serving as shortcuts across the universe or across time (thus allowing for time travel), potentially bringing into contact vastly different "people" and places that are unknown to one another. They are, in physics and science fiction alike, tragically unstable, capable of collapsing at any moment, and would-be travelers are always looking for a way to stabilize them. Although they may be usable, they are also fundamentally destabilizing "places," as spaces and times are thrown together in them in a disorienting way. In my use of the concept, I also have in mind the wormhole as a liminal space-time in which the hierarchies that exist in the ordinary world are temporarily suspended and potentially reversed (MPC's employees see the Bagandula's ancestors—the puny "insects of the forest"—then fall into the hole and die; MPC's employees gets shot at by soldiers who are actually being paid by higher-level diggers; the forest becomes the city, etc.).

In short, the vision of the future that emerges out of artisanal mining is one of mobility, borderless movement, liminality/reversal, and multidimensional collaboration—themes embraced by those involved in mining and by contemporary postcolonial social theorists alike. For example, Achille Mbembe concludes *Necropolitics*, his "genealogy of the contemporary world" from the vantage of the postcolony, by positing a vision of the future based on a kind of radical mobility of the dispossessed rather than of capital (or what he describes as the "unification of the world as part of capitalism's limitless expansion"). As he succinctly puts it, "Let us be content to observe that the future will necessarily be about passage, crossing, and movement" in contrast to walls, borders, and exclusion (2019, 188).[2]

At the same time, the organizational forms that emerge around artisanal mines are also interesting, in part because they parallel the state but also sometimes seem capable of absorbing or sequestering it, at least for a while. The social and political order that emerges around these minerals is very different from the normative model of political order that grew up in the twentieth century. This model emerged around oil, and nowadays it is threatened because oil reserves are running out and because of the ecological and social destruction that this form of extraction has unleashed upon the world. In the twenty-first century, that oil-based model of politics, political economy, and the future—which, though recent, was imagined to be eternal—is clearly coming into view as a temporary situation brought about by a complex of equally temporary social, technical, and political arrangements that emerged around the extraction and distribution of this resource. As Mitchell summarizes in his *Carbon Democracy*:

> The forms of democracy that emerged in leading industrialized countries by the middle decades of the twentieth century were enabled and shaped by the extraordinary concentration of energy obtained from the world's limited stores of hydrocarbons, and by the socio-technical arrangements required for extracting and distributing the energy they contained. . . . Democratic politics developed, thanks to oil, with a particular orientation towards the future: the future was a limitless horizon of growth. This horizon was not some natural reflection of a time of plenty. It was the result of a particular way of organizing expert knowledge and its objects, in terms of a novel world called the "economy." Innovations in methods of calculation, the use of money, the measurement of transactions, and the compiling of national statistics made it possible to imagine the central object of politics [again, the economy] as an object that could expand without any form of ultimate material constraint. (2011, 253)

In its fashioning of itself, the postcolonial Congolese state sometimes taps into this very same oil-based vision of the future, mirroring the social and political relations, modes of knowledge production, and concepts and practices of space and time that emerged around oil's extraction—as if, by doing so, it could catch up to "developed" industrial societies whose economies are based on a source of growth that is already disappearing. Take, for example, one of the major promotional billboards for former President Joseph Kabila, which could be seen in many rural locations during the time of my fieldwork. It showed then President Kabila's smiling face as he gazed up and to the right, where the future is. In it, progress was depicted through three photos of progressively larger and expensive maritime vehicles, some situated on what appeared to be the Congo River; the sign read, "Five Projects [Kabila's Slogan]: The Revolution of Modernity on the Move" (*Revolution de la Modernité en Marche*). To the far left was a photo of pirogues with a few people in each one, underneath which was written the word "Yesterday" (*Jana*), sans exclamation point, in Swahili, despite the fact that the scene was clearly contemporary. To its right was a photo of speedboats (ironically, they were all huddled up next to each other, apparently grounded), accompanied by the more enthusiastically delivered word, "Today!" (*Leo!*). To the far right, the single word "Tomorrow!" (*Kesho!*) welcomed a giant luxury cruise ship. On the surface, the photos represented state-driven progress, or development, over time in terms of speed, size (hence growth), grandeur, modernity, and "cleanliness" (*usafi*), in some cases with the backdrop of the iconic waterway that is the life-blood of the nation that it also symbolizes.

On a very basic level, the billboard harnessed an idea of growth emerging from highly concentrated and centralized energy that is monopolized by a few, which is why these representations of maritime transport and circulation were not actually representing movement in the way that diggers and traders understood the term but rather enclosure and attendant inequality—the opposite of mobility. (The pirogue could have multiple owners and renters, whereas the cruise ship in all likelihood does not, and tickets for the speedboat and cruise ship are sure to be more expensive, well outside the range of "yesterday's" users of pirogues.) This oil-based model of development thus entails a vision of linear movement toward the future in which progress is understood to be infinite and inexorable and in which those who fail to keep up are relegated to the past (the users of pirogues, who actually exist in the present in great numbers, are discursively incarcerated in the past, as if they no longer exist—their actual existence in the present is therefore cast as a temporal anomaly, a kind of mistake). We are, in short, left to imagine a world in which more wealth is in the hands of ever-smaller groups of (probably international) people, while humans are sacrificed for an image of grandeur and global belonging (see also De Boeck and Baloji 2017).

If we were to imagine the modes of political organization that might potentially emerge around artisanally mined minerals if they were not in the process of being crushed by corporations and the state, it's clear that they would be very different from the oil-based vision that has dominated Western visions of the future. They would likely allow for multiple actors and entities—what scholars influenced by Deleuze like to refer to as assemblages—with different relations to and understandings of earth (as opposed to land, which is a concept that is less easy, although not impossible, to dissociate from that of property; Parr 2010). All these approaches, however different, would operate on a more expansive understanding of "development," the future, the social, and earth than the exclusive, or monadic, relations to land exercised by states and companies. Such systems would be comparatively diffuse and democratic—above all, they would be opposed to private property and enclosure. They would also not reiterate the old Western humanist polity based on rational control of nature by the mind for the betterment of society, because they would be forged through a multidimensional collaboration between people and spirits. The worlds that might emerge from artisanal mining are also not in harmony with the idea of a limitless and linear future for a number of reasons, one of which is the fact that, as we have seen, artisanal mining depends on the recuperation of discarded things and places (waste, for lack of a better word) and also involves direct engagement with emissaries of the past. Of course, the worlds that are made through these supple and shifting collaborations and accommodations would not be constructed solely from the volition of the human actors involved. An assemblage of forces comes together in and through the earth, as diggers and others fight to manage their relationship to these forces in real time. This was well known to everyone who worked in the mining trade, who were painfully aware of the paradox of trying to build incremental futures on what they often referred to as sand.

I am suggesting, in short, that the emerging polities and politics of artisanal miners are one among many possible assemblages (not the only one or even necessarily the most privileged or unique) that might have the capacity to eclipse (not usurp) the postcolonial state and the global networks that feed and are fed by it—and which Congolese have a history of analyzing and critiquing through their stories and speculations surrounding war. Still, to the extent that these forms of collaboration coalesce around mineral resources, they cannot continue indefinitely or permanently, certainly not without further destruction to ecologies that are already beyond the brink (see, e.g., Musah-Surugu et al. 2017; many authors also suggest that artisanal miners could "transition" into other activities, but agriculture is also a main driver of deforestation in Congo). Though 3Ts mining is no more ecologically destructive than other forms of mining (unlike gold, it does not require the use of mercury, for example), it does often require the redirection of waterways, and silt deposits from mining can also ruin local water supplies, making agriculture at least temporarily impracticable where mining is taking place. Any practice that extracts minerals from the forest and ground is destructive, as are all the other forms of labor and consumption that depend on such work, including that of immaterial laborers in Silicon Valley (again, the work of mining is not actually more destructive to the earth than the work of writing code; it is simply that miners are involved in an aspect of the overall work that code writers, and everyone else, don't get to see).

It's important to keep in mind that the overall problem here is not the creative mobility of the people involved in digging but their exclusion in the face of enclosure, which the movement of artisanal mining creatively responds to. If those involved in artisanal mining are pushed out of mining, as forest land is grabbed by corporations operating in combination with the state, they will continue to pioneer other sites for companies to exploit later. Ecological collapse is therefore fueled by the privatization and enclosure of space, combined with the displacement of populations, whether in the name of extraction or conservation. Meanwhile, corporations can cloak themselves in the signs of cleanliness, civility, and peace, while those they push out are made to be synonymous with darkness and savagery once again.

Because it's based on private enclosure rather than collaborative networks, industrial mining removes diverse people and entities from the process of deciding how to live within ecologies and of defining what ecologies consist of and what they mean. The fewer people and "ways of knowing" involved in decision, concept, and world making around minerals, the more dangerous the ecological problems surrounding extraction become; meanwhile, conservation practice operates as the other side of the process of enclosure—sharing more in common with privatized extraction than we have been conditioned to think.[3] In short, the collaborative model that emerges around artisanal mining, in which indigenous populations are directly involved in what happens at mines and decide whether or not to do the work of "opening" the holes for extraction, is surely preferable to giving more or less total authority of a piece of land and its products to a company. Because it isn't generally based on exclusive private ownership of land, artisanal mining leaves open the possibility of collective decision-making about the uses and futures of earth and its products, as well as divergent understandings of what that earth entails and includes. Moreover, the conflicts, collaborations, and knowledges that emerge through multiple people working through multiple understandings of earth and forest might give rise to a bottom-up world-building project involving the use and management of "resources." Such a practice would have to be different from the top-down knowledge—what Mitchell calls the "rule of experts"—that has characterized other forms of extraction (most notably, that of oil) and the social, political, and cosmological orders that have emerged from them.

Let us close this with a final image: the idea, commonly held in Congo, that coltan is used to make the bullets that others kill them with. On the one hand, this is a straight-up rendering, in an extreme mode, of the alienation experienced by those at the bottom of a commodity chain in a war-ravaged part of the world—that they have totally lost control over the outcome of their work, such that it actually comes back in processed form to kill them. But on another level, I think this story was also an incitement to a different kind of world, built on "social" relationships that they were forging in the moment. Just as Mai Mai insurgents drew upon their relationships with a host of forces, spiritual and otherwise, to transform hard bullets into soft, flowing water, so too did those in the trade, insurgents in their own way, try to turn the substance that they imagined to be in these bullets into relationships and futures that flowed expansively like water, enabling them to move in a world that had otherwise dispensed with them.

ACKNOWLEDGMENTS

I would like to thank all of my interlocutors in North and South Kivu and the Maniema provinces for their generosity, enthusiasm, and support while I was carrying out fieldwork. This research was carried out with two generous awards from the National Science Foundation (NSF) and supplementary support from UC Davis. As discussed in the prologue, Jeff Mantz (currently director of the Cultural Anthropology Program at the NSF) had a major role in the original conceptualization of the project, carrying out preliminary research early on and acquiring funding, and he has been an immeasurably helpful friend and colleague along the way. Joshua Walker, an anthropologist working on artisanal diamond mining in Mbuji-Mayi (currently director of the Congo Research Group at NYU), helped me to think through my research material in the early stages of this research. I would also like to acknowledge all those who commented on and encouraged me at the various academic venues in the United States and Europe where drafts of parts of this book were presented. Though there are too many to mention here, I would especially like to thank the UC Berkeley Anthropology Colloquium, the UC Santa Cruz Anthropology Colloquium, the Temporalities Research Cluster at UC Davis, the History Department Colloquium at UC Davis, the Approaches to Capitalism Seminar at Stanford University, the UCLA African Studies Workshop, the Emory University African Studies Workshop, the University of Chicago African Studies Workshop, the Institute for African Studies at Columbia University, the Amherst College Anthropology Department, and the organizers of the two Harvard University conferences where some of this work was presented. I learned a great deal about diverse scholarly approaches to mining from participating in the conference on conflict minerals in Congo in Zurich in 2014 (organized by Timothy Raeymaekers and Christoph Vogel) and the Mining in

Comparative Perspective Conference in Ghent in 2017 (organized by Jeroen Cuvelier and others).

This project would not have been possible without the collaboration and support of Raymond Mwafrika and Joseph Nyembo, whose contributions are also discussed more fully in the prologue. Raymond and Joseph also read late drafts of the book and fact-checked what they could; any remaining mistakes are ultimately my own. Major Patrick Paigba, an Azande FARDC major attached to the UN in Walikale, North Kivu, was a great friend and crucial informant in Walikale. My Kenyan friend Ngeti Mwadime played a major role in the early stages of the research, when I was first navigating how to move around and conduct research in the mining areas of the Kivus. Professor Bosco Muchukiwa, former Director of the Institut Superieur de Developpment Rural (ISDR), provided support, encouragement, and introduction to a network of Congolese academics. I would like to thank ISDR for providing academic affiliation during much of the research period. Kubisa Muzenende Sousthene, former president of civil society of Luhwindja and Bukavu, provided important logistical support over the course of the research.

At UC Davis, my graduate student advisees, Jane Saffitz and Matt Nesvet, worked for me as graduate student researchers, helping to sort through fieldwork data and conducting related research early in the writing process. They have both gone on to do exciting ethnographic research with artisanal miners in Tanzania and South Africa, respectively. Alex Robins, Jeff Kahn, Jane Saffitz, Jeremy Jones, Justin Haruyama, and Matt Nesvet read this book, or parts of it, very closely at different stages and offered invaluable comments and suggestions. My wife, Bekah, has been present at every step, reading whatever I've written, and she always seems to know what works and what doesn't. It goes without saying that she has been immensely helpful. Jean and John Comaroff have been brilliant mentors throughout my career, and without their support and encouragement, none of my work would have been possible. Finally, I owe a debt to the reviewers and editors of this manuscript, whose insights have been invaluable and inspiring.

NOTES

PROLOGUE

1. The concept of posthumanism is more complicated than I may seem to be suggesting here. While, for some, it implies evolution beyond the limits of the human, for most critical theorists it entails a critique of the conceptual foundations of Western humanism and particularly the privileging of humans above other forms of life and nonlife.

2. Mantz's research on upper-level entrepreneurs (mainly, high-level négociants and comptoirs) has resulted in a number of excellent publications (Mantz 2011, 2018 a and b).

3. See our coauthored book, Smith and Mwadime 2014.

4. Belgian colonial administrators recognized Hutu and Tutsi as different ethnicities, even races, though they were not (Mamdani 2002). What came to be called the Hamitic hypothesis was a nineteenth-century concept which had it that centralized, hierarchical states had been brought to this region of Africa by more "advanced" pastoralists from the North, who were generally understood to be a race separate from and superior to other Africans. They were seen as responsible for African "development" in the form of states, a political form widely assumed, to this day, the measure of development, despite a long history of anthropological critiques of this (see, e.g., Scott 2009; Graeber 2004). The Hamitic hypothesis was instrumentalized in colonial Rwanda and Burundi, becoming the basis of colonial rule and administration; Europeans governed through the Tutsi and established an apartheid system with separate school, salaries, and opportunities for Tutsi and Hutu. At the same time—in late colonial Rwanda and Burundi—colonial authorities sometimes railed against the aristocratic, "feudal" Tutsi that they had propped up and empowered, and encouraged "democratization" from below in the form of Hutu empowerment. In late colonial and postcolonial Rwanda and Burundi, anticolonial independence and democracy movements took the form of violent revolts and genocides against the Tutsi ruling elite.

5. While mining existed in some regions of precolonial Congo (see Matori 2017 on precolonial copper mining in Katanga), "3Ts" mining came with colonialism; also, everyone I spoke with about gold mining said that the demand for and knowledge of this mineral came to this part of the eastern Congo with the Belgians, but I'm less confident about that.

6. For overviews, see Harden 2001; IPIS 2002; UN Security Council 2002; Human Rights Watch 2009; Enough Project 2009; Global Witness 2009; Seay 2012. For an overview and critique of the crafting of the NGO conflict minerals argument, see Nest 2011.

7. See, e.g., Bafilemba et al. 2014; Bleischwitz et al. 2012; Cuvelier et al. 2014; Draginis 2016; Geenen 2012; Geenen 2017; Johnson 2013; King 2014; Marshall 2017; Maystadt et al. 2014; Parker and Vadheim 2017; Radley and Vogel 2015; Raeymaekers 2013; Raghavan 2014; Seay 2012; Seitzinger 2015; Stoop and van der Windt 2018, 2018b; Vogel and Raeymaekers 2016; Wolfe 2015; Worstall 2011.

8. There were actually a number of competing supply chain tracking initiatives, but in the region where I was working, ITRI's ITSCI soon emerged as the dominant one and remained so during my fieldwork (for a detailed and insightful overview of how this came to be, see Cuvelier et al. 2014).

CHAPTER ONE

1. In a different vein, Michael Taussig famously demonstrated how competing worldviews and ideas about the good life came into conflict through mining (Taussig 2010). He plumbed South American miners' depictions of extractive capitalism as a pact with the devil, focusing on the moral dilemmas, interpretations, and inventions that were unearthed through mining. Taussig's work also made it abundantly clear that there was more to mining than the mere extraction of resources, and that potent moral, spiritual, and ontological forces were at work in the conversion of one form of value into another.

2. Some anthropologists have examined contemporary mining conflicts as particularly fruitful entry points for grasping the globalization of capital and the dispossession of indigenous peoples (F. Li 2009, 2015; T. Li, 2003; Valentine 2012). A growing number have begun to take seriously the different meanings and value potentials of minerals to the people who extract them (most relevant for this work is De Boeck 1998, concerning Congolese diamond miners; for more recent work, see eds. Ferry, Vallard, and Walsh 2019; Coyle 2020; D'Angelo and Pijpers 2018; Ferry and Limbert 2008; High 2017; Walsh 2012). A number have focused on the rhetoric and practice of corporate responsibility, drawing attention to the close connection between corporations and international humanitarian organizations, including how the discourse of corporate responsibility dovetails with the project of resource extraction (Coumans 2011; Kapelus 2002; Kirsch 2014; Rajak 2011, 2011b; Sharp 2006; Welker 2014; this book also contributes to these discussions). And one particularly original ethnography examines how mines and the people who live near them co-create and co-constitute one another, becoming real, and even monolithic sociopolitical entities, in dynamic interaction with one another over time (Golub 2014).

3. Not that all ethnographies of industrial mining are simplistically dualistic. Most notably, Alex Golub shows how indigenous groups and mining companies co-constitute one another in Papua New Guinea (2014).

4. For an excellent genealogy of this dynamic in a different but fundamentally similar context, see Burke 1996.

5. This definition and usage of qualisign is not mine but derives from Nancy Munn's use of Charles Peirce's concept of a quality that acts as a sign but can only do so when it's embodied (it is eloquently and concisely explained in Chu 2010; see Munn 1986; Peirce 1998). In elaborating on the concept, Julie Chu describes how

what she characterizes as the qualisign of mobility (a concept very similar to Congolese understandings of movement) only becomes sensible through the "multiplicity of its materializations in various embodied forms" (Chu 2010). As she puts it in her discussion of Fuzhounese migrants:

> For like other qualities that are also signs, mobility can do little on its own until it is materialized through people, objects, words, and other embodied forms. Yet once actualized through a particular thing, it also inevitably becomes entangled with the other features of whatever material forms it inhabits. When embodied by a passenger on a plane, for instance, mobility cannot help being bundled with other qualities like speed, lightness, or cosmopolitan privilege, just as it cannot avoid insinuations of inefficiency, danger, or deprivation when embodied by a stowaway traveling across the ocean in a shipping container. (Chu 2010, 15)

6. In a related vein, Sanchez (2017) has argued that, because of its material qualities, coltan has been more likely to generate peace and prosperity, as well as state provisioning, than gold mining.

7. Others have drawn attention to how wartime actors in Africa perpetrate acts of violence and expropriation "just in time" to the tune of global capitalism (Hoffman 2011).

8. This comment is not intended to naturalize these "boundaries" (e.g., human/nonhuman or country/city) or to suggest there is anything good or inevitable about them, but simply to draw attention to shifts that were taking place, or preexisting processes that were exacerbated.

9. For a nuanced analysis of the relationship between mining and agriculture in Africa, see Pijpers 2014.

10. Raymond has insisted that I add that a goat that cost $20 in Goma was $120 in the mining town of Kalima.

11. According to Schouten and others, Swahili, colonial, and postcolonial authorities simply supplanted these political cartographies that were based on extracting value from movement, rather than producing entirely new systems of governance *ex nihilo*.

12. While the closure of mobility has often taken the form of organized violence, "the social relations of taxation" are often "complex, ambiguous, ordered, and reciprocal," and so irreducible to violence (Hoffman, Vlassenroot, and Marchais 2017 in Lund and Eilenberg 2017, 235; see also Raeymaekers 2010).

13. As Hoffman puts it in reference to colonial chiefs, "Violence was an omnipresent companion of colonial resource extraction in the Congo Free State. Rape, torture, whipping, hostage taking, mutilation, surprise raids and summary executions underpinned the *prestations* [the giving of tribute to colonial chiefs] system" (Hoffman, Vlassenroot, and Marchais 2016, 239).

14. The foreclosure of movement in the Belgian Congo took many forms, including vaccination campaigns (Hunt 2009, 2016).

15. To quote Long, "Congo's land became a mixture of crown lands, belonging to Leopold; concessions, frequently vast, were licensed to private, usually Belgian, companies (with the state—meaning Leopold—owning 50% of the shares); private land titles; and small areas of land subject to 'customary rights'" (Long 2011, 4). Congolese people were allowed to occupy land but only Europeans could own land.

16. In the immediate postwar era, the World Bank encouraged the unelected transi-

tional government to assert control over mineral and forest resources through the development of a Mining Code and a Forest Code, interventions they saw as key to "relaunching" the economy (Malele 2007; Trefon 2007; Long 2011).

17. In recent years, a number of contemporary scholars—mainly cultural geographers, anthropologists, and historians—have pointed out that what Marx called primitive accumulation doesn't belong to an earlier "stage" of capitalist development, but is an ongoing process that continues to feed the growth of capitalism, sometimes compensating for what Marx referred to as crises of capitalist accumulation (Harvey 2004; Elyachar 2005; Federici 2004; Glassman 2006; Kelly 2011; Hall 2013; Sassen 2014; West 2012).

18. Structural adjustment programs and neoliberalism broadly construed have also been interpreted as key causes of the rise of artisanal mining throughout the world (Banchirigah 2006; Dorner et al. 2012; Hentschel et al. 2002; Hilson and Potter 2005; Jønsson et al. 2009).

19. See, e.g., the Spanish novelist Vásquez-Figueroa's economically titled war thriller, *Coltan* (2009).

20. For example, contemporary journalists and NGOs deplore the ways armed groups and companies have exploited Congolese *and* lament that Congo's political leaders have not yet developed the consciousness necessary to administer a bureaucratic state, an institutional form that is still widely seen as the necessary precondition for peace and prosperity (hence the implied need for interventions designed to bolster the state, regardless of what Congolese think about it or how they experience it; see, e.g., Autesserre 2010). For most commentators, Congo remains the very definition of the Hobbesian State of Nature, in which the state does "not yet" exist, even though it very clearly does (I have been present at more than one talk where an academic working in Congo declared that there was no state, and that Congo was in fact the "state of nature." In one case, it turned out that the researcher's Congolese "fixer" had paid the state officials to go away before the researcher showed up, so that they wouldn't bother the American about papers and stamps; hence the researcher's impression that there was no state at all in nonurban areas).

21. For excellent ethnographies of how these virtual worlds are in fact embodied and the motivations and meaning-making activities of avatars, see Boellstorf 2015 and Golub 2007; for an excellent review of this literature, see Golub 2014.

22. We could go back, further, to the 1960s counterculture, but that would be a different book; see, e.g., Spencer 2018.

23. See also Bukatman 1993: "Cyberspace is a celebration of spirit, as the disembodied consciousness leaps and dances with unparalleled freedom. It is a reality in which the mind is freed from bodily limitations, a place for the return of the omnipotence of thoughts . . . the return of the animistic view of the universe within the scientific paradigm" quoted in Hergert 2017.

24. A rather large portion of these *Star Trek* episodes concern the potential dangers and displeasures of the "inevitable" evolution toward disembodiment (e.g., sad futuristic beings that have evolved out of their own bodies and now seek to possess the bodies of humans in order to experience sensory and sensual pleasures). In fact, the central character (Kirk) seems to exist mainly to reassure people that certain forms of embodied masculinity (fist-fighting, for example) will still be important in a highly computerized and technological future. He also spends a lot of time arguing with overly rule-bound computers until they blow up in frustration. Beyond *Star Trek*, in popular thought about extraterrestrials, embodiment is also a desire or

need that is projected onto supposedly advanced entities (the most frequent among them, the greys, are all head and no genitals, often depicted as requiring humans to procreate; their arms are definitely too skinny to dig for long in the ground; Battaglia 2006).

25. Mining does emerge fairly often in science fiction, from Metropolis to the work of Ursula K. Le Guin and more recent literature on moon and asteroid mining, where it tends to speak to the destruction and cruelty that makes "progress" possible.

26. There is a long line of critical thinking about the Western Idea of Africa. See Mudimbe 1989, 1994.

27. For a nuanced analysis of African appropriation of mobile phones, see McIntosh 2007.

28. When eastern Congolese in mining areas got hold of digital technologies, they often used them to reveal hidden and sometimes horrible things. They posted pictures of the violence committed by militias, including gut-wrenching images of victims of mass rape and assault, on their Facebook pages. They argued over photoshopped images of alleged witches supposedly caught in the act (one popular image showed a group of people in "traditional" dress sitting on the ground, alleged to be passengers of an invisible witch plane that had crashed). Some people even took their lives in their own hands by texting images of secret boys' initiation rites in the forest, including photos of the forest deities that preside over some ethnicities' versions of these rituals. At times this urge to reveal "the truth" was malicious and criminal: in the remote, impoverished mining town of Kalima, a man allegedly had sex with multiple married women for one hundred dollars each, then he posted their names and pictures on Facebook, apparently in an effort to reveal "the truth" about Kalima's women to others. Two of the women were said to have died by suicide, while several others escaped to the city to avoid their neighbors, spouses, and kin.

CHAPTER TWO

1. This approach builds on the work of others who have argued that war is not only about destruction and violence, but shapes worldview and perception (Hoffman 2011).

2. This section on ropes (up to the next section) was published in a somewhat different form in Smith 2015.

3. See, for example, Hochschild 1998, Forbath 1991, Folsom 2016.

4. For example, when I was doing fieldwork in Punia, Maniema, in 2013, the British NGO that was subsidizing the dispensary and sanitizing the water there suddenly disappeared one day without informing anyone. Nearly everyone I knew, including myself, became violently ill from water we had grown accustomed to drinking; in just a few days, mortality rates skyrocketed and malaria claimed my friend's only child.

5. Other meaningful referents emerge in these rumors, like the reference to medical technologies and interventions (hypodermic needles) and, in a somewhat different vein, other more contemporary and even futuristic technology—mainly, computers that can read one's emotional state and suggest solutions to problems. This evokes a longer history—for example, biomedicine, and particularly the hypodermic needle, was an important instrument and symbol in the Belgian Congo because it was through vaccination campaigns that the colonial regime first sought to control the movement of African bodies in space, radically restricting movement with the ostensible aim of stopping the spread of disease in a region that colonial authorities viewed as synonymous with disease.

6. Because this research was carried out in the Kivus and Maniema, I did not hear de-
    tailed stories about the Ugandan-backed MLC, or Movement for the Liberation of
    Congo, which was focused in Ituri and other areas further north.
7. The extent to which this happened was never entirely clear (the militia leader Sheka
    in Walikale was held out as exemplifying this strategy), but many soldiers told me
    that this was a strategy they had seen their superiors employ, and a few offered that
    they might employ such a strategy if they got the opportunity (especially earlier in
    my fieldwork, before 2011).
8. By the time I started doing fieldwork, the problem of the child witch had already
    become an object of humanitarian intervention, as there was an NGO in Goma that
    served as an orphanage for accused child witches and worked to assimilate them
    back into society (I visited there several times in 2011, toward the end of the "time
    of child witches," as some people called it). The NGO, and other NGOs, formally
    blamed Pentecostal churches for spreading and profiting from rumors of child
    witches; one poster that could be found in many NGO, school, and government
    offices for some years depicted a man intervening between a preacher and a young
    child whose parents were trying to protect him: "No! Zackie is not a witch!" ran
    the caption. The Congolese-run NGO in Goma was divided by two narratives that
    its personnel had a hard time reconciling—some employees stuck to the line that
    these children were not witches and were therefore victims who had to be protected,
    while others felt they were in fact witches and needed to be "saved" through reli-
    gious intervention.
9. See also the Institute for Developing Economies Japan External Trade Organiza-
    tion website: https://www.ide.go.jp/English/Data/Africa_file/Company/drc01.html.
    While the general outlines of the story were widely known in the region, most of the
    details of this story, such as the contract percentages, came from the management of
    the partly government-owned company Sakima, stationed in Maniema. Sakima had
    purchased Sominki's 3Ts concessions and knew the details of the Canadian gold-
    mining company Banro's claims on the gold mines.
10. I am adopting the phrase "web of life" from Moore 2015.
11. Failure to keep these observances led directly to their death. Indeed, one of the main
    reasons that former Mai Mai insurgents gave for their frequent deaths and losses
    was failure to follow prohibitions on sex and entering into contact with women, as
    well as violations of food taboos.
12. Some of these rituals resemble precolonial rituals of sovereignty in Congo's various
    kingdoms (see Graeber and Sahlins 2017), in which kings produced extraordinary
    power and created sovereignty through performing acts that were otherwise taboo.
13. One man finished his discussion about how Mai Mai never stole or raped by re-
    membering the time a group of Mai Mai stole his minerals from him. "But you just
    said Mai Mai weren't bad, that they never stole," I said. "Well, on that day they were
    bad," he retorted.

CHAPTER THREE

1. Schouten used this phrase while presenting drafts of his forthcoming book
    (Schouten 2021) at a workshop I participated in at UC Berkeley in 2018. I'm not
    sure if it appears in the book or not, since it is not yet published.
2. It was hard to say how many mines were run like this but educated estimates from
    my interlocuters put the overall percentage at less than half, maybe a third.
3. The vast majority of women I spoke to about this claimed they had no interest in

arguing with men about this issue because they didn't want to, or felt unable to perform, the difficult work of digging in holes; they claimed to be more or less satisfied so long as they weren't excluded from washing and sorting minerals, and from selling commodities to diggers.

4.   There is a recent, but growing, literature on the experiences of women in artisanal mining communities, and their contribution to artisanal mining. For DR Congo, see Bashwira and Cuvelier 2019; for a wider, global context, see Moretti 2006; Yakovleva 2007; Lahiri-Dutt 2006, 2012, 2015.

5.   The way it was explained to me by Congolese who were involved in the early implementation of cooperatives is this (I have no idea if this accurately reflects the intentions of the World Bank or the IMF, but it was my interlocutors' mediated perspective on what was happening): Cooperatives were first foisted onto Congo by the World Bank and the IMF, and written into the Mining Code of 2002, but they have since morphed in Congolese hands, becoming an instrument of collaboration and collusion across differences. The original idea was that these international organizations would fund cooperatives of diggers that would "develop" into companies, ultimately making négociants and foreign comptoirs unnecessary. The alleged intention was the development of a local entrepreneurial class that was relatively self-sustaining and generative. As a result of these interventions, cooperatives became a legal requirement for mines, included in the Mining Code of 2002 (although not all mines, even properly regulated ones, have cooperatives, despite the official requirement that they exist). But, according to this interpretation (which may or may not be true—I'm not sure, but it does sound believable to me), the money for these cooperatives never came through from the World Bank or the IMF because high-level Congolese state actors preferred the money to pass through state institutions, while the international lenders wanted the money to be funneled through local NGOs, in keeping with the neoliberal, anti-statist ideology of the "development apparatus" at the time. Without this funding, the cooperatives were never able to take off and become self-sufficient.

6.   In mining concessions that are outside of a ZEA (zone of artisanal extraction), the owners of the concession (a company that has bought a contract from a preceding company) have representatives on site; they allow diggers to dig and charge fees to others (négociants and comptoirs) who come in to buy. They also charge the PDG for use of the hole.

7.   To complicate matters even more, the Congolese government inscribed the system of négociants and comptoirs into the Mining Code, so that the two systems—one intended to make négociants and comptoirs obsolete, and the other requiring them as part of the system of regulation—today sit alongside one another in that legal code. Over time, the cooperatives became a complex instrument with multiple cross-cutting purposes that were never exactly the same from place to place. For diggers, effective cooperatives organize labor, maintain security, resolve conflicts, and even try the majority of minor cases of conflict between diggers. They also provide diggers with loans, tools, and machines purchased through dues or money forwarded by nondigger members of the cooperatives. When they work, the cooperatives eclipse state authorities—that is, they do the work the small-scale mining authority, SAESSCAM (another institution set up through the Mining Code of 2002), is supposed to do but never does. They also take over from the mining police and the courts, or work in combination with them, doing whatever they can to keep armed actors away from the mines.

8. At that point, in 2013, new government regulations mandated that comptoirs be located in provincial capitals and that middlepersons sell to the comptoirs in the provinces from which they bought minerals, in order to facilitate taxation. Under this system, taxes were collected from the comptoir, and sent to Kinshasa, after which a portion (the retrocession) was supposed to be returned to the province, though this rarely happened.

9. For example, drivers were often surprised by the different document regimes that emerged as they crossed from one province to another, which was made more overwhelming by the fact that these documents might apply to the driver, the car, or even the accompanying passengers.

10. I have a rather straightforward, though twofold, understanding of what commodity fetishism means in the context of capitalism. First, I mean commodities becoming more important, deserving of human concern/care, valuable, powerful (in the sense of directing or orienting human affairs), and decisive than people. Second, I mean the systematic loss, or displacement, of knowledge and awareness about the total process through which commodities come into being and come to have value (which is not a process limited or reducible to wage labor). The assumption that commodity fetishism is mainly a habit of mind or way of thinking (which it partly is)—rather than, say, the stock market, the act of bailing out banks instead of people, or a ride on a Congolese cargo plane—has led to a lot of confusion, including the idea that commodity fetishism is an intellectual mistake that might be resolved intellectually, which is simply not the case. The term itself doesn't help, because it was originally used by Marx to draw attention to what he regarded as the magical thinking of economists, which he was equating with his nineteenth-century understanding of non-European "animists," transmitted to him via mostly eighteenth-century Western authors.

CHAPTER FOUR

1. In a similar vein, Makori 2017 also shows how artisanal miners and artisanal mining in the copperbelt region of Katanga draw upon and reproduce the pasts of mining in the present.

2. See Ferguson's (1999) insightful analysis of the multiple responses to decline and reversal in the Zambian copperbelt following the collapse of modernist optimism that accompanied the "African Industrial Revolution" (see also Apter 2005).

3. An earlier version of this segment can be found in Smith 2015. Lagome has said that he prefers that I use the name he goes by, Lagome, rather than concocting new names or place names.

4. Punia, a colonial chief, is remembered as having been an old man who met the Belgians when everyone else ran away, and whose great-grandson now collects the rent for Sakima.

5. In 2010, new regulations in mining mandated that minerals had to pass through the provincial capitals so that tax could be collected there.

CHAPTER FIVE

1. Major Patrick has indicated that he prefers I use his actual name.

2. De Boeck discusses how women in artisanal diamond–mining towns on the Angolan border were referred to as "dogs breaking their leash," meaning having been freed from patriarchal control.

CHAPTER SIX

1. In addition, they chose the river as a "natural" territorial boundary dividing the two claims so that Bassa henceforth had a partial territorial claim to Bisie in addition to a tributary claim over the part of Bisie over which they did not exercise a territorial claim.

CHAPTER SEVEN

1. Under Manzi, who was in command when I went to Bisie, the system of tolls never went away, and the army continued to collect revenue from Bisie, allegedly funneling that money to Rwanda and Rwandan-backed militias. After the closure of Bisie, Manzi left the army and joined the rebel militia M23, which would temporarily seize the city of Goma in 2011.

CHAPTER EIGHT

1. The videos on the Alphamin website seem to change periodically. The videos I discuss here were recorded and transcribed by me and my then graduate student, Jane Saffitz (currently Assistant Professor of Anthropology at Denison University), in July and August of 2018.

2. Another example was Matamba, a small village with an adjacent mine: the government chose it for validation because it was near the administrative center of Walikale and more or less equidistant to every nonvalidated site. That meant that the minerals from those sites, including Bisie, could be "illegally" cleaned there, receiving a Matamba tag even when they were minerals from somewhere else. This would "open the ways" of diggers, négociants, and state workers, even though this "cleaning" of minerals was in complete violation of how the system was supposed to work. But in terms of avoiding violence, Matamba was a bad choice: it was right on the road, in a small village with no military presence. Any armed group with the will to do so could just march in and "invade" (*kuvamia*) that place.

CHAPTER NINE

1. Being more interested in how they understood it than how planning offices imagined it, I use "ITRI/Pact" here because those on the ground tended to know the word *ITRI* and not *Pact*, even though it was actually Pact and the government doing the implementing.

2. I mostly talked to diggers and négociants about their experiences of the project, rather than seeking out ITRI representatives. But when I did run into ITRI representatives and converse with them, a few days or weeks would go by and I'd receive an email from the ITRI office in England asking me to fill out a form specifying my funding source and asking me how much money I had. In each case, I was in the forest and unable to download the form. There was never any sense that the form was a prelude to us having a conversation, and I was inclined to infer that they felt I needed this form just to exist in the spaces they were in. I would respond asking to meet or talk with them, but they wouldn't respond to that aspect of the message; rather, they would repeat their request that I fill in that line I left blank on the form regarding what my budget was. (I eventually did have a nice phone interview with someone from Pact in DC, who seemed apologetic about the secretiveness of their partner, ITRI.)

3. For an excellent, and more exhaustive, overview, see Cuvelier et al. 2014.

4. For industrial mining companies like Banro and Alphamin, tagging was a mixed bag: on the one hand, it legitimated artisanal mining and made it possible for the Mining Code to be rewritten in 2017, so that artisanal mining and industrial mining were allowed to coexist in the same site (the Mining Code now calls this the "zone of tolerance"). On the other hand, so long as the mines to which they held contracts remained unvalidated, companies could use the fact that they were "red," or bloody, to have the government relocate artisanal miners to other, validated sites (as happened at Bisie). Or, as one NGO representative put it, "the government can say this place is now a place of blood minerals and no one can dig there, so the companies can use the process of traceability to get their rights."

5. It's interesting in this regard that one of the early tracking initiative proposals, before ITRI/Pact became the dominant player, was a conservation NGO that tagged and tracked gorillas (Cuvelier et al. 2014). This suggests that the homology miners saw between conservation and conflict free minerals tracking, as sharing in a cosmology of enclosure, were prescient and not spurious.

CONCLUSION

1. To drive home this exponential growth, I quote from a 2018 report of the Intergovernmental Forum on Mining, Minerals, and Sustainable Development: "Artisanal and small-scale mining (ASM) has experienced explosive growth in recent years due to the rising value of mineral prices and the increasing difficulty of earning a living from agriculture and other rural activities. An estimated 40.5 million people were directly engaged in ASM in 2017, up from 30 million in 2014, 13 million in 1999 and 6 million in 1993. That compares with only 7 million people working in industrial mining in 2013." The authors then go on to wring their hands about the need for state regulation. https://www.iisd.org/publications/global-trends-artisanal-and -small-scale-mining-asm-review-key-numbers-and-issues/.

2. Displacement and dispossession on a global scale, he argues, have brought about a new kind of increasingly common, if not universal, human being that surpasses the limits of humanism—one existing in a condition of "lost citizenship." Having "lost everything," this new person is "always nowhere" and is therefore also always in passing (passant, as he puts it). Mbembe posits that the condition of being passant, or in passing, is different from that of being a migrant, a refugee, or, on the other end of the spectrum, a citizen or native. Rather, the "project" and "fate" of the person in passing lies in "learning to pass constantly from one place to another" all the while carrying the torch of an "idea of the earth that is common to us." Mbembe's voice here joins that of many others who have insisted that the right to mobility in an age of exclusion and expulsion is among the key political motifs of our time.

3. This is clear in the violent expulsion of people from forest to protect nature (Gauthier and Pravettoni 2016; Longo 2019, 2020; Vidal 2020; Zaitchik 2018), but dispossession through conservation happens in more subtle ways as well. See, e.g., Kelly 2011; West 2016.

BIBLIOGRAPHY

Achebe, Chinua. "An Image of Africa: Racism in Joseph Conrad's *Heart of Darkness.*" *Massachusetts Review* 18 (1977): 77.

Adunbe, Omolade. *Oil Wealth and Insurgency in Nigeria.* Bloomington: Indiana University Press, 2015.

Aneesh, Aneesh. *Virtual Migration.* Durham, NC: Duke University Press, 2006.

Autesserre, Severine. "Dangerous Tales: Dominant Narratives on the Congo and Their Unintended Consequences." *African Affairs* 111, no. 443 (2012): 202–22.

Autesserre, Severine. *The Trouble with the Congo: Local Violence and the Failure of International Peacebuilding.* New York: Cambridge University Press, 2010.

Bafilemba, Fidel, Timo Mueller, and Sasah Leshnev. *The Impact of Dodd-Frank and Conflict Minerals Reform on Eastern Congo's Conflict.* Enough Project report, June 2014.

Bales, Kevin. *Blood and Earth: Modern Slavery, Ecocide, and the Secret to Saving the World.* New York: Spiegel and Grau, 2016.

Ballard, Chris, and Glenn Banks. "Resource Wars: The Anthropology of Mining." *Annual Review of Anthropology* 32 (2003): 287–313.

Banchirigah, Sadia Mohammed. "How Have Reforms Fuelled the Expansion of Artisanal Mining? Evidence from sub-Saharan Africa." *Resource Policy* 31 (2006): 165–71.

Banner, Olivia. *Communicative Biocapitalism: The Voice of the Patient in Digital Health and the Health Humanities.* Ann Arbor: University of Michigan Press, 2017.

Bashwira, Marie-Rose, and Jeroen Cuvelier. "Women, Mining and Power in Southeastern Democratic Republic of Congo: The Case of Kisengo." *Extractive Industries and Society* 6, no. 3 (2019): 960–67.

Besley, Tina, and Michael Peters. "Enterprise Culture and the Rise of the Entrepreneurial Self." *Counterpoints* 303 (2007): 155–74.

Biletshi, Raoul. "Killing the Dollar Softly in the Democratic Republic of Congo." *World Crunch,* January 9, 2013. https://worldcrunch.com/world-affairs/killing-the-dollar-softly-in-the-democratic-republic-of-congo.

Bleischwitz, Raimund, Monika Dittrich, and Chiara Pierdicca. "Coltan from Central Africa, International Trade and Implications for Any Certification." *Resource Policy* 37 (2012): 19–29.

Blomley, Tom, Dilys Roe, Fred Nelson, and Fiona Flinton. *"Land-grabbing": Is Conservation Part of the Problem or the Solution?* IIED, London, September 2013. https://pubs.iied.org/sites/default/files/pdfs/migrate/17166IIED.pdf.

Blum, Andrew. *Tubes: A Journey to the Center of the Internet*. New York: Ecco, 2013.

Blunt, Robert. *For Money and Elders: Ritual, Sovereignty, and the Sacred in Kenya*. Chicago: University of Chicago Press, 2019.

Boellstorf, Thomas. *Coming of Age in Second Life: An Anthropologist Explores the Virtually Human*. Princeton, NJ: Princeton University Press, 2015.

Burke, Timothy. *Lifebuoy Men, Lux Women: Commodification, Consumption, and Cleanliness in Modern Zimbabwe*. Durham, NC: Duke University Press, 1996.

Buscher, Karen, and Koen Vlassenroot. "Humanitarian Presence and Urban Development: New Opportunities and Contrasts in Goma, DRC." Special issue, *Disasters* 34, no. S2 (March 11, 2010): S256–S273.

Chu, Julie. *Cosmologies of Credit: Transnational Mobilities and the Politics of Destination in China*. Durham, NC: Duke University Press, 2010.

Clark, John Frank, ed. *The African Stakes of the Congo War*. New York: Palgrave Macmillan, 2002.

Comaroff, John, and Jean Comaroff. *Theory from the South: How Europe Is Evolving Toward Africa*. New York: Routledge, 2012.

Conrad, Joseph. *Heart of Darkness*. 1902. Reprint, Mineola, NY: Dover Thrift Editions, 1990.

Coumans, Catherine. "Occupying Spaces Created by Conflict: Anthropologists, Development NGOs, Responsible Investment, and Mining." Supplement, *Current Anthropology* 52, no. S3 (2011): S29–S43.

Covington-Ward, Yolanda. *Gesture and Power: Religion, Nationalism, and Everyday Performance in Congo*. Durham, NC: Duke University Press, 2016.

Coyle, Lauren. *Fires of Gold: Law, Spirit, and Sacrificial Labor in Ghana*. Oakland: University of California Press, 2020.

Cuvelier, Jeroen, ed. "The Complex Conflict Dynamics in Kalehe's Nyabibwe Mine." In *The Complexity of Resource Governance in a Context of State Fragility: The Case of Eastern DRC*. IPIS, University of Ghent, November 2010.

Cuvelier, Jeroen. "Men, Mines and Masculinities: The Lives and Practices of Artisanal Miners in Lwambo (Katanga Province, DR Congo)." PhD diss., KU Leuven (Belgium), 2011.

Cuvelier, Jeroen. "Money, Migration and Masculinity among Artisanal Miners in Katanga (DR Congo)." *Review of African Political Economy* 44 (2016): 204–19.

Cuvelier, Jeroen, Jose Diemel, and Koen Vlassenroot. "Digging Deeper: The Politics of Conflict Minerals in the Eastern Democratic Republic of Congo." *Global Policy* 4, no. 4 (2013): 449–51.

Cuvelier, Jeroen, Steven Van Bockstael, Koen Vlassenroot, and Claude Iguma. *Analyzing the Impact of the Dodd-Frank Act on Congolese Livelihoods*. New York: Social Science Research Council, 2014.

D'Angelo, Lorenzo, and Robert Pijpers. "Mining Temporalities: An Overview." *Extractive Industries and Society* 5, no. 2 (2018): 215–22.

Day, Sophie, Akis Papataxiarchis, and Michael Steward. *Lilies of the Field: Marginal People Who Live for the Moment*. Studies in the Ethnographic Imagination. New York: Westview, 1998.

De Boeck, Filip. "Dogs Breaking Their Leash: Globalization and Shifting Gender Categories in the Diamond Traffic between Angola and DR Congo (1984–1997)." In *Changements au féminin en Afrique noire: Anthropologie et Littérature* Vol. 1, 87–144. Tervuren: Musée Royale de l'Afrique Centrale, 1999.

De Boeck, Filip. "Domesticating Diamonds and Dollars: Identity, Expenditure and Sharing in Southwestern Zaire (1984–1997)." *Development and Change* 29, no. 4 (1998): 777–810.

De Boeck, Filip. "Inhabiting Ocular Ground: Kinshasa's Future in the Light of Congo's Spectral Urban Politics." *Cultural Anthropology* 26, no. 2 (2011): 263–86.

De Boeck, Filip. "On Being Shege in Kinshasa: Children, the Occult, and the Street." In *A Reader in the Anthropology of Religion*, edited by Michael Lambek, 495–506. Oxford: Blackwell, 2004.

De Boeck, Filip, and Sammy Baloji. "Positing the Polis: Topography as a Way to De-center Urban Thinking." *Urbanisation* 2, no. 2 (2017): 142–54.

De Boeck, Filip, and Sammy Baloji. *Suturing the City: Living Together in Congo's Urban Worlds.* London: Autograph ABP, 2016.

De Boeck, Filip, and Francois Plessart. *Kinshasa: Tales of the Invisible City.* Leuven, Belg.: Leuven University Press, 2014.

De Certeau, Michel. *The Practice of Everyday Life.* Oakland: University of California Press, 1988.

Devlin, Larry. *Chief of Station Congo: Fighting the Cold War in a Hot Zone.* New York: PublicAffairs, 2008.

Diemel, J. A., and J. Cuvelier. "Explaining the Uneven Distribution of Conflict-Mineral Policy Implementation in the Democratic Republic of Congo: The Role of the Katanga Policy Network (2009–2011)." *Resource Policy*, 46, no. 2 (2015): 151–160.

Diemel, J. A., and D. J. M. Hilhorst. "Unintended Consequences or Ambivalent Policy Objectives? Conflict Minerals and Mining Reform in the Democratic Republic of Congo." *Development Policy Review* (Overseas Development Institute) 37, no. 4 (2018): 453–69.

Dorner, Ulricke, Gudrun Franken, Maren Liedtke, and Henrike Sievers. "Artisanal and Small Scale Mining." Working paper, POLINARES, 2012, 19.

Douglas, Mary. *Purity and Danger.* New York: Routledge, 1966.

Draginis, H. *Point of Origin: Status Report on the Impact of Dodd-Frank 1502 in Congo.* Enough Project report, February 2016.

Elyachar, Julia. *Markets of Dispossession: NGOs, Economic Development, and the State in Cairo.* Durham, NC: Duke University Press, 2005.

Enough Project Team. *A Comprehensive Approach to Congo's Conflict Minerals.* Enough Project report, April 2009.

Fabian, Johannes. *Time and the Other: How Anthropology Constructs Its Object.* Chicago: University of Chicago Press, 1986.

Fairhead, James. "Paths of Authority: Roads, the State and the Market in Eastern Zaire." *European Journal of Development Research* 4 (1992): 17–35.

Federici, Silvia. *Caliban and the Witch: Women, the Body, and Primitive Accumulation* Brooklyn, NY: Autonomedia, 2004.

Ferguson, James. *Expectations of Modernity: Myths and Meanings of Urban Life on the Zambia Copperbelt.* Berkeley: University of California Press, 1999.

Ferguson, James. *Global Shadows: Africa in the Neoliberal World Order.* Durham, NC: Duke University Press, 2006.

Ferme, Mariane. *Out of War: Violence, Trauma, and the Political Imagination in Sierra Leone.* Oakland: University of California Press, 2018.

Ferme, Mariane. *The Underneath of Things: Violence, History, and the Everyday in Sierra Leone.* Berkeley: University of California Press, 2001.

Ferry, Elizabeth E., and Mandana E. Limbert, eds. *Timely Assets: The Politics of Resources and Their Temporalities.* Santa Fe, NM: School for Advanced Research Press, 2008.

Ferry, Elizabeth, Annabel Vallard, and Andrew Walsh. *The Anthropology of Precious Minerals*. Toronto: University of Toronto Press, 2019.

Folsom, Jenny. "Antwerp's Appetite for Foreign Hands." *Contexts* 15, no. 4 (2016): 65–67.

Forbath, Peter. *The River Congo: The Discovery, Exploration, and Exploitation of the World's Most Dramatic River*. Boston: Houghton Mifflin, 1991.

Foucault, Michel. *The History of Sexuality*. New York: Vintage, 1990.

Garrett, Nicholas, and Harrison Mitchell. *Trading Conflict for Development: Utilizing the Trade in Minerals from Eastern DR Congo for Development*. Resource Consulting Service, April 2009.

Gauthier, Marine, and Riccardo Ravettoni. "'We Have Nothing but Our Reindeer': Conservation Threatens Ruination for Mongolia's Dukha." *Guardian*, August 28, 2016. https://www.theguardian.com/global-development/2016/aug/28/reindeer -conservation-threatens-ruination-mongolia-dukha.

Geenen, Sara. "A Dangerous Bet: The Challenges of Formalizing Artisanal Mining in the Democratic Republic of Congo." *Resources Policy* 37, no. 3 (2012): 322–30.

Geenen, Sara. "Trump Is Right on Congo's Minerals, but for All the Wrong Reasons." *The Conversation*, February 22, 2017.

Ginsburg, Faye. "Rethinking the Digital Age." *Flow: A Critical Forum on Television and Film* 4, no. 1, January 21, 2005.

Glassman, Jim. "Primitive Accumulation, Accumulation by Dispossession, Accumulation by Extra-economic Means." *Progress in Human Geography* 30, no. 5 (2006): 608–25.

Global Witness. *"Faced with a Gun, What Can You Do?": War and the Militarization of Mining in Eastern Congo*. London: Global Witness, 2009.

Golub, Alex. *Leviathans at the Gold Mine: Creating Indigenous and Corporate Actors in Papua New Guinea*. Durham, NC: Duke University Press, 2014.

Guyer, Jane. *Marginal Gains: Monetary Transactions in Atlantic Africa*. Chicago: University of Chicago Press, 2004.

Guyer, Jane. "Prophecy and the Near Future: Thoughts on Macroeconomic, Evangelical and Punctuated Time." *American Ethnologist* 34, no. 3 (2008): 409–21.

The Hague Centre for Strategic Studies. *Coltan, Congo & Conflict: POLINARES Case Study*. HCSS Report no. 21/05/13. The Hague: HCSS, 2013.

Hall, Derek. "Primitive Accumulation, Accumulation by Dispossession, and the Global Land Grab." *Third World Quarterly* 34, no. 9 (2013): 1582–1604.

Harden, Blaine. "A Black Mud from Africa Helps Power the New Economy." *New York Times*, August 12, 2001.

Hardt, Michael, and Antonio Negri. *Empire*. Cambridge, MA: Harvard University Press, 2001.

Hardt, Michael, and Antonio Negri. *Multitude*. New York: Penguin, 2004.

Hartsock, Nancy. "Globalization and Primitive Accumulation: The Contributions of David Harvey's Dialectical Materialism." In *David Harvey: A Critical Reader*, edited by Noel Castree and Derek Gregory, 167–90. Malden, MA: Blackwell, 2006.

Harvey, David. *The Condition of Postmodernity: An Enquiry into the Origins of Cultural Change*. Malden, MA: Blackwell, 1991.

Harvey, David. *The New Imperialism: Accumulation by Dispossession*. Oxford: Oxford University Press, 2003.

Haufler, Virginia. "Disclosure as Governance: The Extractive Industries Transparency Initiative and Resource Management in the Developing World." *Global Environmental Politics* 10, no. 3 (2010): 53–73.

Hayes, K. "Small-Scale Mining and Livelihoods in Africa." Paper prepared for Common

Fund for Commodities, International Seminar on Small-Scale Mining and Livelihoods in Africa. Zanzibar: CFC, 2008.

Hayles, N. Katherine. *How We Became Posthuman: Virtual Bodies in Cybernetics, Literature, and Informatics.* Chicago: University of Chicago Press, 1999.

Heemskerk, Marieke. "Self-Employment and Poverty Alleviation: Women's Work in Artisanal Gold Mines." *Human Organization* 62, no. 1 (2003): 62–73.

Hegel, Georg Wilhelm Friedrich. *The Philosophy of History.* Mineola, NY: Dover Philosophical Classics, 2004. First published in 1899 by Colonial Press.

Helliker, Kirk, and Tendai Murisa, eds. *Land Struggles and Civil Society in Southern Africa.* Trenton, NJ: Africa World Press, 2011.

Hentschel, Thomas, Felix Hruschka, and Michael Priester. *Global Report on Artisanal & Small-Scale Mining.* London: IIED, 2002.

Hergert, Paol. "How Tangible Is Cyberspace?" *Digital Culturalist,* May 15, 2017. https:// digitalculturist.com/how-tangible-is-cyberspace-dce550c52248.

High, Mette. *Fear and Fortune: Spirit Worlds and Emerging Economies in the Mongolian Gold Rush.* Ithaca, NY, and London: Cornell University Press, 2017.

Hilson, Gavin. "Championing the Rhetoric? 'Corporate Social Responsibility' in Ghana's Mining Sector." *Greener Management International* 53, no. 4 (2006): 3–56.

Hilson, Gavin. "'Constructing' Ethical Mineral Supply Chains in Sub-Saharan Africa: The Case of Malawian Fair Trade Rubies." *Development and Change* 45, no. 1 (2014): 53–78.

Hilson, Gavin. "Small-Scale Mining and Its Socio-economic Impact in Developing Countries." *Natural Resources Forum* 26, no. 1 (2002): 3–13.

Hilson, Gavin. "Small-Scale Mining, Poverty and Economic Development in Sub-Saharan Africa: An Overview." *Resources Policy* 34 (2009): 1–5.

Hilson, Gavin M., ed. *The Socio-Economic Impacts of Artisanal and Small-Scale Mining in Developing Countries.* Exton, PA: Balkema, 2005.

Hilson, Gavin, and James McQuilken. "Four Decades of Support for Artisanal and Small-Scale Mining in Sub-Saharan Africa: A Critical Review." *Extractive Industries and Society* 1, no. 1 (2014): 104–18.

Hilson, Gavin, and Clive Potter. "Structural Adjustment and Subsistence Industry: Artisanal Gold Mining in Ghana." *Development and Change* 36, no. 1 (2005): 103–31.

Hilson, Gavin, Titus Sauerwein, and John Owen. "Large and Artisanal Scale Mine Development: The Case for Autonomous Co-existence." *World Development* 130 (June 2020).

Hochschild, Adam. *King Leopold's Ghost: A Story of Greed, Terror and Heroism in Colonial Africa.* Boston: Mariner, 1998.

Hoffman, Danny. "Violence, Just in Time: Work and War in Contemporary West Africa." *Cultural Anthropology* 26, no. 1 (2011): 34–57.

Hoffman, Danny. *The War Machines: Young Men and Violence in Sierra Leone and Liberia.* Durham, NC: Duke University Press, 2011.

Hoffman, Kasper, Koen Vlassenroot, and Gauthier Marchais. "Taxation, Stateness, and Armed Groups: Public Authority and Resource Extraction in Eastern Congo." *Development and Change* 47, no. 6 (2016): 1434–56.

Human Rights Watch. *"You Will Be Punished": Attacks on Civilians in the Eastern Congo.* New York: Human Rights Watch, 2009.

Hunt, Nancy Rose. "An Acoustic Register, Tenacious Images, and Congolese Scenes of Rape and Repetition." *Cultural Anthropology* 23, no. 2 (2008): 220–53.

Hunt, Nancy Rose. *A Nervous State: Violence, Remedies, and Reverie in Colonial Congo.* Durham, NC: Duke University Press, 2016.

Ingold, Tim. "Toward an Ecology of Materials." *Annual Review of Anthropology* 41 (2012): 427–42.

Intergovernmental Forum on Mining, Minerals, Metals and Sustainable Development (IGF). *Global Trends in Artisanal and Small-Scale Mining (ASM): A Review of Key Numbers and Issues.* Winnipeg: IISD, 2017.

International Peace Information Service (IPIS). *Supporting the War Economy in the DRC: European Companies and the Coltan Trade.* IPIS Report, 2002.

Isaacson, Walter. *The Innovators: How a Group of Hackers, Geniuses, and Geeks Created the Internet Revolution.* New York: Simon and Schuster, 2014.

Jacka, Jerry K. "The Anthropology of Mining: The Social and Environmental Impacts of Resource Extraction in the Mineral Age." *Annual Review of Anthropology* 47 (2018): 61–77.

Jackson, Stephen. "Fortunes of War: The Coltan Trade in the Kivus." In *Power, Livelihoods and Conflict: Case Studies in Political Economy Analysis for Humanitarian Action.* ODI Humanitarian Policy Group Report 13, edited by Sarah Collinson, 21–35. London: Overseas Development Institute, 2003.

Jackson, Stephen. "Making a Killing: Criminality and Coping in the Kivu War Economy." *Review of African Political Economy* 93–94 (2002): 517–36.

Jalbert, Kirk, Anna Willow, David Casagrande, and Stephanie Paladino, eds. *ExtrACTION: Impacts, Engagements, and Alternative Futures.* New York: Routledge, 2017.

Jasanoff, Maya. "With Conrad on the Congo River." *New York Times,* August 18, 2017.

Johnson, Dominic. *No Kivu, No Conflict? The Misguided Struggle against 'Conflict Minerals' in the DR Congo.* Goma, DRC: Pole Institute, April 2013.

Johnson, Dominic, and Aloys Tegera. *Digging Deeper: How the DR Congo's Mining Policy Is Failing the Country.* Goma, DRC: Pole Institute, 2005.

Jones, Jeremy. "'Nothing Is Straight in Zimbabwe': The Rise of the Kukiya-Kiya Economy, 2000–2008." *Journal of Southern African Studies* 36, no. 2 (2010): 285–99.

Jønsson, Jesper Bosse, and Deborah Fahy Bryceson. "Rushing for Gold: Mobility and Small-Scale Mining in East Africa." *Development and Change* 40, no. 2 (2009): 249–79.

Kapelus, Paul. "Mining, Corporate Social Responsibility and the 'Community': The Case of Rio Tinto, Richards Bay Minerals and the Mbonambi." *Journal of Business Ethics* 39 (2002): 275–96.

Kelly, Alice. "Conservation Practice as Primitive Accumulation." *Journal of Peasant Studies* 38, no. 4 (2011): 683–701.

Kergel, David. "The History of the Internet: Between Utopian Resistance and Neoliberal Government." In *Handbook of Theory and Research in Cultural Studies,* edited by P. Trifonas. Cham, Switzerland: Springer, 2020.

Kettle, Martin. "President 'Ordered Murder' of Congo Leader." *Guardian,* August 9, 2000.

King, Ian. "The Conflict over Conflict-Free Minerals." *Bloomberg Businessweek,* June 5, 2014.

Kirsch, Stuart. *Mining Capitalism: The Relationship between Corporations and Their Critics.* Oakland: University of California Press, 2014.

Lahiri-Dutt, Kuntala. "Digging Women: Towards a New Agenda for Feminist Critiques of Mining." *Gender, Place and Culture* 19, no. 2 (2012): 193–212.

Lahiri-Dutt, Kuntala. "The Feminisation of Mining." *Geography Compass* 9, no. 9 (2015): 523–41.

Lahiri-Dutt, Kuntala, and Martha Macintyre, eds. *Women Miners in Developing Countries: Pit Women and Others.* Burlington, VT: Ashgate, 2006.

Le Carré, John. *The Mission Song.* New York: Little, Brown, 2008.

Le Guin, Ursula K. *The Word for World Is Forest*. New York: Tor Books, 2010.

Leshnev, Sasha, and Fidel Bafilemba. "Another Conflict-Free Mining Project Launches in Eastern Congo." Enough Project blog, 2014. https://enoughproject.org/blog/another -conflict-free-mining-project-launches-eastern-congo.

Lewis, Tanya. "'Mind Uploading' and Digital Immortality May Be Reality by 2045, Futurists Say." *Huffington Post*, June 18, 2013.

Li, Fabiana. "Documenting Accountability: Environmental Impact Assessment in a Peruvian Mining Project." *PoLAR* 32, no. 2 (2009): 218–36.

Li, Fabiana. *Unearthing Conflict: Corporate Mining, Activism, and Expertise in Peru*. Durham, NC: Duke University Press, 2015.

Li, Tanya. "Situating Resource Struggles: Concepts for Empirical Analysis." *Economic and Political Weekly* 38, no. 48 (2003): 5120–28.

Long, Cath. "Land Rights in the Democratic Republic of Congo: A New Model of Rights for Forest-Dependent Communities?" In *Land Struggles and Civil Society in Southern Africa*, edited by Kirk Helliker and Tendai Murisa. Trenton, NJ: Africa World Press, 2011.

Longo, Fiore. "Colonial Conservation: A Cycle of Impunity." *Ecologist*, February 14, 2020.

Longo, Fiore. "A Colonialist Landgrab is Happening Right Now in Congo." Editorial. *Common Dreams*, May 14, 2019. https://www.commondreams.org/views/2019/05/24/ colonialist-land-grab-happening-right-now-congo.

Luning, Sabine. "The Future of Artisanal Miners from a Large-Scale Perspective: From Valued Pathfinders to Disposable Illegals?" *Futures* 62 (2014): 67–74.

Magnowski, Daniel. "Tin Spike Shows Congo's Growing Origin Role." *Reuters*, October 30, 2008.

Makori, Timothy. "Mobilizing the Past: Creuseurs, Precarity, and the Colonizing Structure in the Congo Copperbelt." *Africa* 87, no. 7 (2017): 780–805.

Malkki, Lisa. *Purity and Exile*. Chicago: University of Chicago Press, 1995.

Mamdani, Mahmood. *When Victims become Killers: Colonialism, Nativism, and the Genocide in Rwanda*. Princeton, NJ: Princeton University Press, 2002.

Mantz, Jeffrey W. "From Digital Divides to Creative Destruction: Epistemological Encounters in the Regulation of the 'Blood Mineral' Trade in the Congo." *Anthropological Quarterly* 91, no. 2 (2018b): 525–50.

Mantz, Jeffrey W. "Improvisational Economies: Coltan Production in the Eastern Congo." *Social Anthropology* 16, no. 1 (2008): 34–50.

Mantz, Jeffrey W. "The Slow Road to Tartarus: Technological Fetishism, Materiality, and the Trafficking of 'Conflict Minerals' in the Eastern DR Congo." In *Linguistic and Material Intimacies of Cell Phones*, edited by Joshua Bell and Joel Kuipers. New York: Routledge, 2018a.

Marchal, Jules. *Forced Labor in the Gold and Copper Mines: A History of Congo under Belgian Rule, 1910–1945*. Translated by Ayi Kwei Armah. Popenguine, Senegal: Per Ankh, 2003.

Marshall, Bruce G., and Marcello M. Veiga. "Formalization of Artisanal Miners: Stop the Train, We Need to Get Off!" *Extractive Industry and Society* 4 (2017): 300–303.

Marx, Karl. *Capital: A Critique of Political Economy*, Vol. 1. New York: Vintage, 1977.

Massumi, Brian. *Politics of Affect*. Cambridge, UK: Polity, 2015.

Matthysen, Ken, and Andres Zaragoza Montejano. *"Conflict Minerals" Initiatives in DR Congo: Perceptions of Local Mining Communities*. Antwerp: IPIS, 2014.

Maystadt, Jean-François, Giacomo De Luca, Petros G. Sekeris, and John Ulimwengu. "Mineral Resources and Conflicts in DRC: A Case of Ecological Fallacy?" *Oxford Economic Papers* 66 (2014): 721–49.

Mbembe, Achille. "At the Edge of the World: Boundaries, Territoriality, and Sovereignty in Africa." *Public Culture* 12 (2000): 259–84.

Mbembe, Achille. *Necropolitics.* Durham, NC: Duke University Press, 2019.

McIntosh, Janet. "Mobile Phones and Mipoho's Prophecy: The Powers and Dangers of Flying Language." *American Ethnologist* 37, no. 2 (2010): 33–53.

Mintz, Sidney W. *Sweetness and Power: The Place of Sugar in Modern History.* New York: Penguin, 1985.

Mitchell, Timothy. *Carbon Democracy: Political Power in the Age of Oil.* New York: Verso, 2011.

Mitchell, William. *City of Bits: Space, Place, and the Infobahn.* Cambridge, MA: MIT Press, 1996.

Moore, Jason. *Capitalism in the Web of Life: Ecology and the Accumulation of Capital.* New York: Verso, 2015.

More, Max, and Natasha Vita-More, eds. *The Transhumanist Reader.* Malden, MA: Wiley-Blackwell, 2013.

Moretti, Daniele. "The Gender of the Gold: An Ethnographic and Historical Account of Women's Involvement in Artisanal and Small-Scale Mining in Mount Kaindi, Papua New Guinea." *Oceania* 76, no. 2 (2006): 133–49.

Morozov, Evgeny. *To Save Everything, Click Here: The Folly of Technological Solutionism.* New York: Perseus Books, 2013.

Mudimbe, V. Y. *The Idea of Africa.* Bloomington: Indiana University Press, 1994.

Mudimbe, V. Y. *The Invention of Africa.* Bloomington: Indiana University Press, 1987.

Munn, Nancy. *The Fame of Gawa.* Durham, NC: Duke University Press, 1986.

Nash, June. *We Eat the Mines and the Mines Eat Us: Dependency and Exploitation in Bolivian Tin Mines.* New York: Columbia University Press, 1993.

Navaro-Yashin, Yael. "Affective Spaces, Melancholic Objects: Ruination and the Production of Anthropological Knowledge." *Journal of the Royal Anthropological Institute* 15, no. 1 (2009): 1–18.

Navaro-Yashin, Yael. *The Make-Believe Space: Affective Geography in a Postwar Polity.* Durham, NC: Duke University Press, 2012.

Nest, Michael. *Coltan.* Cambridge, UK: Polity, 2011.

Neubauer, Robert. "Neoliberalism in the Information Age, or Vice Versa? Global Citizenship, Technology, and Hegemonic Ideology." *Triple C (Communication, Capitalism, and Critique)* 9, no. 2 (2011). https://doi.org/10.31269/triplec.v9i2.238.

Norwegian Refugee Council. "DR Congo Shelters 1 in 10 of the World's Internally Displaced People." May 5, 2020. https://www.nrc.no/news/2020/may/dr-congo-shelters-1-in-10-of-the-worlds-internally-displaced-people/.

Nzongola-Ntalaja, Georges. *The Congo from Leopold to Kabila: A People's History.* London: Zed Books, 2002.

Pact Inc. *PROMINES Study: Artisanal Mining in the Democratic Republic of Congo.* Washington, DC: Pact Inc., June 2010.

Parker, Dominic P., and Bryan Vadheim. "Resource Cursed or Policy Cursed? US Regulation of Conflict Minerals and Violence in the Congo." *Journal of the Association of Environmental and Resource Economists* 4, no. 1 (2017): 1–49.

Parks, Lisa, ed. *Signal Traffic: Critical Studies of Media Infrastructure.* Champaign: University of Illinois Press, 2015.

Parr, Adrian, ed. *The Deleuze Dictionary.* Edinburgh: Edinburgh University Press, 2010.

Perelman, Michael. *The Invention of Capitalism: Classical Political Economy and the Secret History of Primitive Accumulation.* Durham, NC: Duke University Press, 2000.

Pijpers, Robert J. "Crops and Carats: Exploring the Interconnectedness of Mining and Agriculture in sub-Saharan Africa." *Futures* 62 (2014): 32–39.

Pistilli, Melissa. "Conflict Minerals: ITRI Supply Chain Initiative Fails to Address Major Issues." *Tantalum Investing News*, April 1, 2010.

Poggiali, Lisa. "Seeing from Digital Peripheries: Technology and Transparency in Kenya's Silicon Valley." *Cultural Anthropology* 31, no. 3 (2016): 387–411.

Povinelli, Elizabeth. *Geontologies: A Requiem for Late Liberalism*. Durham, NC: Duke University Press, 2016.

Prunier, Gerard. *Africa's World War: Congo, the Rwandan Genocide, and the Making of a Continental Catastrophe*. Oxford: Oxford University Press, 2011.

Radley, Ben, and Christoph Vogel. "Fighting Windmills in Eastern Congo? The Ambiguous Impact of the 'Conflict Minerals' Movement." *Extractive Industries and Society* 2, no. 3 (2015): 406–10.

Raeymaekers, Timothy. "Protection for Sale? War and the Transformation of Regulation on the Congo-Ugandan Border." *Development and Change* 41, no. 4 (2010): 563–87.

Raeymaekers, Timothy. "Postwar Conflict and the Market for Protection: The Challenges to Congo's Hybrid Peace." *International Peacekeeping* 20, no. 5 (2013): 1–18.

Raghavan, Sudarsan. "How a Well-Intentioned U.S. Law Left Congolese Miners Jobless." *Washington Post*, November 30, 2014.

Rajak, Dinah. *In Good Company: An Anatomy of Corporate Social Responsibility*. Stanford: Stanford University Press, 2011a.

Rajak, Dinah. "Theatres of Virtue: Collaboration, Consensus, and the Social Life of Corporate Social Responsibility." *Focaal* 60 (2011b): 9–20.

Reno, William. *Warlord Politics and African States*. Boulder, CO: Lynne Rienner, 1999.

Richardson, Tanya, and Gisa Weszkalnys. "Introduction: Resource Materialities." *Anthropological Quarterly* 87, no. 1 (2014): 5–30.

Ring, Laura. *Zenana: Everyday Peace in a Karachi Apartment Building*. Bloomington: Indiana University Press, 2006.

Rist, Gilbert. *The History of Development: From Western Origins to Global Faith*. London: Zed Books, 2008.

Roitman, Janet. *Fiscal Disobedience: An Anthropology of Economic Regulation in Central Africa*. Princeton, NJ: Princeton University Press, 2004.

Ross, Aaron, and Barbara Lewis. "Congo Mine Deploys Digital Weapons in Fight against Conflict Minerals." *Commodities News*, September 30, 2019.

Ross, Michael. "The Political Economy of the Resource Curse." *World Politics* 51, no. 2 (1999): 297–322.

Sassen, Saskia. *Expulsions: Brutality and Complexity in the Global Economy*. Cambridge, MA: Harvard University Press, 2014.

Schouten, Peer. "Roadblock Politics in Central Africa." *Environment and Planning D: Society and Space* 37, no. 5 (2019): 924–41.

Seay, Laura E. "What's Wrong with Dodd-Frank 1502? Conflict Minerals, Civilian Livelihoods, and the Unintended Consequences of Western Advocacy." Working paper, Center for Global Development, 2012, 284.

SEC. "Reconsideration of Conflict Minerals Rule Implementation." Public statement, US Securities and Exchange Commission, January 31, 2017.

Seitzinger, Michael V., and Kathleen Ann Ruane. *Conflict Minerals and Resource Extraction: Dodd-Frank, SEC Regulations, and Legal Challenges*. Congressional Research Service Report 7-5700. 2015.

Sharp, John. "Corporate Social Responsibility and Development: An Anthropological Perspective." *Development Southern Africa* 23, no. 2 (2006): 213–22.

Sheller, Mimi. *Mobility Justice: The Politics of Movements in an Age of Extremes.* London: Verso, 2018.

Simone, Abdou Maliq. "On the Worlding of African Cities." *African Studies Review* 44, no. 2 (2001): 15–41.

Simone, Abdou Maliq. "People as Infrastructure: Intersecting Fragments in Johannesburg." *Public Culture* 16, no. 3 (2004): 407–29.

Smith, James. *Bewitching Development.* Chicago: University of Chicago Press, 2008.

Smith, James. "'May It Never End': Price, Networks, and Temporality in the '3Ts' Mining Trade in the Eastern Congo." *HAU: Journal of Ethnographic Theory* 5, no. 1 (2015): 1–34.

Smith, James. "Tantalus in the Digital Age: Coltan Ore, Temporal Dispossession, and 'Movement' in the Eastern Democratic Republic of Congo." *American Ethnologist* 38, no. 1 (2011): 17–35.

Smith, James H., and Jeffrey W. Mantz. "Do Cellular Phones Dream of Civil War? The Mystification of Production and the Consequences of Technology Fetishism in the Eastern Congo." In *Inclusion and Exclusion in the Global Arena,* edited by Max Kirsch and June Nash, 71–93. New York: Routledge, 2006.

Smith, James, and Ngeti Mwadime. *Email from Ngeti.* Oakland: University of California Press, 2014.

Spencer, Keith. *A People's History of Silicon Valley.* London: Eyewear, 2018.

Starosielski, Nicole. *The Undersea Network: Sign, Storage, Transmission.* Durham, NC: Duke University Press, 2015.

Stearns, Jason. "Causality and Conflict: Tracing the Origins of Armed Groups in the Eastern Congo." *Peacebuilding* 2, no. 2 (2014): 157–71.

Stearns, Jason. *Dancing in the Glory of Monsters: The Collapse of the Congo and the Great War of Africa.* New York: PublicAffairs, 2012.

Stearns, Jason. *From CNDP to M23: The Evolution of an Armed Movement in Eastern Congo.* Nairobi: Rift Valley Institute and Usalama Project, 2012.

Stoler, Anne, ed. *Imperial Debris: On Ruins and Ruination.* Durham, NC: Duke University Press, 2011.

Stoop, Nik, Marijke Verpoorten, and Peter van der Windt. "More Legislation, More Violence? The Impact of Dodd-Frank in the DRC." *PLoS ONE* 13, no. 8 (2018a): e0201783.

Stoop, Nik, Marijke Verpoorten, and Peter van der Windt. "Trump Threatened to Suspend the 'Conflict Minerals' Provision of Dodd-Frank. That Might Actually Be Good for Congo." *Washington Post,* September 27, 2018b.

Strathern, Marilyn. *Audit Culture: Anthropological Studies in Accountability, Ethics, and the Academy.* New York: Routledge, 2000.

Streeter, Thomas. *The Net Effect: Romanticism, Capitalism, and the Internet.* New York: New York University Press, 2010.

Sunstein, Cass. *Infotopia: How Many Minds Produce Knowledge.* New York: Oxford University Press, 2008.

Taussig, Michael. *The Devil and Commodity Fetishism in South America.* Chapel Hill: University of North Carolina Press, 1980.

Templeton, Graham. "Inside Intel's Phenomenal Efforts to End Its Use of Conflict Minerals." *Extreme Tech,* February 11, 2016.

Terranova, Tiziana. "Post-Human Unbounded: Artificial Evolution and High-Tech Subcultures." In *The Cybercultures Reader*, edited by David Bell and Barbara M. Kennedy, 268–79. London: Routledge, 2000.

Thiong'o, Ngugi wa. "The Contradictions of Joseph Conrad." Review of *The Dawn Watch: Joseph Conrad in a Global World*, by Maya Jasanoff. *New York Times*, November 21, 2017.

Trefon, Theodore. *Congo's Environmental Paradox: Potential and Predation in a Land of Plenty*. London: Zed Books, 2016.

Tsing, Anna. *Friction: An Ethnography of Global Connection*. Princeton, NJ: Princeton University Press, 2004.

Tsing, Anna. *The Mushroom at the End of the World: On the Possibility of Life in Capitalist Ruins*. Princeton, NJ: Princeton University Press, 2017.

Tsing, Anna. "Supply Chains and the Human Condition." *Rethinking Marxism* 21, no. 2 (2009): 148–76.

Turner, Thomas. *Congo*. Cambridge, UK: Polity, 2013.

Turner, Thomas. *Congo Wars: Conflict, Myth, Reality*. London: Zed Books, 2007.

United Nations Security Council. *Final Report of the Panel of Experts on the Illegal Exploitation of Natural Resources and Other Forms of Wealth of the Democratic Republic of the Congo*. New York: United Nations Security Council, 2002.

Van Reybrouck, David. *Congo: The Epic History of a People*. New York: Ecco, 2015.

Vázquez-Figueroa, Alberto. *Coltan*. Winchester, UK: O Books, 2008.

Vidal, John. "Armed Ecoguards Funded by WWF 'Beat Up Congo Tribespeople.'" *Guardian*, February 7, 2020. https://www.theguardian.com/global-development/2020/feb/07/armed-ecoguards-funded-by-wwf-beat-up-congo-tribespeople.

Vlassenroot, Koen. "Magic as Identity Maker: Conflict and Militia Formation in Eastern Congo." In *Displacing the State: Religion and Conflict in Neoliberal Africa*, edited by James Howard Smith and Rosalind I. J. Hackett. Notre Dame, IN: University of Notre Dame Press, 2011.

Vogel, Christoph, and Timothy Raeymaekers. "Terr(it)or(ies) of Peace? The Congolese Mining Frontier and the Fight against 'Conflict Minerals.'" *Antipode* 48, no. 4 (2016): 1102–21.

Walker, Joshua. "Torn Dollars and War Wounded Francs: Money Fetishism in the Democratic Republic of Congo." *American Ethnologist* 48, no. 2 (2017): 288–99.

Weissman, Stephen. "What Really Happened in Congo: The CIA, the Murder of Lumumba, and the Rise of Mobutu." *Foreign Affairs* 93, no. 4 (2014): 14–24.

West, Paige. *Dispossession and the Environment: Rhetoric and Inequality in Papua New Guinea*. New York: Columbia University Press, 2016.

West, Paige. "Environmental Conservation and Mining: Between Experience and Expectation in the Eastern Highlands of Papua New Guinea." *Contemporary Pacific* 18, no. 2 (2006): 295–313.

Wild, Franz, Michael Kavanagh, and Jonathan Ferzier. "Gertler Earns Billions as Mine Deals Fail to Enrich Congo." *Bloomberg*, December 5, 2012.

Wilson, Thomas. "Mountain of Tin Draws Investors to World's Biggest Untapped Site: Alphamin Braves Armed Groups and Remote Location in Democratic Republic of Congo." *Bloomberg*, May 3, 2016. https://www.bloomberg.com/news/articles/2016-05-03/mountain-of-tin-draws-investors-to-world-s-biggest-untapped-site.

Wolfe, Lauren. "How Dodd-Frank Is Failing Congo." *Foreign Policy*, February 2, 2015.

World Bank. *Mining Together: Large-Scale Mining Meets Artisanal Mining: A Guide for Action*. New York: World Bank, 2009.

Worstall, Tim. "How Enough Project and Global Witness Make Life Tougher in Congo." *Forbes*, November 2, 2011.

Yakovleva, Natalia. "Perspectives on Female Participation in Artisanal and Small-Scale Mining: A Case Study of Birim North District of Ghana." *Resource Policy* 32, nos. 1–2 (2007): 29–41.

Zaitchik, Alexander. "How Conservation Became Colonialism." *Foreign Policy*, July 2018.

Zuboff, Shoshana. *Surveillance Capitalism: The Fight for a Human Future at the New Frontier of Power.* London: Profile Books, 2019.

INDEX

adulthood, 74, 177, 259

AFDL (Alliance of Democratic Forces for the Liberation of Congo), 17–18, 22, 74, 92, 235

Africa: anthropology of, 46, 59, 62; and development, 59; as Other, 55; as producer of resources, 54, 95; scholars of, 46; the state in, 47, 323n4; in the Western imagination, 55–56, 61–64, 327n26; violence in, 325n7

Africans, 262, 323n4; and development, 62; relationship with whites, 166; in the Western imagination, 55–56, 59, 110, 323n4; as workers, 48

Afrofuturism, 99–100, 201

Agence Nationale de Renseignements (ANR), 23, 133, 135, 181, 193–94, 203, 240, 242, 244, 269

agriculture: as alternative work, 161, 275, 319, 332n1; as collective identity, 239; conflicts with mining, 17, 20, 44, 121, 166–67, 283, 288, 318; as inferior to mining, 166, 278, 318, 325n9; and war, 20, 51

Alphamin Resources, 41, 217, 219, 234, 253–54, 256–57, 261–63, 267–68, 286, 308, 332n4

Al Qaeda, 104, 214

Amnesty International, 85

amputation, 167

ancestors, 26, 32, 34–35, 39, 45–46, 52, 75–76, 82, 86, 96–103, 109–11, 117,

121, 134–35, 142–54, 156, 168–69, 172–73, 177, 186–87, 196–97, 210, 219–20, 222, 227–28, 233, 259, 264–65, 302, 306–7, 313–15

Aneesh, Aneesh, 59

Anglo American, 19

Angola, 16–17, 330n2 (chap. 5)

Apple, 30, 253–54, 270, 300

Arabs, 103–4, 172, 214, 228

artisanal mining: and commodities, 43; economic dependence on, 3, 20, 205, 303; global growth of, 58, 302–3, 310, 326n18; and labor, 60, 110, 158, 163, 276, 303; legal status of, 12, 16, 29–31, 39, 41, 48, 65, 112, 251, 255, 257, 269, 273, 275, 277, 284, 296, 302–3, 332n4; and movement, 34, 42, 67, 158, 197, 305–6, 308, 315, 318–19; research on, 9, 32–33, 38, 71, 202; significance of, 73, 158, 179–80, 205, 261, 276, 298, 315; as a system, 162, 182, 188, 217, 276, 298, 302–3, 319, 329n6; and value, 154, 197; visibility of, 37; and war, 71–72, 89–90, 92; wealth of, 28, 197, 303, 310

bag-and-tag, 13, 30, 67, 261, 263, 268–69, 274

Bagandula, 12, 142–43, 208, 210, 212, 216, 219–33, 238–42, 244–45, 252, 258, 260–62, 264–65

Bales, Kevin, 217–18, 249

350 / Index

Sodexmin, 224
soldiers, 6–7, 14, 17, 21–22, 24–25, 29, 42,
45, 64–65, 72–76, 78, 81–86, 88–89,
103, 110, 116, 120, 130–31, 133, 136–
37, 152, 154, 185, 187, 189, 191, 193,
203–4, 206, 208, 213, 218, 229, 232,
245–49, 254, 260–61, 263, 267, 269,
281–82, 286–89, 291–92, 315, 328n7
Sominco, 93–94
Sominki, 90, 92–94, 157–58, 161, 163,
165–69, 178, 182, 192, 211, 223, 267,
305, 328n9
South Africa, 20, 205, 234, 284
South Kivu, 9, 12, 17–18, 52, 77–79, 83,
87, 90, 93, 111, 140, 186, 193, 211, 247,
267, 269, 272–73, 275, 281–83, 286
spatiotemporality, 26, 44, 48, 50, 72, 81–82,
89, 96, 135, 144, 154, 165, 172, 177–
78, 180, 230–31, 265, 268, 303, 313
stability, 42, 45, 74, 106, 146, 213, 217,
289, 305, 315
Standard Chartered, 19
Stanley, Henry, 165
Stanleyville, 98, 108, 185
storytelling, 72, 81, 204, 217
Sudan, 95
suicide, 77, 79, 81, 327n28
Sun City Agreement, 20, 241, 284
supply chain, 6, 30, 37, 47, 49, 51, 62, 68,
109, 117, 125, 137–38, 142, 144, 153–
54, 156, 162–63, 197, 270–71, 274,
300–308, 310, 324n8
surveillance, 26, 48, 66–67, 140, 241, 276,
312
Swahili, 6, 14, 24, 33–34, 143, 188, 204,
228, 284, 305, 312, 317, 325n11
Symetain, 161, 163, 165, 167–69, 178

tantalum, 4, 30, 35–36, 40, 109, 126, 141,
158, 187, 299, 310
Tanzania, 17, 214
taxation, 112, 275, 288, 325n12, 330n8
technology, 4, 18, 20, 23, 25–27, 58, 60–
63, 66–67, 74, 77–79, 86, 89, 95–99,
101–2, 126, 146, 151–54, 169, 252,
292, 299, 310, 327n5
technosolutionism, 4, 295
techno-utopianism, 4–5
tempo-politics, 256

temporality, 32, 38, 54, 56, 59, 74, 79, 90,
95–96, 103, 107, 114, 135, 140, 154,
159, 161, 163, 165–66, 170, 174, 177–
78, 196, 204, 234, 254, 259, 265, 299,
312–13, 315, 317
3T minerals, 4, 30, 36, 75, 90, 93, 117, 127,
133–34, 187, 263, 272, 319, 323n5
time (politics of), 56
tin, 4, 13–15, 18, 30, 40, 116, 129, 158,
187, 211, 220, 248, 253–54, 261, 267–
68, 272, 274, 308
touching, 30, 32, 41, 46–47, 50, 66, 73, 88,
104, 117–18, 270, 293, 312
transcendence, 52, 54, 58–59, 63, 315
transparency, 32, 37, 54, 61–62, 64, 67,
231, 241, 252, 265, 273, 288, 295, 299,
310, 312
trust, 33–34, 38, 41, 106, 138, 154, 206,
216, 247, 262, 272, 285, 313–14
tungsten, 4, 30, 187
Tutsi, 17–18, 22, 24, 206, 224, 282–83,
286–87, 289, 323n4
Twitter, 62

Uganda, 17–18, 65, 85, 92, 95, 146, 284,
328n6
United Nations, 17, 19, 67, 74–75, 80, 95,
113, 151, 192–93, 196, 204, 214, 244,
247, 289
United States (country), 16, 19, 21, 24, 30,
37, 46, 58–59, 94, 150, 251–52, 284–
85, 295
United States of America (mine), 170–71,
182, 195
uranium, 14, 164, 190
urban, 5, 40, 42, 88, 95, 114, 132, 139,
148, 153, 155, 212, 239, 296, 311–12
uses of a body, 75, 81

value, 20, 27, 31, 35–38, 41–42, 45, 47, 51–
52, 59, 61, 63, 75, 78, 89–90, 92, 94,
102–4, 106, 108–9, 111, 114, 117, 121,
126–30, 139, 141–44, 146–47, 150–51,
153–55, 163–64, 166, 170, 173, 180,
184, 186, 195, 206, 209, 216, 219, 224,
232–33, 236, 248–49, 253, 256, 265,
275, 285, 293, 295, 312–13, 324nn1–2,
325n11, 330n10, 332n1
violence, 11, 14, 17, 22, 27, 29–30, 32, 35,